Aspects of Signal Processing

Part 1

NATO ADVANCED STUDY INSTITUTES SERIES

Proceedings of the Advanced Study Institute Programme, which aims at the dissemination of advanced knowledge and the formation of contacts among scientists from different countries

The series is published by an international board of publishers in conjunction with NATO Scientific Affairs Division

A	Life Sciences	Plenum Publishing Corporation
B	Physics	London and New York
C	Mathematical and Physical Sciences	D. Reidel Publishing Company Dordrecht and Boston
D	Behavioral and Social Sciences	Sijthoff International Publishing Company Leiden
E	Applied Sciences	Noordhoff International Publishing Leiden

Series C – Mathematical and Physical Sciences

Volume 33 – Aspects of Signal Processing
Part 1

Aspects of Signal Processing

With Emphasis on Underwater Acoustics

Part 1

Proceedings of the NATO Advanced Study Institute held at Portovenere, La Spezia, Italy 30 August–11 September 1976

edited by

G. TACCONI
University of Genoa, Italy

D. Reidel Publishing Company
Dordrecht-Holland / Boston-U.S.A.

Published in cooperation with NATO Scientific Affairs Division

Library of Congress Cataloging in Publication Data

Nato Advanced Study Institute on Signal Processing with
Emphasis on Underwater Acoustics, Portovenere, Italy, 19
Aspects of signal processing with emphasis on
underwater acoustics.

(NATO advanced study institutes series : Series C,
Mathematical and physical sciences ; v. 33, pts. 1-2)
Sponsored by NATO Division of Scientific Affairs and
Selenia s. p. a., Rome.
Includes bibliographical references and index.
1. Signal processing--Addresses, essays, lectures.
2. Sonar--Addresses, essays, lectures. 3. Underwater
acoustics--Addresses, essays, lectures. I. Tacconi, G.,
1925- II. North Atlantic Treaty Organization.
Division of Scientific Affairs. III. Selenia s. p. a.
IV. Title. V. Series.
TK5102.5.N36 1976 621.38'043 77-3238
ISBN 90-277-0798-7 set of two parts
ISBN 90-277-0799-5 part 1 ISBN 90-277-0800-2 part 2

Published by D. Reidel Publishing Company
P.O. Box 17, Dordrecht, Holland

Sold and distributed in the U.S.A., Canada, and Mexico
by D. Reidel Publishing Company, Inc.
Lincoln Building, 160 Old Derby Street, Hingham, Mass. 02043, U.S.A.

Printed in The Netherlands

TABLE OF CONTENTS

SUBJECT 3 - TIME-SPACE PROCESSING, ADAPTIVE PROCESSING AND NORMALIZATION, QUANTIZATION METHODS

CONTENTS OF PART 2

SUBJECT 3 (continued).

+ Author unable to attend; paper not presented at the Institute but included in these proceedings.

ASPECTS OF SIGNAL PROCESSING

Proceedings of the NATO Advanced Study Institute on
Signal Processing with emphasis on Underwater Acoustics

held at

Portovenere, La Spezia, Italy
30 August to 11 September 1976

Sponsored by

NATO SCIENTIFIC AFFAIRS DIVISION
SELENIA S.p.A., Rome
THE ITALIAN NATIONAL RESEARCH COUNCIL, C.N.R. Rome [+]

With the support of

MARINA MILITARE ITALIANA, La Spezia
SACLANT ASW RESEARCH CENTRE, La Spezia

Director

G. Tacconi
Istituto di Elettrotecnica
University of Genoa, Italy

Scientific Committee

J.W.R. Griffiths   H. Urban
H. Mermoz   C. Van Schooneveld
P.L. Stocklin   G.C. Vettori

Secretary

A. Plaisant

[+] C.N.R. Contract 206072/07/67084.

LIST OF PARTICIPANTS

* *Contributors* ** *Tutorial Lecturers*

W.G.L. Adaway
Ferranti Limited
Western Road
Bracknell
Berkshire, U.K.

Ø. Andersen
Det Norske Veritas
Division for Marine Technology
P.O. Box 6060
Oslo 6, Norway

H.V. Asselt
Holland Signaal Appl.
Hegelo (0)
The Netherlands

C. Atzeni *
IROE
Università di Firenze
Via Panciatichi 56
Firenze, Italy

A. Aytun
TBTAK Structural & Seismic Engineering
Building Research Institute
Ankara, Turkey

K. Aytun (Mrs)
Turkish Petroleum Co.,
T.P.A.O.
Ankara, Turkey

N. Bache
SACLANT ASW Research Centre
Viale San Bartolomeo 400
La Spezia 19026
Italy

A.B. Baggeroer *
M.I.T.
Cambridge
Mass.
U.S.A.

A. Bakkers
Holland Signaal Appl.
Hengelo (0)
The Netherlands

K.J. Beauchamp
University of Lancaster
Bailrigg
Lancaster, LAI 4YW, U.K.

A. Bevilacqua
SELENIA
Via Tiburtina
00131 Roma
Italy

G. Bienvenu *
Thomson - CSF
Chemin des Travails
06802 Cagnes-sur-Mer
France

L. Bjørnø **
Technical University of Denmark
Department of Fluid Mechanics
DK-2800 Lyngby, Denmark

B. Blixhavn
The Norwegian Institute of Cosmic Physics
P.O. Box 1038
Oslo 3, Norway

H. Block
LEOK
Haarlemmerstraatweg 7
Oegstgeest
The Netherlands

I.S. Blumental
SACLANT ASW Research Centre
Viale San Bartolomeo 400
La Spezia 19026
Italy

K. Bom *
Erasmus Universiteit Rotterdam
Thorax Technology
Postbus 1738, Rotterdam
The Netherlands

M. Boninu
Istituto Elettro-Acustica
MARIPERMAN
Marina Militare Italiana
La Spezia, Italy

C. Braccini **
Istituto di Elettrotecnica
Università di Genova
Viale Causa 13
Genoa, Italy

L. Brock-Nannestad
Danish Defence Research Establishment
Osterbrogades Kaserne
Copenhagen, Denmark

H.J. Bückmann
Forschungsinstitut für Hochfrequenzphysik
D-5307 Wachtberg-Werthhoven
Germany

W. Bühring
FFM
Werthoven
Germany

D. Butler
Plessey Company
Templecombe
Somerset
U.K.

P.G. Cable *
Naval Underwater Systems Center
New London
Conn. 06320
U.S.A.

V. Cappellini **
Istituto di Elettronica
Università di Firenze
Via Santa Marta
Firenze, Italy

G.C. Carter *
Naval Underwater Systems Centre
New London
Conn. 06320
U.S.A.

V. Chander
Naval Physical & Oceanographic Laboratory
DO No. PL/930/VC, Naval Base
Chochin 682004, S. India.

D.L. Chenoweth
University of Louisville
Louisville
Kentucky 40208
U.S.A.

T. Cheston
SACLANT ASW Research Centre
Viale San Bartolomeo 400
La Spezia 19026
Italy

R.H. Clarke *
Imperial College
Exhibition Road
London S.W.7 2BT, U.K.

J.E. Cole
Tufts University
Anderson Hall
Medford
Mass. 02155, U.S.A.

B.W. Conolly *
Chelsea College
London University
Manresa Road
London S.W.3 6LX, U.K.

H. Debart *
Centre de Recherches de la
Comp. Générale d'Electricit'
91460 Marcoussis
France

P. de Heering **
SACLANT ASW Research Centre
Viale San Bartolomeo 400
La Spezia 19026
Italy

P.A. Devijver **
MBLE (Philips) Research Lab.
Av. Emile van Bacelaere 2
Brussels 1170
Belgium

E. Diamanti
Istituto Elettro-Acustica
MARIPERMAN
Marina Militare Italiana
La Spezia, Italy

R. Diehl
Krupp-Atlas Elektronik
Postfach 448545
Bremen 44
Germany

J. Doutt
SACLANT ASW Research Centre
Viale San Bartolomeo 400
La Spezia 19026
Italy

A.D. Dunsiger
Memorial University of Newfoundland
St. Johns
Newfoundland, Canada

T.S. Durrani *
Strathclyde University
Royal College
204 George Street
Glasgow G.1 1XW, U.K.

C. Fimerelis
Ministry of Defence
Hellenic Navy Command
Geten, Cholargos
Athens, Greece

A.H.C. Frazer
Royal Naval Engineering College
Manadon
Plymouth PL5 3AQ
U.K.

O.L. Frost **
Argosystems Inc.
2332 South Court
Palo Alto
Cal. 94301, U.S.A.

G. Gambardella **
Istituto di Elettrotecnica
Università di Genova
Viale Causa 13
Genoa, Italy

O.B. Gammelsaeter
Norwegian Defence Research Establishment
P.O. Box 115
Horten N-2191, Norway

C. Giraudon *
CIT - Alcatel
1 Avenue Aristide Briand
94 Arcueil
France

J.W.R. Griffiths **
University of Technology
Depart. of Electronic Engineer.
LOUGHBOROUGH
Leicestershire LEII 3TU, U.K.

L.J. Griffiths *
University of Colorado
Depart. of Electrical Engineer.
Boulder
Colorado 80302, U.S.A.

H. Gudat
Institut für Informationsverbeitung in Technik & Biologi
Sebastian-Kneipp 12-14
D-7500 Karlsruhe, Germany

E. Helmer
Forschungsanstalt der Bundeswehr für Wasserschall & Geophysik
Klausdorfer Weg 2-24
23 Kiel, Germany

G. Hemmie
BWB
Koblenz
Germany

G. Hermstrüwer
Krupp-Atlas Elektronik
Postfach 448545
Bremen 44
Germany

W.S. Hodgkiss *
Naval Undersea Center
San Diego
California 92132
U.S.A.

F. Hoering
ELAC
Kiel
Germany

J.M. Hovem
ELAB
Trondheim
Norway

J.E. Hudson **
University of Technology
Depart of Electronic Engineer.
LOUGHBOROUGH
Leicestershire LEII, 3TU, U.K.

E. Hug
SACLANT ASW Research Centre
Viale San Bartolomeo 400
La Spezia 19026
Italy

M.L. Hull
University of California
5144 Etcheverry Hall
Berkeley
California 94720, U.S.A.

F. Jensen
SACLANT ASW Research Centre
Viale San Bartolomeo 400
La Spezia 19026
Italy

J.K. Johnsen
Norwegian Defence Research Establishment
P.O. Box 115
N-2191 Horten, Norway

A. Kracht
Krupp-Atlas Elektronik
Postfach 448545
Bremen 44
Germany

T. Kuyama *
Defence Academy
3-2-12 Kakinokizaka
Meguroku, Japan

R.T. Lacoss **
M.I.T. Lincoln Laboratory
Lexington
Mass.
U.S.A.

J.L. Lacoume *
University of Grenoble
ENS D'Electrotechnique et de Genie Physique
38040 Grenoble, France

R. Laval *
Societe AERO
3 Av. de l'Opera
75001 Paris
France

R. Leisterer
Krupp-Atlas Elektronik
Postfach 448545
Bremen 44
Germany

G. Loubet
University of Grenoble
ENS D'Electrotechnique et de Genie Physique
38040 Grenoble, France

O. Løvhaugen
Central Institute for Industrial Research
Forskningsveien 1
Oslo 3, Norway

M. di Lullo
Scientific Affairs Division
NATO Headquarters
Brussels
Belgium

E.B. Lunde *
Norwegian Defence Research Establishment
P.O. Box 115
N-2191 Horten, Norway

R. MacKinnon
Department of National Defence
Victoria
Canada

C.J. MacLeod
University of Strathclyde
Royal College
204 George Street
Glasgow G1 1XW, U.K.

J.N. Maksym *
Defence Research Establishment Atlantic
P.O. Box 1012
Dartmouth, N.S. B2Y, 3Z7, Canada

A. Marchioni
Laboratoire de Mécanique et d'Acoustique
31 Chemin Joseph Aiguier
13 274 Marseille, France

L. Masotti **
IROE
Via Panciatichi 56
Firenze
Italy

J. McCool *
Naval Undersea Center
Point Lomo
San Diego
California 92132, U.S.A.

L. Meier *
SACLANT ASW Research Centre
Viale San Bartolomeo 400
La Spezia 19026
Italy

H. Mermoz *
Laboratoire de Detection Sous Marine
DCAN
Toulon, France

L. Montefusco
Università di Firenze
Istituto di Idraulica
Via S. Marta
Firenze, Italy

D. Muster
University of Huston
Department of Mech. Engineer.
Houston
Texas 77004, U.S.A.

U. Nickel
FFM
Werthhoven
Germany

K. Ossenberg
IITB
Sebastian-Kneipp Str. 12-14
D75 Karlsruhe
Germany

N.L. Owsley *
Naval Underwater Systems Center
New London
Conn. 06320
U.S.A.

S. Pardini
Selenia S.p.A.
Via Tiburtina
Roma 00131
Italy

R.A. Pearce
Plessey Co.
Templecombe
Somerset
U.K.

L. Pescatori
Selenia S.p.A.
Via Tiburtina
Roma 00131
Italy

G. Pettersen
Norwegian Defence Research Establishment
P.O. Box 115
Horten, Norway

B. Picinbono **
Laboratoire d'Etude des Phénomènes Aléatoires
Université de Paris-Sud
91405 Orsay, France

D.G. Pincock
University of New Brunswick
Fredricton
New Brunswick,
Canada

G. Pranzo-Zaccaria
SACLANT ASW Research Centre
Viale San Bartolomeo 400
La Spezia 19026
Italy

A.R. Pratt *
University of Technology
Depart. of Electronic Engineering
Loughborough
Leicestershire LEII 3TU, U.K.

P. Prinsen **
RVO-TNO, Physics Laboratory
Vlakte van Waalsdorp
The Hague
The Netherlands

C.N. Pryor *
Naval Underwater Systems Center
New London
Conn. 06320
U.S.A.

M. Rabinovitz
Tadiran Electronics
P.O. Box 648
Tel-Aviv
Israel

P.J.W. Rayner **
University Engineering Department
Trumpington Street
Cambridge CB2 1PZ
U.K.

H.A. Reeder *
Tracor Inc
6500 Tracor Lane
Austin
Texas 78721, U.S.A.

M.L. Retter *
University of Oxford
Department of Nuclear Physics
Keble Road
Oxford OX1 3RH, U.K.

D. Ross
SACLANT ASW Research Centre
Viale San Bartolomeo 400
La Spezia 19026
Italy

O. Roth
Brüel & Kjaer
23 Linde allé
DK-2850 Naerum
Denmark

U. Rupe
SACLANT ASW Research Centre
Viale San Bartolomeo 400
La Spezia 19026
Italy

D. Saxton
Admiralty Research Laboratory
Queens Road
Teddington
Middlesex, U.K.

H. Schachter
Polytechnic Institute of New York
Farmingdale
New York 11735, U.S.A.

B. Scholz *
Forschungsanstalt der Bundeswehr für Wasserschall und Geophysik
Klausdorfer Weg 2-24
2300 Kiel, Germany

C. van Schooneveld **
RVO-TNO, Physics Laboratory
Vlakte van Waalsdorp
The Hague
The Netherlands

P. Schultheiss **
Yale University
525 Becton Center
New Haven
Conn. 06520, U.S.A.

M. Siegel
Krupp-Atlas Elektronik
Postfach 448545
2800 Bremen
Germany

D. Solimini
Università di Roma
Istituto di Elettronica
Via Eudossiana 18
Rome, Italy

K. Søstrand
Norwegian Defence Research Establishment
P.O. Box 115
N-3191 Horten, Norway

R.G. Taylor *
Admiralty Surface Weapons Establishment
Portsmouth
Hants., U.K.

R. Thiele *
Forschungsanstalt der Bundeswehr für Wasserschall und Geophysik
Klausdorfer Weg 2-24
2300 Kiel, Germany

G.L. Tinelli
Compagnia Generale di Elettricità
Dipartimento Sistemi Difes
Via Montefeltro 8
20156 Milano, Italy

P. Tournois
Thomson CSF
Cagnes-sur-Mer 06802
France

C. Turcat (Miss)
University of Grenoble
ENS D'Electrotechnique et de Genie Physique
38040 Grenoble, France

H. Urban
Krupp-Atlas Elektronik
Postfach 448545
2800 Bremen
Germany

R.J. Urick **
Tracor Sciences & Systems
1601 Research Blvd.
Rockville
Maryland 20850, U.S.A.

J. Viale
ECAN
Saint Tropez
France

J.L. Vernet
Thomson CSF
Cagnes-sur-Mer 06802
France

G. Vettori
SACLANT ASW Research Centre
Viale San Bartolomeo 400
La Spezia 19026
Italy

W. Von Winkle
Naval Underwater Systems Centre
New London
Conn. 96321
U.S.A.

A. Wasiljeff **
Universität Bremen
BF Electrotechnik
Bremen, Germany

M. Weber
Forschungsinstitut für Hochfrequenzphysik
D-5307 Wachtberg-Werthhoven
Germany

V. Westerlin
Försvarets Forskningsanstalt
Stockholm
Sweden

H. Whitehouse **
Naval Undersea Center
San Diego
California 92132
U.S.A.

B. Widrow **
Stanford University
Department of Electrical Engin.
Stanford
California, U.S.A.

R. Zoppoli *
Università di Genova
Facoltà di Ingegneria
Viale Cambiaso 6
Genoa, Italy

PREFACE

The summer school held in Portovenere followed a tutorial format with the purpose of familiarizing postdoctoral or postgraduate students in the basic theories and up-to-date applications of present knowledge. Although, from a teaching point of view, a certain amount of overlapping is always useful, in order to avoid excessive duplication direct contact between lecturers expert in the same subject was encouraged during the preparation phase.

In recent years computer facilities and theoretical implementation have considerably increased the possibility of solving problems relating to signal detection in noise. Any type of communication may take advantage of signal processing principles, including any type of physical measurement that can be considered as a non-semantic and/or quasi-semantic communication. Since signal processing techniques are common to many branches of science (telecommunications, radar, sonar, seismology, geophysics, nuclear research, space research and others), the advanced and sophisticated levels reached singularly in any one of them could be used to the advantage of the others. In particular, underwater acoustics is a discipline which, to some extent, represents a practical general model that has permitted the development of signal processing techniques suitable to meet data reduction and interpretation needs of other branches of science. This ASI consequently underlined the inter-disciplinarity of signal processing in order that the principles of outstanding methods developed in one field may be adapted to others.

Among the various aspects of signal processing treated during the school, that regarding the auto-regressive systems or maximum entropy power spectral analysis was recognized as a very useful tool for many different applications, and the tutorial lectures supplied the audience with valuable basic elements for new applications. The same can also be said for adaptive beamforming, time-space variant channel, detection estimation, and processor architecture.

The present proceedings include the tutorial lectures and the complementary advanced research papers by contributing students on topics presented under the following six subjects:

Subject 1 Properties of Time-Space Variant Transmission Channel and Underwater Communications

Subject 2 Detection Estimation and Tracking Techniques

Subject 3 Time-Space Processing, Adaptive Processing and Normalization, Quantization Methods

Subject 4 Displays, Pattern Recognition, Human Decision

Subject 5 Relevant Inputs from Other Fields

Subject 6 Modern Processor Architecture and Techniques

Towards the end of the Institute, attendees were invited to suggest topics for discussion and, consequently, the following four workshops were held:

Properties of Time/Space Variant Channel
Chaired by Prof A. Wasiljeff

Autoregressive Spectral Analysis
Chaired by Dr O.L. Frost and Prof L. Griffiths

Adaptive Beamforming
Chaired by Dr N.L. Owsley

Modern Processor Architecture+
Chaired by Dr H.J. Whitehouse

Summaries on these workshops are also published in these proceedings.

The Institute was sponsored and financed by the NATO Scientific Affairs Division and Selenia S.p.A., Rome. Much valuable support was also given by the Italian Navy (MMI), La Spezia; SACLANT ASW Research Centre, La Spezia; U.S. Office of Naval Research, Washington; Italian National Research Council (CNR), Rome; National Science Foundation, U.S.A.; The North Italy Section of IEEE.

I should like to express my gratitude for the advice and cooperation given by all members of the Scientific Committee and Mr A. Plaisant, and to thank Mrs J. Barbieri for her assistance.

GIORGIO TACCONI

+ Not printed in these proceedings.

INAUGURAL ADDRESSES

I am most pleased to welcome you all to Italy and to this NATO ADVANCED STUDY INSTITUTE ON SIGNAL PROCESSING WITH EMPHASIS ON UNDERWATER ACOUSTICS, and I trust that your stay here will be pleasant and fruitful.

I should like to express my gratitude to Dr R. Cappellini, Prefect of La Spezia, as representative of the Italian Government, and to Admiral G. Oriana, Commander-in-Chief of the Upper Tyrrhenian Naval District, who have honoured the opening of this Institute with their presence. The Mayor of Portovenere, who is unfortunately sick, requested that I also extend his welcome and his regrets for not being present today.

And now, on declaring this Advanced Study Institute open, I have pleasure in introducing you to Dr M. Di Lullo of the Scientific Affairs Division of NATO, Brussels.

GIORGIO TACCONI

Dr. Capellini, Admiral Oriana, Ladies and Gentlemen,

It is an honour and a pleasure for me to welcome you, on behalf of the Assistant Secretary General for Scientific and Environmental Affairs of NATO.

This is but one of the 39 Advanced Study Institutes to be sponsored by the NATO Science Committee in 1976. The Science Committee, which has an annual budget of approximately $7 million, also sponsors fellowships, research grants and administers a few specialised programmes in selected areas, for example, Marine, Ecological and Systems Sciences. The NATO Science Committee was created in 1958 in recognition of the fact that to be powerful and effective an alliance must look beyond a resolve to unite in mutual military defence; it must also determine to maintain and strengthen the power of its science and technology. The standard of scientific excellence which characterises the western democracies has been largely achieved through freedom of movement and exchange of ideas among professional scientists.

I would like to give you some details on these activities not only for your information, but also in case you might like to take advantage of the opportunities offered by the Programmes in the member countries.

THE SCIENCE FELLOWSHIPS PROGRAMME

The NATO Science Fellowships Programme's main purpose is to stimulate the international exchange of post-graduate and post-doctoral students of the pure and applied sciences. Each year about 600 NATO Science Fellowships are awarded. The total number to date, well over 10,000, means that an important part of post-graduate scientific education in the West has been accomplished through the NATO Science Fellowships Programme.

THE RESEARCH GRANTS PROGRAMME

The purpose of the NATO Research Grants Programme is to stimulate, encourage and facilitate scientific research to be carried out in collaboration among scientists working in different member countries of the Alliance, and thus to promote the flow of ideas and of experimental and theoretical methods across frontiers. So far, several thousand applications have been considered and 1,200 proposals have been funded. By supporting only those projects for which international collaboration is demonstrably necessary, and by careful consideration of interdisciplinary approaches, the Programme successfully enhances and does not duplicate national research activities. In this respect the Programme is unique.

THE ADVANCED STUDY INSTITUTES PROGRAMME

The purpose of the Advanced Study Institutes Programme is to contribute to the dissemination of advanced knowledge and the formation of contacts among scientists from different countries. To date, about 750 ASIs have been held, in almost all areas of science, over 30,000 scientists have participated, of which over 10% have come from non-member countries. The proceedings of approximately two-thirds of the ASIs appear in book form. These have generally been recognized as authoritative surveys of their subjects, and have reached a very large audience, both inside and outside the North Atlantic Alliance.

An Advanced Study Institute which fulfils its aim is quite different from an ordinary scientific meeting. A summer school has a clear tutorial nature and enables scientists, particularly young people establishing themselves, to get together without the usual haste of a normal conference. They get the opportunity of hearing leading research workers lecture in depth on their subjects and of contributing to a major reappraisal of a field themselves. In 1976, thirty-nine applications out of eighty-three were accepted for support, several of those rejected were worthy of support and were excluded solely through lack of funds.

I cannot anticipate the outcome of this Advanced Study Institute because it is now entirely in your hands. All possible effort has been made to create the right atmosphere for its success both from a scientific and practical point of view. Outstanding lecturers have been invited and the students have been carefully selected on the basis of their qualifications. Now, we expect the lecturers and students alike to work together for the duration of the course so that the dissemination of the most up-to-date knowledge will be the result of an interaction characterised by great dynamism. This opportunity to have frequent and informal exchanges of views, the participants being together in an isolated but delightful location, and the using of one common language are all factors which have contributed to the success of NATO ASIs in the past.

We hope that by the bringing together of multinational scientists of various ages and experience from related but not identical disciplines, will lead to an enthusiastic and novel approach to the problem areas and allow a critical assessment of current deficiences in the field as a whole.

What we expect to achieve are:

(a) in the sphere of teaching: an improved assessment of the present state-of-the-art so that a coherent presentation of the subject could be made to those who wish to work in this specific field;

(b) in the sphere of research: we hope that through stimulating discussions and accurate teaching of what is currently known about the most promising avenues of research, a large part of the participants will be encouraged to undertake new and challenging investigations;

(c) in the sphere of documentation: we hope that the book which will be published on this summer school will constitute valuable reference material for the future.

I would like to thank the Italian authorities for having honoured us with their presence in the opening session of this Advanced Study Institute and for having assisted in its organization. Particular thanks are extended to the Director and his Organizing Committee for their tireless efforts in preparing this course. I am grateful to you too, the lecturers and students, for having accepted to come here, and I wish you all success in your scientific activities and a pleasant stay in this wonderful corner of Italy.

MARIO DI LULLO

S U B J E C T 1

PROPERTIES OF TIME/SPACE VARIANT TRANSMISSION CHANNEL AND UNDERWATER COMMUNICATIONS

MULTIPATH PROPAGATION AND ITS EFFECTS ON SONAR DESIGN AND PERFORMANCE IN THE REAL OCEAN

R. J. Urick

formerly, U.S. Naval Surface Weapons Center,
Silver Spring, Maryland 20910*

INTRODUCTION

Traditionally, the most important military use of underwater sound has been the detection of submarines at as great a distance as technology and costs will allow. Nowadays, these distances are as great as hundreds of miles for passive sonars, utilizing the noise radiated by the target, and tens of miles for active sonars, utilizing echoes from their targets. Detection ranges of these magnitudes require the maximum state-of-the-art employment of techniques in signal processing, array designs and displays.

The purpose of this paper is to show that these techniques are limited--in their possible gains and their degree of sophistication--by the peculiar and disadvantageous peculiarities of the ocean medium. These characteristics are the result of the presence of multipaths in the ocean--that is, the fact that there is always more than one ray path between source and receiver. In this paper, some of the consequences of multipath propagation in deep water will be pointed out, and the limitations imposed on processing and the gain of arrays will be mentioned. A shortened version of the present paper under the title, "Sonar Design in the Real Ocean: Multipath Limitations on Sonar Performance", was presented at the IEEE International Conference on Acoustics Speech and Signal Processing, April 1976.

---

*present address: Tracor, Inc., 1601 Research Boulevard, Rockville, Maryland 20850

G. Tacconi (ed.), Aspects of Signal Processing, Part 1, 3-18.

MULTIPATHS IN THE SEA

In contrast to radar, sonar systems must reach a long range target by multipath propagation of sound in the sea. These multipaths are caused by the peculiar lens-like structure of the velocity--or index of refraction profile--of the deep ocean, and by the presence of its surface and bottom. A typical ray diagram for a deep source is shown in Fig. 1. The solid lines show the refracted paths from the source, while the dashed line shows one path reflected from the ocean boundaries. At low frequencies, some of these reflected rays are able to penetrate into the bottom and come back up again, either by refraction or reflection from sub-bottom layers. Many other ray paths exist. These are the surface-bottom multipaths, amounting to four in number for each bottom encounter, carrying nearly the same amount of energy, and having different travel times. They tend to fill up the shadow zones of Fig. 1 and greatly complicate the ray-path picture of deep-water sound transmission.

The net result of all of this is that any given point in the sea receives sound that has travelled via a multitude of propagation paths carrying different amounts of sound energy and having differing travel times. These differences associated with the source, target and the sea itself, cause signal distortion, fluctuation, de-correlation and frequency shifting and smearing of a received signal.

DISTORTION

After having travelled a distance in the sea, a short pulse becomes stretched out in time by multipath transmission. A classic instance is a so-called SOFAR signal having a slow build-up and a sudden cessation. Fig. 2 shows an explosive pulse, with its shock wave and bubble pulses, as recorded near the source. It has an overall duration of a few milliseconds. But, when received at a long range, it acquires a duration measured in seconds of time, as shown by the envelopes of Fig. 3, taken from (1). The reason for the extreme time-stretching lies in the differences in travel times (or mean sound velocity) along refracted paths like those of Fig. 1. The smooth envelope is due to the great number of multipaths overlapping in travel time when both source and receiver are on the channel axis. When source or receiver, or both, are off-axis, the envelope becomes irregular, as shown in Fig. 4, where the source was at a depth of 800 ft and the receiver was on-axis at 4000 ft. With this geometry the multipaths become resolved and each blob in the envelope represents a discrete, identifiable multipath arrival.

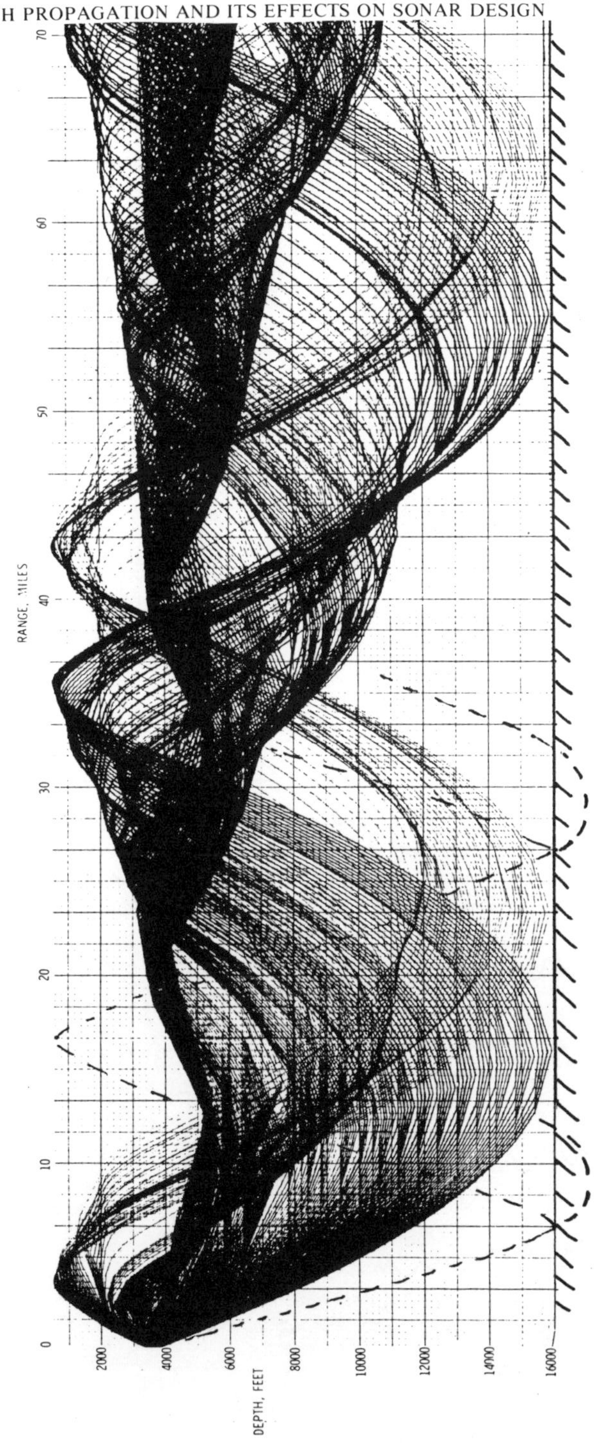

Fig 1. Multipaths in the Deep Ocean (Sofar) sound channel out to a range of 70 miles. Water depth is 16000 feet. Dashed line shows one bottom-surface reflected path; other such paths are not shown to avoid confusion.

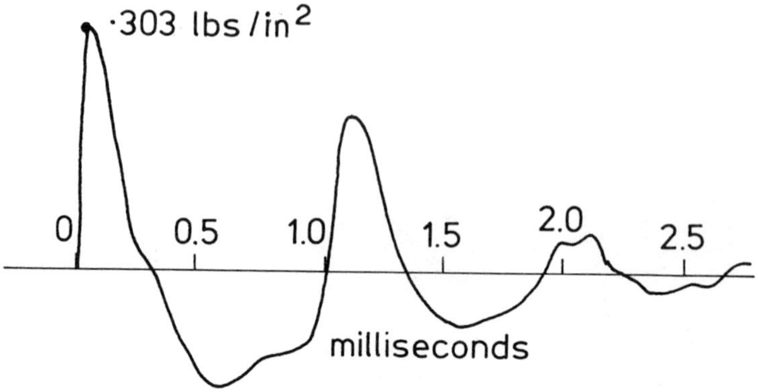

Fig. 2. Pressure signative of a lb charge detonated at 22,000 feet recorded overhead at a shallow depth.

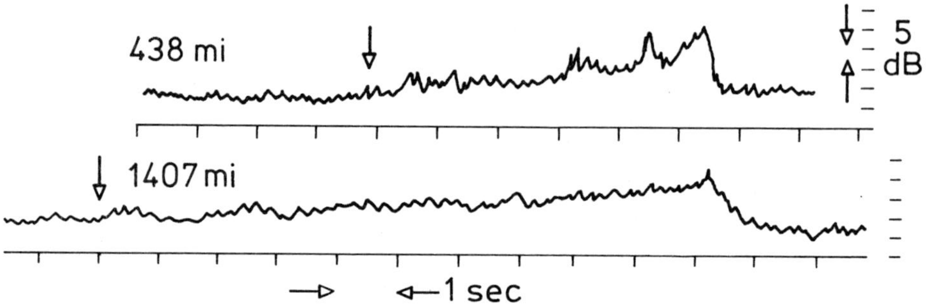

Fig. 3. Envelope of an explosive pulse after travelling through the Deep Ocean (Sofar) sound channel. The arrows show the approximate times of commencement of the received signal.

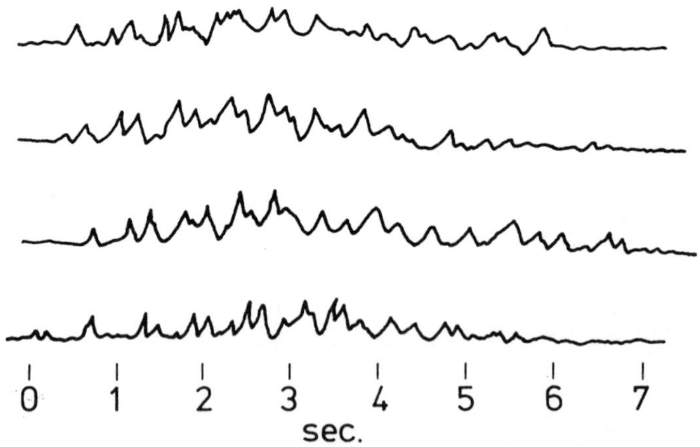

Fig. 4. Envelopes of explosive pulses off axis on a circle at 600 miles 10 degrees apart south of Bermuda.

The target itself causes distortion, even in the absence of propagation effects, because of its extension in range and the fact that scattering and reflection occur to some extent all along the target. Fig. 5 shows two echoes of a 15 ms pulse obtained 1 second apart from a moving submarine at a short range such that propagation effects were likely to be negligible. Time stretching takes place here as well. Moreover, there is little similarity between the shapes of the two echoes, as a result of the small changes in heading made by the sub's helmsman as he tried to steer a steady course.

Explosive echoes from a submarine also show great distortion, even at beam aspect, as illustrated by the echo shown in Fig. 6. Here the responsible culprits are likely to be the propagation multipaths. They cause the classic near-field signature of the explosive source (Fig. 2) to be unrecognizable in the echo.

Because of distortion, replica correlation of the received echo against a stored waveform is seldom an effective type of processing in sonar echo ranging. Pulse shapes other than a flat-topped sinusoid or a linear FM slide rarely result in gains that would make than worthwhile.

FLUCTUATION

The signal from a steady source of sound always is observed to fluctuate at a distant receiver. Fluctuation is the result of 1) multipath transmission and 2) motion of the source, the target, the sea surface, and the currents and turbulence of the body of the sea itself. In a completely frozen-over ocean, there would be no fluctuation. Fig. 7 shows the signal from a steady source as received by hydrophones 5 miles away at depths of 1000 and 8000 feet. The signal at 8000 feet is stronger and more steady than that at 1000 feet because of the presence of a direct path from the source at the deeper depth. The cause of the fluctuation is the moving sea surface, which by reflection and scattering, creates multipath interference and accordingly, a fluctuating signal. The faster fluctuations are caused by sea-surface motion, and the spectrum of the fluctuation duplicates the spectrum of the waves. This is illustrated by Fig. 8, where the solid curve is the measured spectrum of the envelope of a portion of the data of Fig. 9 in the period range 3-30 seconds and the dashed curve is the computed Neuman-Pierson wave spectrum for the wind speed existing at the time the data was recorded.

Slower and larger fluctuations occur in bottom-bounce transmission from a moving source, as illustrated in Fig. 9, where we see the fluctuations from a steady 275 Hz source towed

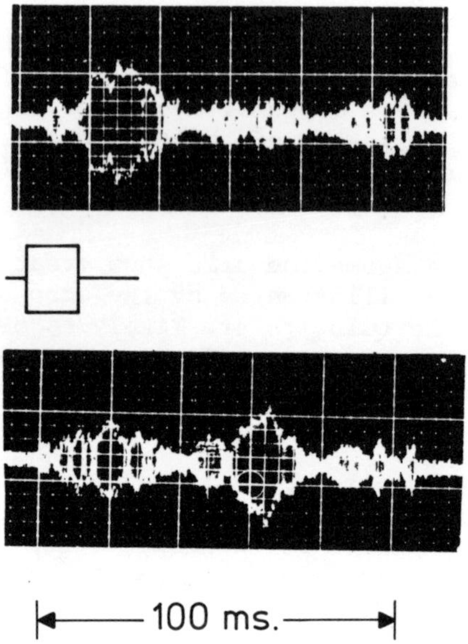

Fig 5. Two echoes photographed 1 second apart from a submarine at a close range. The rectangle between the photos shows the envelope of the pulse sent out by the source.

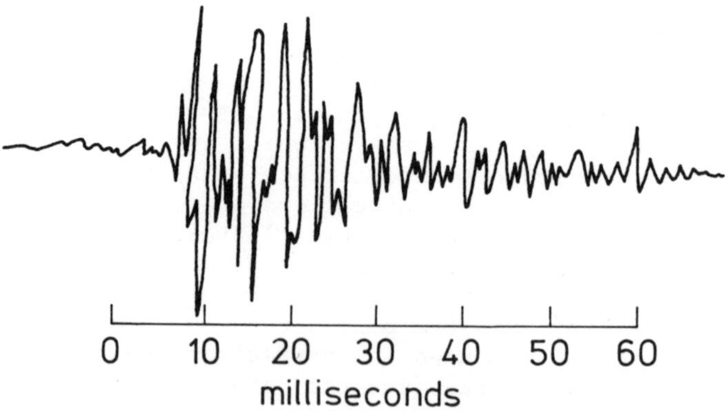

Fig 6. An explosive echo from a submarine near beam aspect.

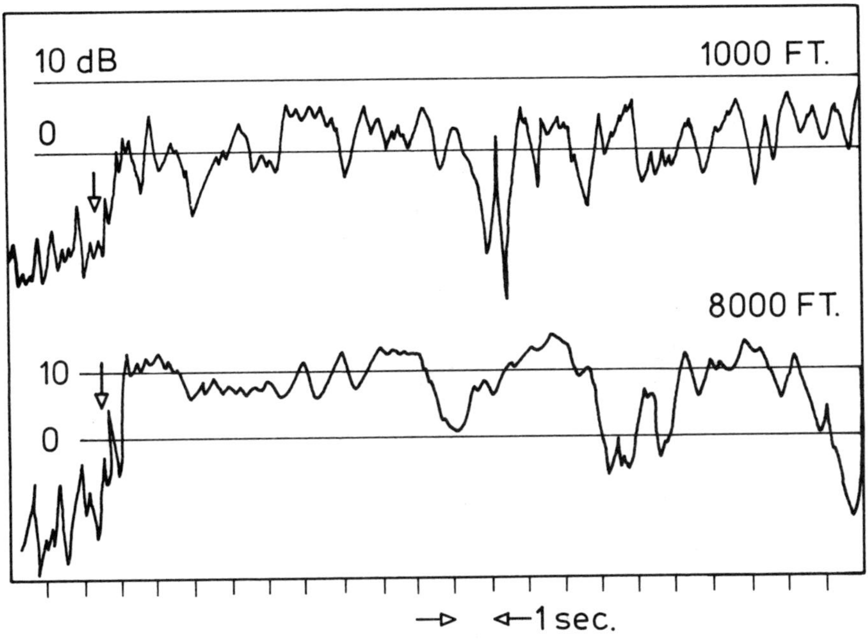

Fig 7. Fluctuation observed at two depths from a steady cw 1120 Hz. source at a distance of 5 miles.

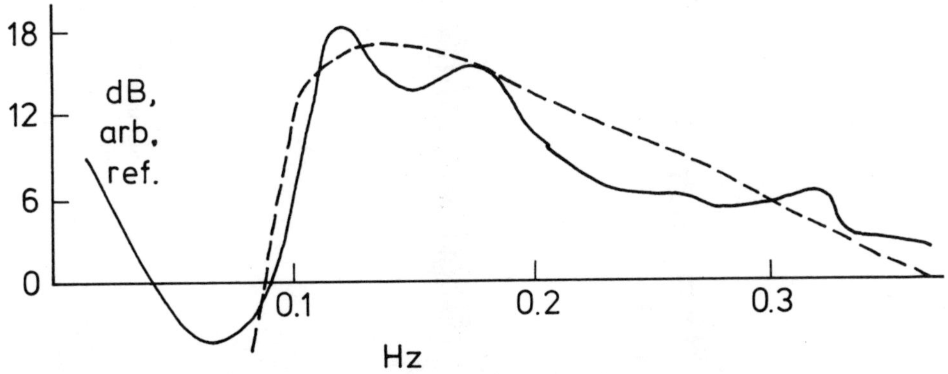

Fig 8. Modulation spectrum of the envelope from a 275 Hz cw source towed at 2.7 knots at a range of 23 miles. Dashed line is Neuman-Pierson wave spectrum for a wind speed of 20 knots. Analysis bandwidth 0.02 Hz.

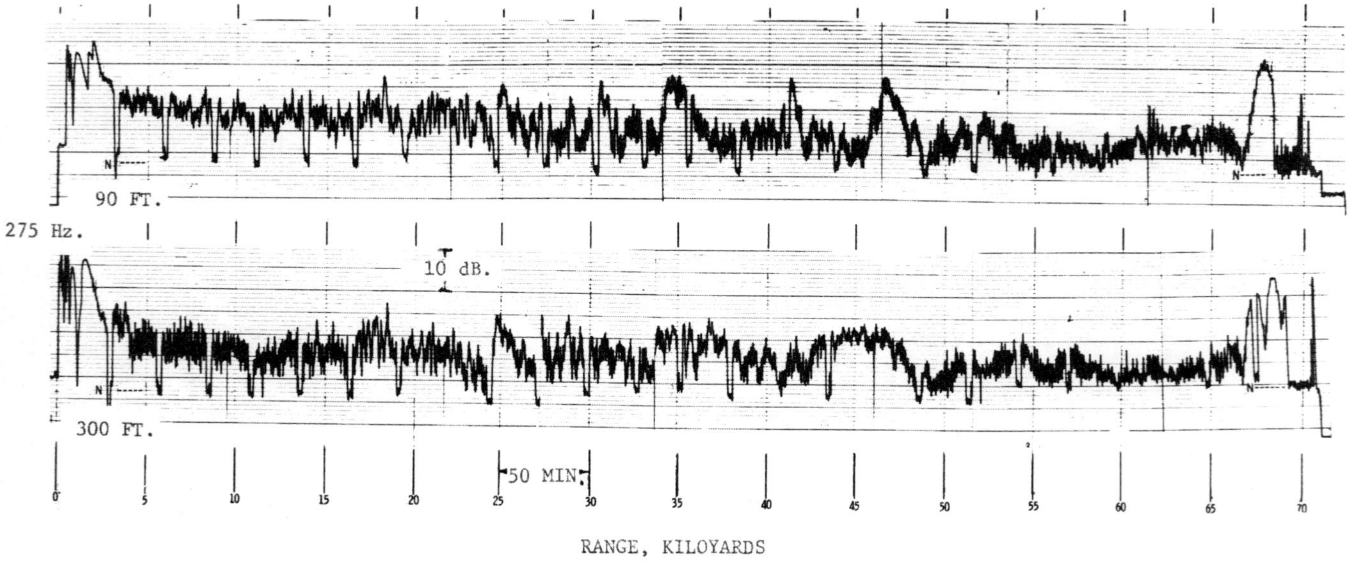

Fig 9. Signal from a 275 Hz cw source towed at 2.7 knots outward in range from hydrophones at 90 and 300 Hz as far as the first convergence zone.

outward in range to the first convergence zone and recorded at two depths (2). The broad fluctuation spectrum is particularly evident. There is little apparent similarity of the fluctuations at the two depths. Here the cause is interference between the surface-bottom multipaths resulting from the motion of the source.

Between a fixed source and receiver, the fluctuation of amplitude and phase resulting from internal waves and other oceanic motions has recently received considerable theoretical (3) and observational (4)(5) attention.

Still and all, whatever the cause of the fluctuations, it appears from a recent study (6) that amplitude fluctuations obey a modified-Rayleigh (or "Rician") distribution, and have a time-scale that depends on the rate of interference of the multipaths (7).

Fluctuations cause targets to be detected during periods of signal surges and to be lost during periods of signal fades. They prevent the use of long integration times for signal detection.

DE-CORRELATION

Multipaths cause the signal received by separated hydrophones to be different--that is, to be de-correlated. Fig. 10 shows the clipped correlation coefficient of the sound from a ship in the octave band 177-354 Hz between two hydrophones 500 feet apart vertically. The ship passed overhead out to a range of 5 miles. The rapid de-correlation with horizontal range is the result of bottom-bounce multipath interference with the direct-path sound from the source. As the range increases, the direct sound becomes weaker and the interference field stronger, causing a greater dissimilarity in waveform at the two receivers (8).

Another example is shown in Fig. 11. Here the level at one hydrophone, and the clipped correlation coefficient between two hydrophones 2.2 ft apart vertically, is shown as a 1120 Hz cw source entered a convergence zone (cz) at 29.5 miles. Within the cz both the level and coherence of the signal is higher than at shorter ranges, due to the emergence of a single, strong, convergent path. Thus, in a cz a vertical array performs better than outside the cz, not only because of a higher signal level, but because of a higher array gain as well.

Signal de-correlation places a limit on the size of arrays used to increase the signal-to-noise--or array gain--for detection.

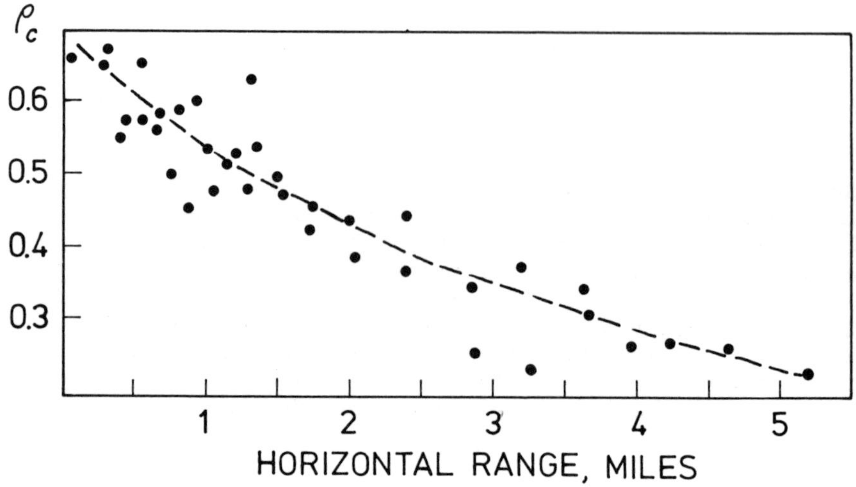

Fig 10. Clipped correlation coefficient of the signal from a shallow broadband cw source in the octave 177-354 Hz between two hydrophones at 4000 and 4500 feet. Water depth 7000 feet.

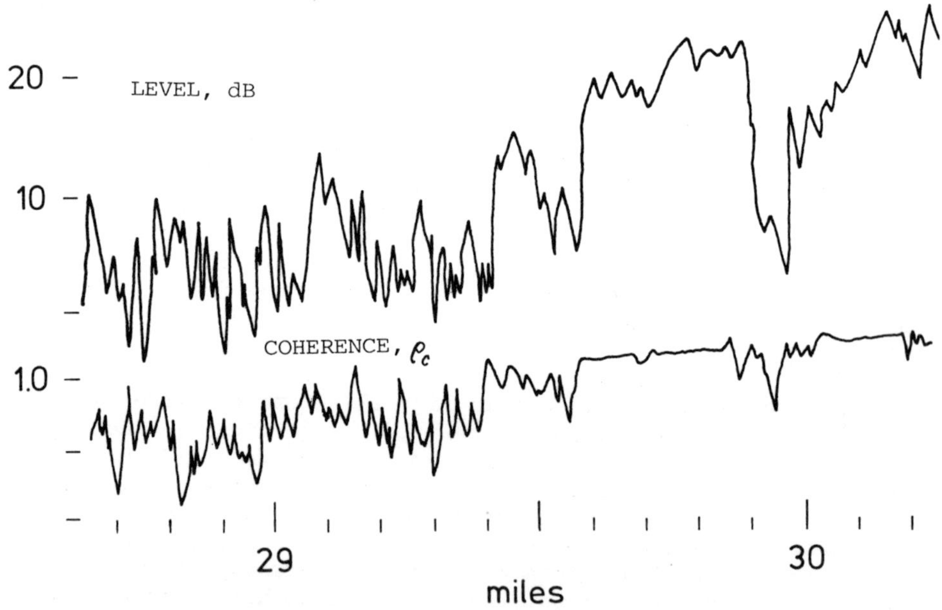

Fig 11. Level and coherence between hydrophones 2.2 ft. apart of the signal from a towed 1120 Hz cw source entering a convergence zone.

Stated another way, an array of n elements has a gain less than the value of 10 log n by the fractional amount of uncorrelated sound in the total received signal.

FREQUENCY EFFECTS

A signal-frequency source acquires side-bands, and is otherwise smeared out in frequency, by multipath transmission. Fig. 12 shows the spectrum of signals received at a distance of 19 miles from a fixed source by a fixed receiver. The sidebands A and A' are caused by the rough, moving sea surface, which modulates the received signal and superposes its own spectrum on that of the carrier frequency (9). In a multipath environment, a moving source will also smear its single-frequency output by producing differing doppler shifts among the different existing multipaths.

In back-scattering, as well as in forward scattering, the sea surface broadens and shifts the spectrum incident upon it. Fig. 13 shows reverberation (back-scattered) spectra of 110 ms pulses at 60 kHz received back from the sea surface at a grazing angle of 30° (10). The spectrum of the source pulse (B) is broadened and doppler-shifted up (C) or down (A) in frequency depending on whether the surface is insonified up-wind or down-wind.

Although they are small, these frequency effects become important in underwater communication and in passive detection, by requiring wider filter bands for signal reception.

SUMMARY

Multipath transmission has the great beneficial effect in producing stronger sound fields than would exist in their absence. Yet these stronger signals are not always utilizable, in that they are smeared in time and in frequency, are unsteady, and tend to become rapidly de-correlated with hydrophone separation in arrays. These effects are not predictable, because of the complexity of the existing multipaths, and are such as to make elaborate processing techniques, such as replica correlation and matched filter designs, useless for signal enhancement. Similarly, arrays become limited in size and their performance degraded by multipath transmissions from the source. As a result, many of the elegant techniques of radar and radio communication are ineffective in long-range sonar because of the multipath environment in which the sonar must operate.

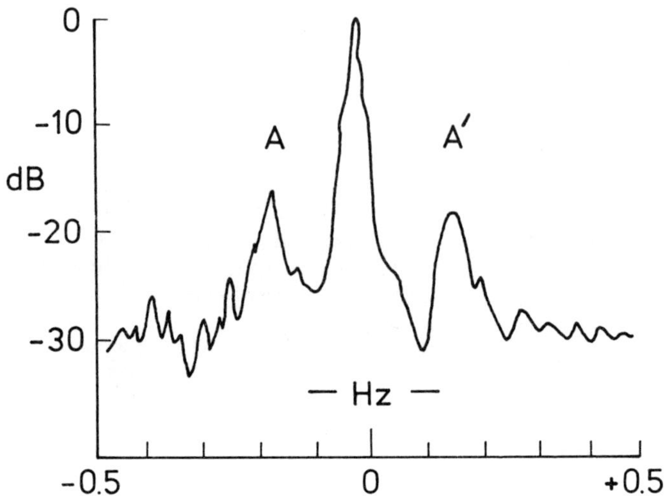

Fig 12. Frequency spread of forward-scattered 100 ms. 1702 Hz pulses over a distance of 19 miles in shallow water; analysis band 0.01 Hz. USN Underwater Systems Center data.

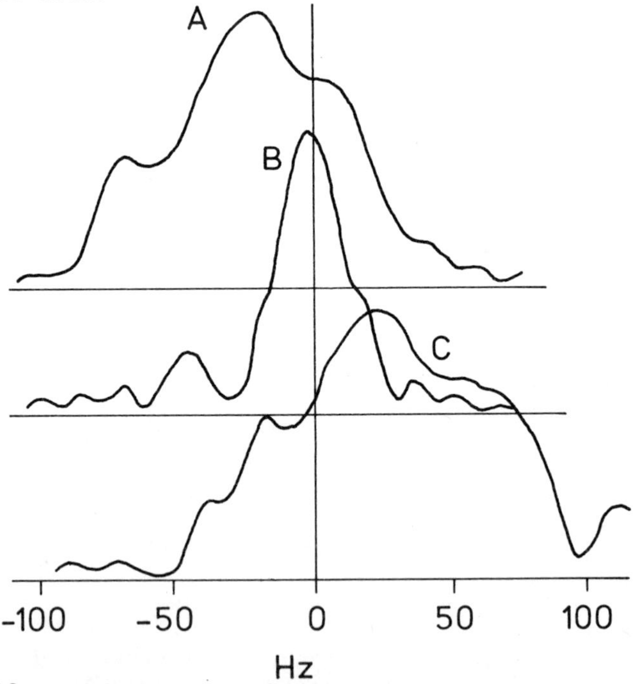

Fig 13. Backscattered (reverberation) spectra of 110 ms pulses at 60 kHz from the sea surface A: downwave; B: source spectrum; C: upwave. 30° grazing angle. Igarashi and Stern data.

REFERENCES

(1) R. J. Urick, Principles of Underwater Sound, 2nd ed., McGraw-Hill Book Co., New York, 1975, Fig. 6.10, traces 6 and 16.

(2) R. J. Urick, Amplitude fluctuations of the sound from a moving source in the deep sea, Naval Ordnance Laboratory Report NOLTR 74-156, 1974.

(3) W. H. Munk and F. Zachariasen, Sound propagation through a fluctuating stratified ocean, J. Acoust. Soc. Am. 59, 818, 1976.

(4) J. G. Clark and M. Kronengold, Long period fluctuations in deep and shallow water, J. Acoust. Soc. Am. 56, 1071, 1974.

(5) R. M. Kennedy, Phase and amplitude fluctuations in propagating through a layered ocean, J. Acoust. Soc. Am. 46, 737, 1969.

(6) R. J. Urick, A statistical model for the fluctuation of sound transmission in the sea, Naval Surface Weapons Center Report NSWC/WOL/TR 75-18, 1975.

(7) R. J. Urick, The time scale of the fluctuation of a bottom-bounce narrow-band signal from a moving source in the sea, Naval Surface Weapons Center Report NSWC/WOL/TR 75-83, 1975.

(8) R. J. Urick, The vertical coherence of sound from a near-surface source in the sea and the effect on the gain of an additive vertical array, J. Acoust. Soc. Am. 54, 115, 1973.

(9) W. I. Roderick and B. F. Cron, Frequency spectra of forward scattered sound from the ocean surface, J. Acoust. Soc. Am. 48, 759, 1970.

(10) Y. Igarashi and R. Stern, Observations of wind-generated doppler shifts in surface reverberation, J. Acoust. Soc. Am. 49, 802, 1971.

DISCUSSION

Comment : K.C. BEAUCHAMP

I am not too familiar with sonar processing, but have experience in the seismic processing field. To what extent are hydrophone arrays used with added delays to confine the detection to a particular direction in order to improve the signal/noise ratio and to simplify the problem? An example in the seismic field is the large LASA array set up in Montana, U.S.A. Is there a similar array for sonar purposes?

Reply : R.J. URICK

The nearest type of device existing in sonar is the flexible, long-line array, similar to the seismic streamer but using time delays at the outputs of the sensor elements to form directional beams. These are formed simultaneously by using the DIMUS technique involving clipping and shift-register time delays.

Comment : P.E. CABLE

To what extent can the underwater medium be considered an uncorrelated scatter medium? Bello has examined this question for a point-to-point tropospheric communication link, and I wonder if the same analysis is also applicable to the direct-path, underwater, acoustic channel. The underwater, surface-scatter channel probably cannot be considered to be an uncorrelated scatter channel in most circumstances.

Reply : R.J. URICK

I do not know the work to which you refer. However, we have found experimentally that the sonar surface duct is in large part a coherent duct. The coherence of transmission may be expected to be dependent on frequency of sea state. Not much is known about the coherence characteristics of signals and of noise in the surface duct.

Comment : L. GRIFFITHS

Have there been any multiple-sensor experiments conducted in which the multipath components were spatially resolved? It would appear that this kind of measurement would be extremely useful for determining whether or not individual multipath components have sufficient spatial and temporal stability to allow combining of these components.

Reply : R.J. URICK

Not that I know of, but I agree that such measurements would be useful. There has been a lot of work with explosive sounds and these have shown that a single multipath has considerable temporal and spatial stability. However, in echo ranging it is difficult to use short enough pulses to resolve the multipaths, and (in passive sonar) to achieve a geometry where only a single multipath exists.

Comment : N.L. OWSLEY

What chance of success do you anticipate for coherent multipath recombination and for what conditions would you make this prediction?

Reply : R.J. URICK

It is a matter of stability in each of the multipaths involved. If the recombination can be implemented within the decorrelation time constraints, it is certainly possible to do coherent recombination. However, a difference in frequency for each arrival caused by a source-receiver relative motion, would be a complicating factor in this process.

Comment : D. SAXTON

Decorrelation of signals received due to multipaths can surely be an advantage in pulse-compressed transmissions, since this would allow separation of multipaths even more easily, i.e. (i) good correlation of direct path; (ii) poor correlation of multipaths, so that correlation peaks separated in time due to pulse compression, amplitudes low due to decorrelation.

Reply : R.J. URICK

In sonar the pulse lengths required to sort out the multipaths are, at long ranges, very short. Short pulses are difficult to process and require broad receiving bandwidths. Instead, in sonar, long pulses are used in long-range work in order to get a better gain over noise and reverberation.

Comment : R.G. TAYLOR

Is there any way in which fades in signal strength due to multipath and fluctuations can be smoothed by a wideband frequency transmission?

Reply : R.J. URICK

Frequency diversity is always a way to smooth out fluctuations. However, fluctuations are desirable rather than undesirable for long range detection, since targets are detected first during signal surges. Also, frequency diversity over broadbands is difficult and expensive in sonar, since multiple sound sources are required.

Comment : C. VAN SCHOONEVELD

In the lecture, the notion of diversity emerged in relation to signal fluctuations, and I should like to make the following remarks: When signal fluctuations occur, the SMR has a probability distribution with a certain width, σ, around an average value, SMR_{AV}. In principle, diversity systems aim at a reduction of σ, while keeping SMR_{AV} approximately constant. This reduction is desirable when SMR_{AV} is large, since it decreases the probability of the occasional missed detection when SMR is accidentally small. [Example: bit detection in data communication systems.] On the other hand, a reduction of σ is undesirable when SMR_{AV} is low. In that case, the reduction of fluctuations suppresses the occasional detection when SMR is accidentally high. [Example: active radar or sonar systems at long range.] To summarize, the desirability of fluctuation suppression depends on the average SMR. I feel that this is the reason for the use of diversity systems in data communication and for its non-use in radar or sonar. (To avoid misunderstanding, it should be added that some diversity systems do not only obtain a reduction of σ but also an increase of SMR_{AV}. Example: space diversity, where a 2nd receiving antenna is added to a 1st one, thus increasing the collecting area. In those cases, the conclusion given must clearly be modified.)

Reply : R.J. URICK

The above remarks concerning the suppression or enhancement of fluctuations are indeed cogent and reasonable. I would add to them the thought that, in addition to frequency diversity and spatial diversity as tools for reducing fluctuations, there is a simpler and better known one, namely, temporal diversity or smoothing, as in a post-detector averager. This, too, should depend on the mean SNR: long averaging times at high SNR's to avoid lost target detections, and short averaging times at low SNR's to take advantage of signal surges. But what is "long" and what is "short" depends on the coherence time, or the lag in the auto-correlation function of SNR required for it to fall to near zero, is not at all well-known under the practical conditions of sonar.

THE INFLUENCE OF TIME- AND SPACE-VARIANT RANDOM FILTERS ON SIGNAL PROCESSING

A. Wasiljeff

Department of Physics and Electrotechnics,
University of Bremen, Germany

ABSTRACT. To analyze signals that have been propagating in random media with rough boundaries the medium between transmitter and receiver may be considered as a time and space variant random filter. A historical outline of the theory of random fields is given with applications to tropospheric scatter links and to underwater acoustics. To discuss the influence of the randomness of the transmitting channel on signal processing procedures, the concept of a multidimensional scattering function is presented. Its effect on the combined ambiguity-power directivity function of signals and on arrayprocessing gain are considered.

1. INTRODUCTION

Underwater systems are not limited to applications in antisubmarine-warfare like active or passive sonar, IFF or target classification. New systems are developed for tracing schools of fish, measuring depth and structure of the sea floor and for discovering minerals like oil below the sea bottom. Each one of these systems transmits or receives information of one kind or another. Therefore all of them can be considered as communication systems. In many situations the description of the system depends on parameters which exhibit random variations in time and space. The signals transmitted through the acoustic channel suffer time delay spread, frequen-

G. Tacconi (ed.), Aspects of Signal Processing, Part 1, 19-32. *All Rights Reserved.*

cy spread and angular spread. This is mainly due to instabilities of the propagation paths of sound energy, the motion of the sea surface and scattering by rough boundaries and inhomogeneities of the volume. Another important field of application of random field theory is the transmission of telephone and television signals well beyond the optical horizon at UHF frequencies which plays an important role in communication techniques. Statistical wave theory describes wave fields from a statistical point of view in describing their average values. This averaging procedure generally chooses the second order moments. They are termed spatial or temporal coherence or correlation function. The theory was first developed in the field of optics but applies equally well to radio waves and to acoustic radiation. In optics it plays an important role in image processing and in the studies of the atmospheric effects on resolution. Similar problems arise in radio astronomy, in applications of RADAR and SONAR systems and in the development of LASERS and holography.

2. HISTORICAL OUTLINE

2.1 Theory of random wave fields

Verdet demonstrated in 1869 that the light from two pinholes in a screen will interfere, if the separation is less than 1/2o mm. This received little interest until 19o7 when von Laue described in a paper [1] the statistical aspects of diffraction. Van Cittert discusses in 1936 the complex correlation in the illuminated plane in terms of the intensity distribution across the source plane. A further development of the theory of partial coherence was given by Zernike in 1938 [1]. In that paper a theory was formulated that relates the light intensity on a plane to an illuminating incoherent source. The lateral coherence function of the far field of an incoherent source is the Fourier transform of the intensity distribution. The Van Cittert/Zernike theorem is the optical analogy to the Wiener-Khinchine theorem of communication theory. In formulating the theorem in terms of modern antenna engineering we may say: The angular power spectrum and the correlation function of the aperture distribution of an array form a Fourier pair. These early authors on random wave fields paid more attention to the statistical behaviour of the source than to the randomness of the propagation medium. Correlation theory of random fields has been applied to acoustic propagation phenomena in 1947 by Pekeris [2]. He calculated the scattering of a plane sound wave when passing through a medium in which the sound velocity fluctuates slightly and irregularly around its mean value. To calculate the scattered field he used the Born approximation to the wave equation. This is an approximation that neglects multiple scattering. He mentioned that the corresponding electromagnetic scattering problem in an optically inhomogeneous medium can be treated along similar lines by referring to a study of scattering problems made by himself in 1942. In 1953 Mintzer [3] published a paper on wave propagation in a randomly inhomogeneous medium where he studied the propagation of sound pulses from a point source in a medium with randomly varying index of refraction. Mintzer compared experiment and theory. He suggested however that motion of source and receiver will distort the correlation and the results found from the data may be more representative of the ship's motion than of the refractive index variations. This remark applies to almost all correlation measurements performed during sea trials. As a consequence Mintzer suggested to obtain fluctuation data from a stationary source and receiver. This remark is as topical today as it was more than twenty years ago. In 195o the theory of Pekeris was applied by Booker and Gordon [4] to radio scattering in the troposphere. They explained microwave propagation beyond the optical horizon as a scattering from inhomogeneities of the refractive index. Further development of the theory is due to Bremmer [5] and Wheelon [6]. According to this theory the correlation between the field strength at different places and or at different times can be derived from a corresponding correlation function for the random fluctuations of the refractive index. An extensive review of literature on tropospheric scatter propagation has been given by Giovanni d'Auria in 197o[7].

2.2 Random communication channels

The propagation physicist studies how the fields are affected by the propagation medium and its boundaries at various points in space. A different approach is taken by the communication engineer. He considers the medium as a black box with random characteristics and studies the input-output relations of the black box. This black box is called a communication channel or filter. Bello [8] and Kennedy [9] tried to combine communication theory and the propagation physics described in the previous section. They developed a channel model for tropospheric scatter links which is based on the scattering function of the random communication channel. The medium is considered as a large collection of point scatterers. Each of this differential elements or blobs is characterized in part by a time delay, a Doppler shift and an average scattering cross section. The scattering function describes the distribution of the reflection cross section in time delay and in Doppler shift. It is assumed that the impulse response of the filter is a sample function of a Gaussian random process. In that case the specification of the channel reduces to the specification of the mean and correlation function of the filter's random impulse response. The propagation is established by single scattering from a large number of independent elements. The same channel model can be developed from the concepts of linear system theory using the equivalent assumption of wide sense stationary uncorrelated scattering channels. It can be shown that uncorrelated scattering is equivalent to wide sense stationarity in frequency domain of the correlation of the transfer function. As far as input-output relations are concerned in the transmission of narrowband signals through random channels like troposcatter links, scattering at different path delays may be regarded as uncorrelated. This concept has been successfully used to calculate intermodulation distortion in analogue frequency modulation systems with frequency division multiplexing [1o] and to calculate the error rate in digital transmission due to multipath spread of the troposcatter channel [11] . The WSSUS channel approach has been applied by Ellinthorpe and Nuttal [12] to model an undersea channel. Since the time varying features allow a multipath structure to change with time in a random way, the generality of the approach is great enough to include one-way and two-way propagation, target back scattering, reverberation and noise. Søstrand applied the concept of scattering function and related Fourier transforms to characterize the underwater channel and to measure coherence and stability of underwater acoustic transmission [13] . Further discussions and measurements of the scattering function are given by Thiele for the shallow water case [14] .

Thus the main historical stream of research in the field of random communication channels ran from acoustics to troposcatter links and came back to the starting point of underwater sound transmission. One should mention, of course, that apart from communication engineering a tremendous amount of research has been carried on concerning the physics of sound propagation in random media since the early papers of Pekeris and Mintzer. A review of the most important developments can be found in the text books of Chernov [15] Tatarski [16] URICK [17] and in the reports of Fortuin [18] .

3. SIGNAL PROCESSING

3.1 Multidimensional Scattering function

To describe the time and space varying channel one can use the mathematical tool of filter theory as has been done by Laval and Hovem [19, 2o] . For the sake of simplicity stationarity in frequency, time and space is assumed. That means that the correlation function of the transfer function depends only on the difference of the different variables and not on their absolute value. We combine frequency f_i, time t_i and the space coordinates x_i and y_i in a plane that is perpendicular to the main direction of propagation to a point vector in the four-dimensional space

$$\vec{p_i} = \begin{pmatrix} f_i \\ t_i \\ x_i \\ y_i \end{pmatrix} \qquad (3.1)$$

We define a vector $\vec{p}$ as a difference vector between two points

$$\vec{p} = \vec{p_i} - \vec{p_k} = \begin{pmatrix} \Delta f \\ \Delta t \\ \Delta x \\ \Delta y \end{pmatrix} \qquad (3.2)$$

The assumption of stationarity in the four variables yields a correlation function of the time and space dependent transfer function, that depends only on the difference vector $\vec{p}$

$$R_H(\vec{p}) = \langle H(\vec{p}_i)H(\vec{p}_k)\rangle = \langle H(f_i,t_i,x_i,y_i)H(f_k,t_k,x_k,y_k)\rangle \quad (3.$$

Let us define a new vector $\vec{q}$ which is formed of the delay spread τ , the doppler spread Φ and the angular spread u and v

$$\vec{q} = \begin{pmatrix} \tau \\ \Phi \\ u \\ v \end{pmatrix} \tag{3.4}$$

The variables u and v are defined as the wave number components in the angular space

$$u = \frac{1}{\lambda}\alpha ; \quad v = \frac{1}{\lambda}\beta \tag{3.5}$$

where λ denotes the wave length and α and β are the direction cosines of the wavenumber vector k with respect to the x and y axis. k is really formed by the three components u, v, w with $w = \frac{1}{\lambda}\gamma$ and γ the direction cosine with respect to the z axis. The third component however is not independent, since it can be calculated from the relation $\alpha^2 + \beta^2 + \gamma^2 = 1$
Therefore it is sufficient to consider only two variables at a time. The antenna engineer is more familiar with the angles, Θ and $\emptyset$ of a spherical coordinate system. The direction cosines are given by the relation

$$\begin{aligned} \alpha &= \sin\Theta \cos\emptyset \\ \beta &= \sin\Theta \sin\emptyset \\ \gamma &= \cos\Theta \end{aligned} \tag{3.6}$$

Let us formally define a four dimensional Fourier integral of the correlation-function $R_H(\vec{p})$ with respect to the variables $\vec{q}$

$$L(\vec{q}) = \int_{-\infty}^{\infty} R_H(\vec{p})\, e^{j2\pi\vec{p}\cdot\vec{q}} dp \tag{3.7}$$

Here $\vec{p}\cdot\vec{q}$ signifies the scalar product between the vectors p and q whereas dp is an abbreviation for the four-dimensional volume element

$$dp = d(\Delta f)\ d(\Delta t)\ d(\Delta x)\ d(\Delta y)$$

The integration limits may be extended to infinity if the correlation function decreases rapidly enough to guarantee the existence of the integral. The inverse Fourier transform yields

$$R_H(\vec{p}) = \int_{-\infty}^{\infty} L(\vec{q})\, e^{-j2\pi\vec{p}\vec{q}}\, dq \tag{3.8}$$

To simplify the discussion of the physical meaning of the function $L(\vec{q})$ we assume that the function $R_H(\vec{p})$

$$R_H(p) = R_H(\Delta f, \Delta t, \Delta x, \Delta y) \tag{3.9}$$

can be factorized into a product

$$R_H(p) = R_{H_1}(\Delta f, \Delta t) \cdot R_{H_2}(\Delta x, \Delta y) \tag{3.10}$$

The function

$$R_{H_2}(\Delta x, \Delta y) = \langle H(x_i, y_i)\ H(x_k, y_k)\rangle \tag{3.11}$$

is called the spatial correlation function of the space dependent transfer function of the medium. It should be determined perpendicular to the mean propagation direction of the field. Factorization in the p domain yields a product of the Fourier integral in the q domain.

$$L(\vec{q}) = L_1 \ (\tau, \phi) \ L_2(u,\ v) \tag{3.12}$$

The function $L_1(\tau, \emptyset)$ is the scattering function of Bello and Kennedy. [8,9]. The function $L_2(u,v)$ is the angular scattering function introduced to underwater acoustics by Hovem and Laval [19, 2o]. The Fourier realtionship between spatial correlation function and angular scattering function

$$R_{H_2}(\Delta x, \Delta y) \circ\!\!-\!\!\bullet\ L_2(u,\ v) \tag{3.13}$$

is the equivalent to the van Cittert-Zernike theorem in the theory of partial coherence in optics as mentioned in the beginning. A more detailed discussion of the angular spectrum representation of waves and its relation to coherence theory can be found in a paper by R. Clarke [21]. It should be emphasized however, that the correlation function R_H was factorized only for convenience and without justifying it physically. The same thing applies to the assumption of stationarity in frequency and space. Only for special cases like narrow band signals and not too large arrays such simplifications may be meaningful.

3.2 Ambiguity functions

The ambiguity function of a signal has been defined by Woodward [22] as the crosscorrelation function of the transmitted signal s(t) with the received echo, that has suffered a doppler spread $\emptyset$ and a time delay

$$\chi(\emptyset, \tau) = \int_{-\infty}^{\infty} s^*(t)\, s(t+\tau)\, e^{j2\pi\phi t}\, dt = \int_{-\infty}^{\infty} S^*(f)\, S(f+\emptyset)\, e^{j2\pi f t}\, df \tag{3.14}$$

It describes the combined resolution properties of the signal in time and frequency. It is a correlation function for time and frequency shift.

When χ is equal or close to unity the resolution in time and frequency is ambiguous, a target with range and velocity (τ_0, ϕ_0) cannot be resolved from a target at $(\tau_0 + \tau, \phi_0 + \phi)$. An ideal ambiguity function which allows perfect resolution in range and velocity would be a thumbtack-ambiguity function. Such a function is smeared, due to the random time-varying communication channel in the delay-time-doppler (=Range-velocity)-space. The output ambiguity function of the random channel can be described as the two dimensional convolution of the input-ambiguity function and the scattering function of the channel [13].

$$|\chi_{out}(\emptyset, \tau)|^2 = \iint |\chi_{in}(\emptyset - \hat{\emptyset}, \tau - \hat{\tau})|^2 \, L(\hat{\emptyset}, \hat{\tau}) \, d\hat{\emptyset}, d\hat{\tau} \qquad (3.15)$$

This relation can be generalized to the multidimensional case, using the scattering function of the last section. Here the input ambiguity function does not only describe the properties of the transmitted signal but also gives information about the power directivity function of the array. Such a generalized ambiguity function has been defined e.g. by Vakman [23] to characterize the angular resolution of an antenna. The word ambiguity may be a bit misleading and therefore it may be better to use the term combined Ambiguity-Powerdirectivity function or the acronym CAP. The output CAP of a random space and time variant channel is related to the input CAP by the multidimensional convolution product

$$|\chi_{out}(\vec{q})|^2 = \int |\chi_{in}(\vec{q} - \vec{\hat{q}})|^2 \, L(\hat{q}) \, d\hat{q} \qquad (3.16)$$

The angular uncertainty of the medium , which is the effective beam width of the scattering function can be measured in the corresponding Fourier domain by means of the spatial correlation function. The effective width of the spatial correlation function, which is also called coherence length, determines its reciprocal value as the angular uncertainty of the medium. Measurements of the coherence length have been reported in the literature [24] but unfortunately the measurements have been performed only along either the horizontal or the vertical axes. Since the factorization of the 4 dimensional scattering function is only an idealized assumption, one should measure the correlation in the whole plane perpendicular to the propagation direction together with frequency and time correlation to determine the four dimensional scattering function of delay-doppler and angular spread as its Fourier transform.

4. SPATIAL FILTERING

4.1 Structure classification

In Radar and Sonar applications one is interested in resolving the structure of a target in range, doppler and angular space. It is insufficient to design signals which are optimal only with respect to range and doppler resolution. One should try to design a four dimensional signal-array-processing system which is matched to the scattering function of the target in the four variables τ, ϕ, u and v. Since the medium smears this four dimensional signal, it is important to determine the function of the medium as well in order to optimize the system design. If we restrict ourselves to the second order moment's approach to random media we can formulate the 4 dimensional CAP function which is received as

$$\chi_{\text{receiver}} = \chi_{\text{transmitter}} * \chi_{\text{medium}} * \chi_{\text{target}} * \chi_{\text{medi}} \quad (4.1)$$

Since we cannot know the scattering function of the medium completely we have to develop some first order approximation models to get some feeling for the limitations of the medium in system design.

4.2 Array gain

Array processing systems are designed to extract one particular signal waveform, which is incident on a receiving array of sensors in the presence of many other signals. These other signals are considered to be interfering noise. The combination of array and processing of the outputs of the individual array elements can be described as a filter in space and frequency. Optimization is usually achieved through a mean square error criterium, that minimizes the distorting noise and therefore maximizes the signal to noise ratio.

For this procedure the space time correlation matrix of the outputs of all sensors has to be known [25] . The randomness of the space and time variant communication channel gives limitations on the performance of array processing. Let us consider as a simple example a linear additive array of K elements of equal sensitivity. According to Urick [17] the array signal to noise power is given as

$$\frac{\overline{S^2}}{\overline{N^2}} = \frac{\overline{s^2}}{\overline{n^2}} \frac{\sum_{n=1}^{k} \sum_{m=1}^{k} (\rho_s)_{nm}}{\sum_{n=1}^{k} \sum_{m=1}^{k} (\rho_N)_{nm}} \tag{4.2}$$

Here $(\rho_s)_{nm}$ are the signal crosscorrelation coefficients between the n-th and m-th elements of the array and $(\rho_n)_{nm}$ are the noise crosscorrelation coefficients. The average signal power across the array is $\overline{S^2}$ and the average noise power is given as $\overline{N^2}$. $\overline{s^2}$ is the signal power and $\overline{n^2}$ the noise power of a single element. The gain of the array may be defined following Urick as the ratio in decibel units of the signal to noise ratio of the array to the signal to noise-ratio of a single sensor element:

$$AG = 10 \log \frac{\overline{S^2}/\overline{N^2}}{\overline{s^2}/\overline{n^2}} = 10 \log \frac{\sum_{n=1}^{k} \sum_{m=1}^{k} (\rho_s)_{nm}}{\sum_{n=1}^{k} \sum_{m=1}^{k} (\rho_N)_{nm}} \tag{4.3}$$

If we know the correlation coefficients ρ_{nm} we can use these relations to calculate the gain.
Let us assume for instance a correlation function ρ_{nk} that contains a constant term that is independent of distance (n-k). This term we call the coherent part of the spatial correlation function ρ_c.

Thus we get $\rho_{nk} = \rho_c + \tilde{\rho}_{nk}$

where ρ_c denotes the coherent and $\tilde{\rho}_{nk}$ the incoherent part.
This yields the gain for the array:

$$AG = 10 \log \left(1 + \frac{2}{k} \sum_{n=1}^{k-1} n(\rho_c + \tilde{\rho}_{nk}\right) = \tag{4.6}$$
$$= 10 \log (1 + (k-1)\rho_c + R)$$

The second term R which is given as the sum of the incoherent contribution

$$R = \frac{2}{k} \sum_{n=1}^{k-1} n \tilde{\rho}_{nk}$$

will in general converge to a constant with increasing k whereas the first term of the formula will continue to increase with k. We have a gain reduction due to the randomness of the medium; in that case we can increase the array gain as much as we want by increasing the length of the array. An important case, that yields a correlation function, which can be split into a coherent and an incoherent part is the spatial correlation function for a plane wave that has suffered one surface reflection from the rough sea [26] . Unfortunately we are very often confronted with the situation (e.g. in Shallow water acoustics) that the coherent part of the correlation function of the medium is so small that it cannot be determined. Then we have to determine the constant term R and we shall get a limiting number for the length of an array to be used for reasonable array processing. Such numbers have been obtained for special cases and reported in the literature [27].

REFERENCES

1. Beran, M., Parrent Jr., G.B. Theory of Partial Coherence, Englewood Cliffs, N.J.: Prentice Hall (1964)
2. Pekeris, C.L., Phys. Rev., 71, 268 (1947)
3. Mintzer, D., J. Acoust. Soc. Am., 25, 186 (1954)
4. Booker, H.G., Gordon, W.E., Proc. IRE, 38, 4o1 (195o)
5. Bremmer, H., J. Res. NBS, 68D, 967 (1964)
6. Wheelon, A.D., J. Res. NBS, 63 D, 2o5 (1959)
7. d'Auria, G., AGARD conf. proc., 7o, paper 25, (19
8. Bello, P.A., IEEE Trans CS, 16, 36o (1963)

9. Kennedy, R.S., Fading Dispersive Communication Channels, New York, Wiley-Interscience (1969)
1o. Wasiljeff, A., AGARD conf. proc. 37, paper 26 (1968) Sandefjord
11. Bello, P.A., Ehrman, L., IEEE Trans. Comm. Tech. 17, 183 (1969)
12. Ellinthorpe, A.W., Nuttal, A.H., IEEE 1st Ann. Comm. Conv. Record, 585, (1965)
13. Søstrand, K.A., Signal Processing ASI-Proc. II paper 25, 26, Enschede (1968)
14. Thiele, R.,SACLANTCEN Conf. Proc., 14, 173 (1974)
15. Chernov, L.A., Wave Propagation in a Turbulent Medium, New York, McGraw Hill, (196o)
16. Tatarskij,V.I., Rasprostranenie Voln v turbulentnoi atmosfere, Izd. Nauka, Moskva (1967)
17. U rick, R.J., Principles of Underwater Sound for Engineers, New York, McGraw Hill (1967)
18. Fortuin, L., SACLANTCEN SR , 7 (1974) AD 922744
19. Laval, R., in Signal Processing, 223, London, Academic Press (1973)
2o. Hovem, J.M., SACLANTCEN TR, 214 (1972)
21. Clarke, R.H., J. Sound a Vibration, 34(3) 1,(1974)
22. Woodward, P.M., Probability and Information Theory with Applications to Radar, New York, Pergamon Press (1957)
23. Vakman, D.E., Sophisticated Signals and the Uncertainty Principle in Radar, Springer Verlag, Berlin-Heidelberg-New York (1968)
24. Wille, P., Thiele, R., J. Acoust. Soc. America, 5o, 348 (1971)
25. Horton, C.W., Signal Processing of Underwater Acoustic Waves, US Gov. Printing Office, Washington (1969)
26. Wijmans, W., SACLANTCEN SM, 26 (1973) AD 77o245
27. Wasiljeff, A., SACLANTCEN SM, 68 (1975)

DISCUSSION

Comment : R.J. URICK

(1) Your formula for array gain apparently imply uncorrelated (i.e., incoherent) noise, is this correct?

(2) What do you mean by "angular uncertainty"?

Reply : A. WASILJEFF

(1) The formula for the array gain of a linear additive array of equi-spaced elements of equal sensitivity is a generalization of the classical array gain definition. It takes into account the cross-correlation coefficients of the signal between different elements and the cross-correlation coefficients of the noise. The noise was assumed to be uncorrelated to simplify the discussion.

(2) The angular uncertainty is the effective beamwidth of the angular scattering function. Since angular scattering function and spatial correlation function form a Fourier pair, the product of angular uncertainty and coherence length has the order of one, similar to the time bandwidth relation in communication theory.

Comment : G. VETTORI

Under what circumstances are you allowed to factorize the scattering function?

Reply : A. WASILJEFF

Factorizing of the scattering function, that is separation into products of functions of the coordinates τ, ϕ, u and v was done for graphical illustration, since we cannot make a draft in five dimensions. To build a processing filter that is matched to the space-time variant random channel one should know the complete unfactorized scattering function, measuring the spatial correlation along one coordinate line gives only a rough estimate of the width of the angular scattering function.

PARAMETRIC ACOUSTIC ARRAYS

L. Bjørnø

Department of Fluid Mechanics, Technical University of Denmark, DK-2800 Lyngby, Denmark

ABSTRACT. An account of the historical background and the fundamental theory of the parametric acoustic array has been given. Low-amplitude wave interactions in absorption and spreading-loss limited parametric transmitting arrays is discussed for continuous and pulsed primaries and for field points outside or inside the interaction region. High-amplitude wave interactions leading to nearfield saturation limited parametric transmitting arrays are further treated for field points outside and inside the interaction region. The parametric receiving array for low- and high-amplitude pump waves is discussed and finally is given an exposition of the possibilities for obtaining an improvement of the parametric conversion efficiency for low- and high-amplitude wave interactions.

Introduction.

The generation of sum- and differens-frequencies by the interference between two finite-amplitude sound waves has been the subject of discussion for more than two hundred years. Helmholtz [1] and Lamb [2] credit the original observation of differens-frequency tones to Sorge (1745) and Tartini (1754). Since then the subject of difference-frequency wave generation has received the attention of several authors, but only the last 15 years have brought a strong development in the practical exploitation of the finite-amplitude wave interaction products, in particular the difference- and sum-frequency waves, while earlier works seem to have considered the effect as an occasional, undesirable nuisance or as a rather

G. Tacconi (ed.), Aspects of Signal Processing, Part 1, 33-59. *All Rights Reserved.*

academic subject. The practical exploitation of the non-linear sum- and differens-frequency wave generation in particular for underwater sound purposes has particularl through the last 5 - 10 years went through a rapid development, and the spectrum of fields of underwater applications now includes parametric transmitting and receiving arrays for echo-ranging, for bottom and subbottom profiling, for marine archeological detection of buried artifacts, for selected mode excitation in shallow water sound propagation and for ultrasonic imaging in medical diagnostics. The development hitherto in the field of parametric acoustic arrays gives no basis at all for an expectation of a future reduced research activity in finding and developing new fields of application of the parametric acoustic arrays and in improving the parametric acoustic arrays aiming at a better adaptation to and signal processing in fields where it already is being used.

I. Historical background and fundamental theory.

The formulation of the parametric acoustic array problem has its roots in a work by Lighthill [3],[4] who transformed the basic equations of fluid mechanics into a form being particularly suited for the study of sound generated aerodynamically. Lighthill's exact equation for arbitrary fluid motion can be written as:

$$\partial^2\rho/\partial t^2 - c_o^2\,\nabla^2\rho = - c_o^2\,\Box^2\rho = \partial^2 T_{ij}/(\partial x_i \partial x_j) \qquad (1)$$

where $T_{ij} = \rho u_i u_j + p_{ij} - c_o^2 \rho \delta_{ij} + D_{ij}$, with D_{ij} comprising the viscous stresses.

Equation (1) formed the starting point for Westervelt's formulation of his theory of "scattering of sound by sound" [5],[6], being one aspect of Lighthill's general theory. A suggestion by Lighthill that the scattering of sound by sound method might be applied to a sound beam led to Westervelt's derivation of his now classical theory for the parametric acoustic array [7]. The following simplifying assumptions and approximations underlie Westerwelt's work:

(a) The equation of motion for an ideal fluid is used and the attenuation effect is introduced in an "ad hoc" way.

(b) The two superimposed, high-frequency, plane primary waves are assumed to form beams so narrow and so perfectly collimated that the volume distribution of sources may adequately be represented by a line distribution located along the axis of the primary waves. The cross-sectional dimensions of the primary wave interaction region are assumed to be small compared with the wavelength at the differens-frequency.

(c) No attenuation of the differens-frequency wave is assumed to occur.
(d) The amplitude attenuation coefficients for each of the two primary waves are equal and assumed to be one or more orders of magnitude less than the wave number of the differens-frequency wave.
(e) Nonlinear attenuation is negligible.

Rewriting (5) on basis of the assumptions (a) - (e) and using a perturbation analysis retaining terms to second order in the field variables only, Westervelt's quasi-linear approach led to the following inhomogeneous wave equation for the pressure amplitude p_s of the differens-frequency wave:

$$\nabla^2 p_s - c_0^{-2}\partial^2 p_s/\partial t^2 = -\rho_0(\partial q/\partial t) \qquad (2)$$

where $q = \beta(\rho_0^2 c_0^4)^{-1}\partial p_i^2/\partial t$ (3)

q is the source strength density responsible for the generation of acoustic energy through the nonlinear interaction of the primary waves in which the instantaneous pressure at a source point is p_i. β is related to the second order nonlinearity ratio B/A [8] of the fluid through: $\beta = 1 + B/2A$, where :

$$A = \rho_0((\partial p/\partial\rho)_s)_{\rho=\rho_0} = \rho_0 c_0^2 \qquad (4)$$

$$B = \rho_0^2((\partial^2 p/\partial\rho^2)_s)_{\rho=\rho_0} \qquad (5)$$

Expression (3) is valid for plane waves travelling in the same direction while a general expression for the source strength density q of a primary field of any configuration may be found in [9].

The general solution to the equation (2) may be written as a volume integral by:

$$p_s(\underline{R},t) = -(\rho_0/4\pi)(\int_V(\partial q/\partial t)(\exp(ik_s|\underline{R}-\underline{r}|)/(|\underline{R}-\underline{r}|))dV \qquad (6)$$

where $\underline{R}$ and $\underline{r}$ denote the position vectors from the origin to the location of the observer (i.e. the field point) and to the differential volume dV of the zone of volume integration V, respectively. The volume integral (6) was used by Westervelt for a derivation of the differens-frequency sound field generated by the nonlinear interaction of the two perfectly collimated, plane, monochromatic, primary waves of equal source amplitude p_o. His expression for the pressure amplitude p_s as a function of the distance R from the projector emitting the primary waves to the observation point and as a function of the angle θ between the observation point and the acoustic axis of the projector may be written as:

$$p_s(R,\theta) = \omega_s^2 p_o^2 S\beta(8\pi\rho_o c_o^4 R\alpha_o)^{-1}(1 + k_s^2/\alpha_o^2(\sin^4(\theta/2)))^{-\frac{1}{2}} \quad ($$

where the time and phase dependences have been omitted. ω_s is the angular frequency of the differens-frequency wave, S denotes the cross-sectional area of the collimated wave region and k_s denotes the wave number of the differens-frequency wave. α_o is the mean absorption coefficient of the primary waves for infinitesimal wave amplitudes.

Westervelt's solution (7) is restricted to the farfield of the scattered wave by the condition: $k_s R > (k_s/\alpha_o)^2$.

The bracket in (7) leads to the half-power beamwidth θ_h of the differens-frequency wave given by:

$$\theta_h \simeq 2(\alpha_o/k_s)^{\frac{1}{2}} \qquad (8)$$

which shows that a narrowing of the beam takes place for a decrease in the primary wave frequency, opposite to what is the case for a conventional linear projector. Further, a narrowing of the beam will follow an increase in the difference-frequency. It should be noted that the influence of the primary frequencies on (7) and (8) is only through the absorption coefficient α_o.

The difference-frequency signal amplitude p_s may be considered to be radiated from an array of sources distributed continuously throughout the interaction region, being bounded by the collimated beams and extending a distance along the acoustic axis determined by the small signal absorption of the carrier waves. This virtual arr was by Westervelt [7] likened to an end-fire parametric array. The parametric array is shaded by virtue of the naturally smooth decay in the conversion of the carrier frequency waves to the difference-frequency wave with increasing distance from the signal source. The shading of the array being due to the carrier wave absorption and diffraction within the interaction region gives rise to a monotonically decaying angular response of the diff rence-frequency wave with increasing θ-values, thus avoiding the undesirable minor lobes that are common in conventional piston type transducers. Due to the small widt of the interaction region compared to its length the par metric array produces a field of radiation much narrower than the one which would be produced by a conventional u derwater sound source operating linearly at the differen frequency. Furthermore, the wide band character of the parametric conversion process enables one to remedy some of the disadvantages of the rather low efficiency of the nonlinear conversion process by the use of wide-band sig nal processing techniques. In spite of the low source level efficiency - ranging from 10% down to 10^{-5}% -

systems employing parametric arrays can be superior to conventional linear systems when the reduction of the beamwidth, the transducer size or the absorption (due to the low difference frequency) are taken into account.

Since Westervelt's publication of his quasilinear approach leading to his asymptotic solution being valid at long ranges from the interaction region, a great deal of theoretical and experimental works has been done in order to improve the understanding of the characteristics of the parametric acoustic arrays. Experimental evidence of (7) was found by Bellin & Beyer [10] using a 1-in-diameter quartz projector radiating primary frequency waves of 13 and 14 MHz. The interaction between such high primary frequencies is effectively confined to the nearfield of the projector.

The aperture effect due to the finite size of the projector, which should be taken into account when the interaction volume is substantially limited to the nearfield of the projector by the rate of absorption of the primary waves was discussed by Naze & Tjøtta [11] and by Berktay [12]. Berktay assumed a rectangular projector of sides 2b and 2d situated in the y-z-plane and he obtained the following expression for the difference-frequency pressure amplitude produced by the nonlinear interaction of two collimated, plane waves of initial pressure amplitudes p_1 and p_2:

$$p_s = p_1 p_2 S \omega_s^2 \beta (\exp(-\alpha_s R) \psi(b,d,k_s,\gamma,\theta)/(4\pi\rho_o c_o^4 R)) \times (\alpha_T^2 + 4k_s^2 \sin^4(\theta/2))^{-\frac{1}{2}} \quad (9)$$

where the aperture effect is represented by the expression:

$$\psi(b,d,k_s,\gamma,\theta) = (\sin(dk_s\cos\gamma)/(dk_s\cos\gamma))(\sin(bk_s\sin\gamma\sin\theta)/(bk_s\sin\gamma\sin\theta)) \quad (10)$$

and where $S = (2b)(2d)$.
α_T denotes an absorption coefficient determined by:

$$\alpha_T = \alpha_1 + \alpha_2 - \alpha_s\cos\theta \simeq \alpha_1 + \alpha_2 - \alpha_s \quad \text{for } \theta \ll 1.$$

α_1, α_2 and α_s are the absorption coefficients of the primary waves and the difference-frequency wave, respectively. γ is the angle between the z-axis and the radial distance R from the center of the projector $(x,y,z = 0,0,0)$ to the observation point.

For a circular projector the $(\sin(N)/(N))$ in (10) will be replaced by $(2J_1(N)/(N))$.

For spherical spreading primary waves confined to a cone of angular width $2\psi_1$ and by assuming a uniform intensity distribution across the cone, Berktay found for the difference-frequency pressure amplitude along the axis of symmetry:

$$p_s(R,0) = \beta p_1 p_2 \omega_s^2 (\exp(-\alpha_s R))(2\rho_o c_o^4 R k_s)^{-1} \times ((\tfrac{1}{2}\ln(1+\psi_h^4))^2 + (\tan^{-1}(\psi_h^2))^2)^{\frac{1}{2}} \quad (11)$$

where ψ_h is given by:

$$\psi_h^4 = (\psi_1/\theta_h)^4 \simeq (k_s/\alpha_T)(1-\cos(\psi_1))^2$$

and where the half-power beamwidth θ_h (for the expressions (9) and (11)) is given by:

$$\theta_h \simeq 2(\alpha_T/2k_s)^{\frac{1}{2}}, \quad \text{for } \alpha_T/2k_s << 1. \quad (12)$$

α_T is for the spherical wave case given by:

$$\alpha_T = \alpha_1 + \alpha_2 - \alpha_s \cos\theta\cos\psi \simeq \alpha_1 + \alpha_2 - \alpha_s$$

Berktay's asymptotic solutions (9) and (11) are valid for the field point at long ranges from the interaction region.

An experimental verification of the plane wave aperture influence was given by Hobæk [13] and later Muir & Blue [14] reported experimental results involving spherical waves by Fraunhofer zone interaction between the primaries. For a more detailed exposition of the experimental and theoretical results of the 1960'es is referred to [9].

All works previously mentioned comprise finite-amplitude primary wave interaction, but finite-amplitude absorption was not included. Some definitions being essential for the understanding of problems related to finite-amplitude absorption, for instance the acoustical saturation concept, shall here be developed through a brief account of the distortion course of finite-amplitude, monochromatic, plane or spherical waves of angular frequency ω (=c
at various distances from the source.

The cumulative distortion of a finite-amplitude propagating wave is due to two sources of nonlinearity, one kinematic due to convection and one thermodynamic due to nonlinearity of the equation of state of the fluid [8]. These sources give rise to a phase velocity of the finite-amplitude wave being given by:

$$(dx/dt) = c_o + \beta u \quad (13)$$

where u and x denote the local particle velocity and the axial distance from the signal source, respectively. (13) shows that a condensed phase travels faster than a rarefaction phase, thus leading to a steepening of the wave which may result in a shock formation. For finite-amplitude wave propagation in a "lossless" fluid the shock formation distance from the source may be given by:

$$\ell = c_o^2/(\beta\omega u_o) = (\beta k\varepsilon)^{-1} \quad (14)$$

for plane waves, and

$$\bar{r} = r_o \exp(\beta k\varepsilon r_o)^{-1} \quad (15)$$

for spherical waves. ε is the acoustic Mach number being given by: $\varepsilon = u_o/c_o$, with u_o denoting the peak particle velocity at the source, and r_o is the radius of the spherical wave source, or the range at which spherical divergence begins. The process of shock formation is frequently characterized in terms of a dimensionless parameter σ given by:

$$\sigma = x/\ell = \beta k\varepsilon x \quad (16)$$

for plane waves and

$$\sigma = \beta k\varepsilon r_o \ln(r/r_o) \quad (17)$$

for spherical waves.

An essential parameter in nonlinear acoustics is the Gol'dberg number Γ representing the relation between the nonlinear and the dissipative effects, thus likened to an "acoustic Reynolds number" and being given by:

$$\Gamma = \beta p_o/(b\omega) \quad (18)$$

where

$$b = 4\eta/3 + \zeta + \kappa(1/(c_v) - 1/(c_p)) \quad (19)$$

with p_o denoting the pressure amplitude at the source. $\Gamma > 1$ will lead to a shock formation [15], which occurs for $\sigma = 1$ [8]. If Γ is high enough a "sawtooth" wave form may be obtained through further distortion of the finite-amplitude wave. The "sawtooth" formation takes place at $\sigma \sim 3$, which for plane and spherical waves, respectively, can be expressed by:

$$\tilde{x} = 3\ell = 3(\beta k\varepsilon)^{-1} \quad (20)$$

and

$$\tilde{r} = r_o \exp(3(\beta k\varepsilon r_o)^{-1}) \quad (21)$$

Due to absorption, the fundamental wave amplitude will have been reduced with several dB at $\sigma = 3$. In the "sawtooth" region, where a stable wave form is maintained over a distance of propagation due to equilibrium of the nonlinear distortion and the dissipative effects, the fundamental component amplitude p_1 will be reduced according to:

$$p_1 = 2\alpha\rho_o c_o^2/(\beta k \sinh(\pi\Delta/2)) \quad (22)$$

where $\pi\Delta/2 = \alpha/(\beta k\varepsilon) + \alpha x$ for plane waves and $\pi\Delta/2 = \alpha r/(\beta k\varepsilon r_o) + \alpha r \ln(r/r_o)$ for spherical waves with $(kr_o >> 1)$. Δ, being a measure of the shock thickness, is given by:

$$\Delta = 2(\pi\Gamma)^{-1}(1 + \sigma)(r/r_o)^n \tag{23}$$

with n = 0 for plane waves and n = 1 for spherical waves [8].

For increasing initial pressure amplitude p_o, (23) and thus (22) will become independent of p_o and an <u>acoustic saturation</u> takes place.

A transition from the "sawtooth" region to "old age" small amplitude wave propagation - due to the increasing relative influence of dissipation leading to an increasing shock thickness - takes place at a source distance approximately given by:

$$x_{max} = 1/\alpha \tag{24}$$

for plane waves and

$$r_{max} = \beta k\varepsilon r_o(\alpha(1 + \beta k\varepsilon r_o \ln(r_{max}/r_o)))^{-1} \tag{25}$$

for spherical waves [16].

(24) corresponds to $\pi\Delta/2 \simeq 1$ (for large Γ-values), which with (22) leads to a fundamental component amplitude at x_{max} given by: $p_1 = 1.7\alpha\rho_o c_o^2/(\beta k)$, which can be used for a calculation of an equivalent source level for $x > x_{max}$

If the source amplitude is increased indefinitely it may be shown that the fundamental frequency pressure amplitude approaches a finite limiting value, which for spherical waves may be expressed as a function of r by [17] :

$$p_{sat} = 2\rho_o c_o^2(\exp(-\alpha(r - R_{max})))(\beta kr\ln(R_{max}/r_o))^{-1} \tag{26}$$

where

$R_{max} = \int_{p_o\to\infty} r_{max} = (\alpha\ln(R_{max}/r_o))^{-1}$ with r_{max} given by (25). Note, that (26) is independent of the initial source amplitude p_o.

The plane and spherical waves hitherto discussed form parts of the wave field arising from a piston source embedded in an infinite, rigid baffle. Some characteristic source distances shall briefly de defined in the following due to the fact that piston sources have most frequently been used for primary wave transmission by parametric acoustic arrays.

For a piston source of area S embedded in am infinite, rigid baffle operating simultaneously at the angular frequencies ω_1 and ω_2 (wave numbers k_1 and k_2) the wave fields can be treated as plane collimated waves within a distance - the Rayleigh distance or the collimation distance forming the nearfield of the primary wave source - expressed by:

$$R_r = k_o S/(2\pi) = S/\lambda_o \qquad (27)$$

where $k_o = \frac{1}{2}(k_1 + k_2)$. At distances greater than R_r the primary wave field can be represented as spherically spreading waves.

The Fresnel distance may analogous be expressed by:

$$R_f = S/(4\lambda_o) \qquad (28)$$

Empirically [17], it has been found that a source radius $R_o = S/(2\lambda_o)$ provides acceptable agreement with experimental results, thus forming an effective radius of a spherical wave source.

Using the down-shift-ratio $H_s = k_o/k_s = \omega_o/\omega_s$, (with $\omega_o = \frac{1}{2}(\omega_1+\omega_2)$), the difference-frequency wave collimation distance may be defined through:

$$R_s = R_r/H_s \qquad (29)$$

The half-power beamwidth θ_h of the difference-frequency wave may under certain circumstances (to be discussed later on) asymptotically approach that of the primary wave at a source distance given by:

$$R_\theta = R_r H_s \qquad (30)$$

The possibility of shock formation in the primaries necessitates the definition of a dimensionless "saturation number" χ [18], being expressed using (14) by:

$$\chi = R_r/\ell \qquad (31)$$

II. Low-amplitude wave interactions in parametric transmitting arrays. Absorption and spreading-loss limited arrays.

In this section are considered primary waves of low amplitudes, i.e. waves whose peak amplitudes are below their respective shock formation threshold by means of which nonlinear absorption can be neglected.

Westervelt's farfield approximation, treated in section I, for the difference-frequency signal generated by nonlinear interaction of infinitely plane, unsaturated primary waves of finite-amplitude only subject to viscous absorption and the extension of Westervelt's theoretical approach to include the difference-frequency signal generation by the nonlinear interaction between two monochromatic, unsaturated spherical waves emitted by a circular piston projector given by Muir [19] and Muir & Willette [20] are examples on low-amplitude wave interaction parametric transmitting arrays.

If the nearfield primary wave absorption loss ($\alpha_T R_r >> 1$ N to ensure that the primary waves are sufficiently absorbed within R_r to such an extent that no further nonlinear interactions takes place beyond R_r, the parametric array is termed: absorption limited, and Westervelt's solution can be used for a determination of the difference-frequency signal. If $\alpha_T R_r$ is very small, the primary wave interactions takes place predominantly beyond R_r and the array is essentially spreading-loss limited, which demands the use of a spherical wave solution for the difference-frequency signal.

While the absorption limited arrays are characterized by a farfield half-power beamwidth of the difference-frequency signal given by expression (12), spreading-loss limited arrays (also including some viscous absorption effects) will show a half-power beamwidth increasing with the source distance r and asymptotically approaching the half-power beamwidth of the product of the primary beam directivity patterns. The two half-power beamwidths may become equal at the source distance R_θ.

The validity of the asymptotic solution derived for the parametric array using Rutherford's scattering formula demands that the field point (point of observer) is far outside the interaction region, but volume integral solutions have been derived for field points in the interaction region. The use of continuous or pulsed primaries will influence the parametric wave generation process Further, the question whether most difference-frequency signal generation takes place in the near- or in the far field of the primary wave projector - or is shared between the two fields - can be replied to when the primary wave frequencies, their source level and the size of the projector are known.

The discussion in the following may therefore appropriately be subdivided into discussion of procedures and resul arising from (a) field point outside or inside the primary wave interaction region, (b) continuous or pulsed primaries and (c) predominant near- or farfield interactions.

1. Field point outside the interaction region.

a. Continuous primaries. The absorption limited, predominant nearfield interaction, arrays are represented by Westervelt's [7] and Berktay's [12] approaches discussed in section I. Another method for predicting the performance of parametric transmitters, published by Mellen &

Moffett [18],[21], uses $\alpha_T R_\theta$ as the basic parameter for absorption/spreading-loss limited arrays. Their model combine the simple plane and spherical solutions by adding the difference-frequency wave contributions from sources in a well-collimated end-fire array ($r<R_\theta$) and from sources in a spherical region ($r>R_\theta$) having a directivity given by the product of the primary directivity functions.
This model has proven to be of considerable use in calculating the parametric efficiency in cases when the field point is sufficiently far from the source for all points in the interaction region for which diffraction is important to be represented as being at the same range. Their general expression for the parametric sonar source level efficiency (i.e. the ratio of the difference-frequency to the primary source levels) can be written as:

$$|Rp_s/(R_r p_o)| = \chi/2(k_s/k_o)^2 \int^{\infty} (dr/R_r)(1 + (k_s/k_o)^2(r/R_r)^2)^{-\frac{1}{2}} \times (\exp(-2\alpha_o r))(1 + ((\chi/2)\sinh^{-1}(r/R_r))^2)^{-1} \quad (32)$$

where it is assumed that the primary waves were of the same initial amplitude p_o. Neglecting finite-amplitude absorption ($\chi < 1$) and assuming that the parametric array is absorption limited at $r < R_\theta$, (32) reduces to the asymptotic case:

$$|Rp_s/(R_r p_o)| \simeq \chi/2(k_s/k_o)^2 \int^{\infty} (dr/R_r)\exp(-2\alpha_o r) = \chi(4\alpha_o R_r)^{-1}(k_s/k_o)^2 \quad (33)$$

which is the Westervelt case. This may easily be verified by insertion of $\chi = \beta p_o k_o R_r (\rho_o c_o^2)^{-1}$ and by the use of (27).

The square dependence of p_s on the down-shift-ratio ($p_s \propto \omega_s^2$) and on the primary wave amplitude p_o for interactions in the collimated beam region should be noted by (33).

Numerous authors have contributed to our knowledge about the spreading-loss limited arrays ($\alpha_T R_r \ll 1$ Np) where farfield interaction predominates. Two ways of handling the geometry of the array have been prevailing. The first is to approximate the interacting signals as one-dimensional propagating waves [22],[23],[24] and the second is to perform a three-dimensional integration [19],[20].

Berktay & Leahy [24] considered the interaction taking place in the farfield of a piston projector. Their geometry with the projector surface in the y-z-plane is shown in Figure 1. The farfield beam patterns of the primary waves of angular frequencies ω_1 and ω_2 can be represented in the form:

$$p_{1,2} = (\bar{p}_{1,2}/r) D_{1,2}(\gamma,\phi) \exp(-(\alpha_{1,2} + ik_{1,2})r) \quad (34)$$

where $D_{1,2}(\gamma,\phi)$ are the normalized directivity functions of the primaries. $\bar{p}_{1,2}$ denote the source amplitudes of the primaries.

The difference-frequency pressure at the field point (R,θ,η) in Figure 1 can now be written as:

$$p_s(R,\theta,\eta) \simeq \omega_s^2 \bar{p}_1 \bar{p}_2 \beta (4\pi\rho_o c_o^4 R)^{-1} (\exp(-(\alpha_s + ik_s)R)) \times \int_{-\pi/2}^{\pi/2}\int D_1(\gamma,\phi) D_2(\gamma,\phi)(\alpha_T + ik_s(1-\nu))^{-1} \cos\gamma \, d\gamma \, d\phi \quad (35)$$

where $\nu \simeq 1 - \frac{1}{2}(\gamma-\theta)^2 - \frac{1}{2}(\phi-\eta)^2$.

The double integral in (35) is the two-dimensional convolution of the product of the primary directivity patterns (weighted by $\cos\gamma$) with directivity function $D(\theta,\eta) = (1+i(k_s/\alpha_T(1-\cos\theta\cdot\cos\eta))^{-1}$, which is the Westervelt directivity function (7) transferred to the coordinate system given in Figure 1. For extremely narrow primary beams and for $D_1() = D_2()$, (35) reduces to the Westervelt case, and when θ_h given by (12) is much smaller than the beamwidth associated with the product $D_1()D_2()$, the difference-frequency beam patterns become the product of the primary-frequency directivity functions.

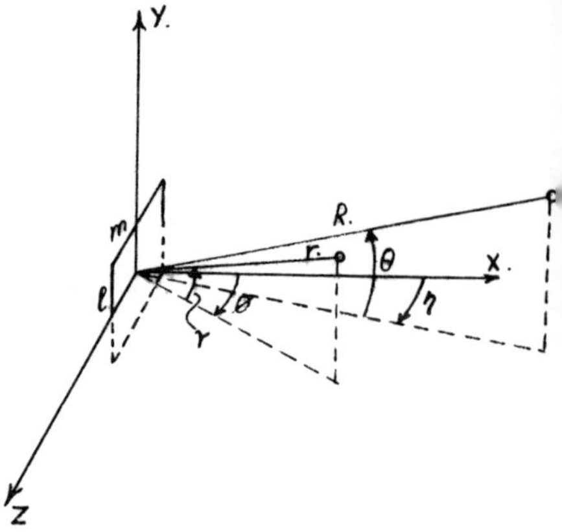

Figure 1. Geometry |24|.

Evaluation of (35) for both a rectangular and a circular projector embedded in an infinite, rigid baffle have been given in [24]. For a rectangular projector with the dimensionless side lengths L and M (normalised by use of λ_o) parallel with the y- and z-axis, respectively (35) may be rewritten as:

$$p_s(R,\theta,\eta) \simeq P_w(R,0) V(\psi_y,\psi_z,\theta',\eta') \quad (36)$$

where:

$$P_w(R,0) = -\omega_s^2 (W_1 W_2)^{\frac{1}{2}} \beta (2\pi c_o^3 R \alpha_T)^{-1} (\exp(-(\alpha_s + ik_s)R)$$

W_1 and W_2 are the acoustic power transmitted at the primary frequencies and where:

$$V() \simeq LM\theta_h^2 \left[\int_{-\pi/2\theta_h}^{\pi/2\theta_h} \int (\sin^2(\sqrt{2}\gamma'/\psi_y))/(\sqrt{2}\gamma'/\psi_y)^2 \times (\sin^2(\sqrt{2}\phi'/\psi_z))/(\sqrt{2}\phi'/\psi_z)^2 \times \right.$$

$$(1 + i((\theta' - \gamma')^2 + (\eta' - \phi')^2))^{-1}d\gamma'd\phi' \qquad (37)$$

All angles are normalized with respect to θ_h and the half-power beamwidths $2\gamma_1$ and $2\phi_1$ of the primary beams in the two planes of symmetry are given by: $\pi L\gamma_1 = \pi M\phi_1 \simeq \sqrt{2}$. $\gamma' = \gamma/\theta_h$, $\phi' = \phi/\theta_h$, $\psi_y = \gamma_1/\theta_h$ and $\psi_z = \phi_1/\theta_h$.

A calculation of (37) along the acoustic axis ($\theta' = \eta' = 0$) has been performed [24] and is given in Figure 2. Using B/A = 5 for water, the RMS source level at the difference-frequency can be calculated by (36) through the expression:

$$SL_- \simeq 137 + 20\cdot\log(f_s) - 40\cdot\log(\theta_h^\circ) + 10\cdot\log(W_1) + 10\cdot\log(W_2) + + 20\cdot\log|V| \quad \text{dB re 1 } \mu\text{Pam} \qquad (38)$$

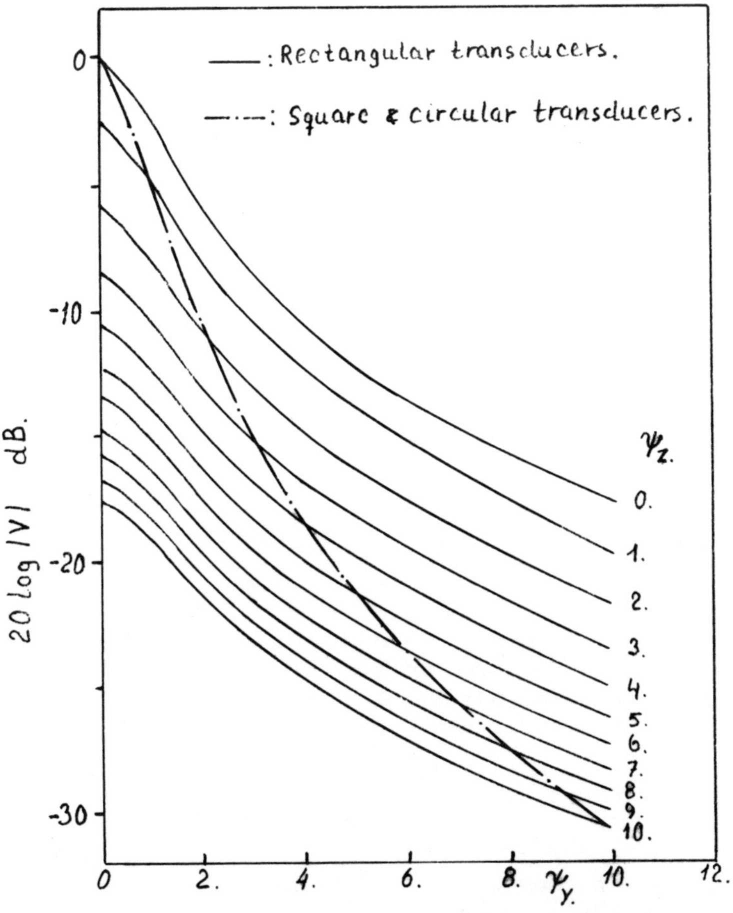

Figure 2. Pressure reduction factor, |V|, for rectangular, squared and circular transducers. $\theta' = \eta' = 0$, [24].

(38) can be used for an evaluation of the farfield response of a parametric transmitter. f_s is the difference-frequency in kHz, θ_h is in degrees, W_1 and W_2 are in watts and $20\cdot\log|V|$ is obtained from Figure 2, where a circular projector of radius a gives: $\psi_d = 92.5/(k_o a\theta_h^\circ)$ and a rectangular projector of sides ℓ and m gives: $\psi_y = 163/(k_o \ell\theta_h^\circ)$ and $\psi_z = 163/(k_o m\theta_h^\circ)$.

A comparison between the farfield difference-frequency amplitudes calculated by means of (32) and by means of (37) can be performed for both circular and rectangular projectors through the use of: $\alpha_T R_\theta = 0.650/\psi_d$ (for circular projectors) and $\alpha_T R_\theta = 0.635/(\psi_y\psi_z)$, (for rectangular projectors)

For $\chi < 1$ and $2\alpha_o R_r < 1/H_s$, i.e. the array is spreading-loss limited in the farfield of the primaries where most of the difference-frequency generation takes place, expression (32) reduces to:

$$|Rp_s/(R_r p_o)| \simeq \chi/2(k_s/k_o)E(2\alpha_o R_\theta) \qquad (39)$$

where E() is the well-known exponential integral function. The linear dependence on the down-shift-ratio, i.e. $p_s \propto \omega_s$, represented by (39) for the farfield interaction should be noted. In the general case, when both near- and farfield interactions contribute to p_s, the exponent of the down-shift-ratio is between 1 and 2, which also has been experimentally verified [14],[25].

If χ is put equal to 1 ($\ell = R_r$), high, but not too high for the continued omission of strong finite-amplitude distortion and absorption effects, initial primary amplitudes occur. The use of B/A = 5, $\rho_o = 10^3$ kg/m^3 and c_o = 1500 m/s will then lead to the maximum obtainable difference-frequency source level given by:

$$SL_{-m.} \simeq 20\cdot\log(f_s) - 40\cdot\log(f_o) + 20\cdot\log(|V|/\alpha_T R_\theta) + 274 \quad \text{dB re 1 μPam} \qquad (40)$$

b. <u>Pulsed primaries.</u> The discussion of the parametric array thus far has comprised the nonlinear interactions between continuous (monochromatic) primary waves, only, but also pulsed carriers have been used.

Berktay [12] considered primary waves which at the source were of the form:

$$p_i(t) = p_o F(t)\cos(\omega t) \qquad (41)$$

where F(t) is an envelope function assumed to vary slowl compared with the cos(ωt)-term, thus covering a relatively narrow band (no components higher than $\omega/3$) in orde to avoid overlap in the frequency spectra of the scattered and the primary waves. The time-domain solution for the scattered wave may for plane waves, along the axis of the array be expressed by:

$$p_s(R,t) \simeq p_o^2 \beta S (16\pi\rho_o c_o^4 R\alpha)^{-1} (\partial^2/\partial t^2) F^2(t') \qquad (42)$$

where viscous absorption of the primary has been introduced. $t' = t - R/c_o$, is the retarded time.

Experiments have shown that pulsed carriers yield a scattered component being about 2 dB greater than in a two-frequency parametric array [25],[26],[27]. A detailed discussion of the transient (pulsed) carrier is given in [19]. The use of pulsed carriers demands that the projector bandwidth must be much wider than the scattered wave frequency, in order to avoid interpulse distortion. (42) stresses the wide band character of the parametric array at the scattered wave frequency.

2. Field point inside the interaction region.

The full consequences of the asymptotic theory for parametric arrays cannot be met with the field point in the interaction region, which is frequently the case for laboratory tank experiments. Two different theoretical approaches have been used in order to calculate the difference-frequency signal level at points in the interaction region. The first approach is based on numerical integration of a volume integral [19],[20] and the second is based on the introduction of a correction factor to the asymptotic theory [28].

Muir & Willette [20], investigated the generation in fresh-water of the sum- and difference-frequency signal of two high-frequency primaries (418 and 482 kHz) transmitted by a 3-in-diameter, circular piston projector. In their theoretical approach they only considered contributions from farfield interactions and they derived the following volume integral expression for the (sum- or) difference-frequency pressure amplitude, valid for $R \gg r$, but approximately valid for the field point in the interaction region:

$$p_s(R,\theta) = 2p_1 p_2 \omega_s^2 \beta R_r^2 (\pi\rho_o c_o^4 k_1 k_2 a^2)^{-1} \times$$
$$\int_{R_r'}^{R} \int_0^{\phi e^{+\theta}} \int_0^{\pi} J_1(k_1 a \sin\sigma) J_1(k_2 a \sin\sigma)/(\sin^2\sigma) \times$$
$$(\exp-(((\alpha_1+\alpha_2) - ik_s)r - (ik_s - \alpha_s)r'))/r' \times$$
$$\sin\phi \, d\psi \, d\phi \, dr \qquad (43)$$

where $\sin\sigma = ((\sin(\theta-\phi) \mp \sin\phi\cos\theta(1-\cos\psi)) + \sin^2\phi\sin^2\psi)^{\frac{1}{2}}$, and $r' = (R^2 + r^2 - 2rR\cos\phi)^{\frac{1}{2}}$. ϕ_e is an effective array angle and $R'_r = 3a^2/4\lambda_o$ with a being the projector radius.

The geometry of expression (43) is shown in Figure 3.

The numerical solution to (43) was compared with experimental results showing a good agreement even for the field point in the interaction region, and an expression like (43) has been used by Bjørnø et al [25] for a calculation of p_s in the interaction region.

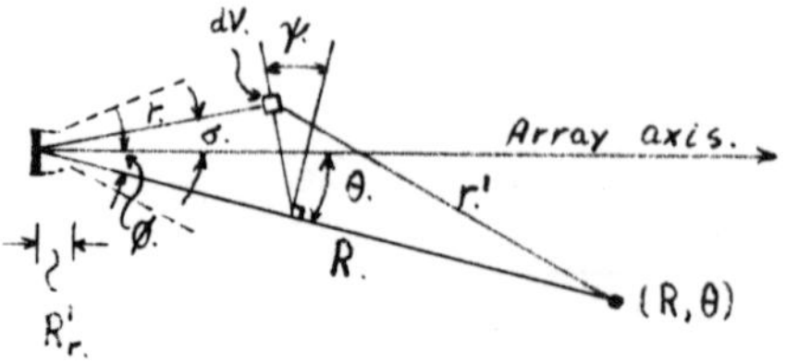

Figure 3. Geometry [20].

Their theoretical and experimental findings [25] are given in Figure 4, showing the applicability of the volume integral method for calculation of p_s in the interaction region.

The half-power beamwidth dependence on distance from the projector is given in Figure 5 for both the sum- and the difference-frequency waves [20] showing that the parametric arrays develop their narrow beam features exponentially and rather early in the nonlinear interaction process, thus making them well suited even in their near-field regions.

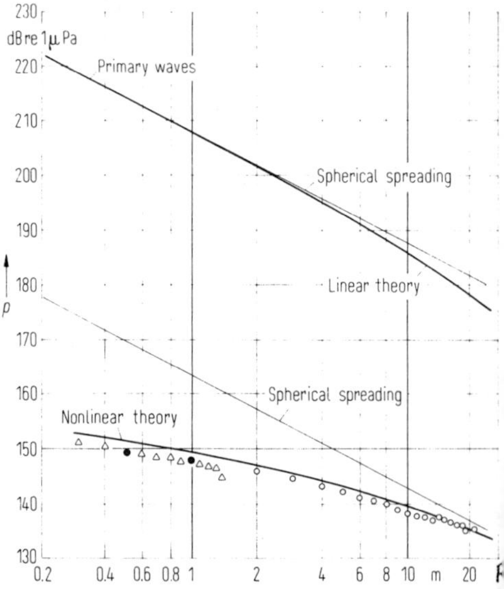

Figure 4. △, • and ○ are experimental results. The curve marked "Nonlinear Theory" arise from the numerical integration of a modified version of (43). [25].

The number of full-scale at-sea tests with parametric arrays have increased considerably during the recent years. The first at-sea test was reported by Walsh [28] using a mean primary frequency of 200 kHz and a difference-frequency of 3.5 - 12 kHz in a 65 ms FM pulse of a bandwidth of 2 kHz. The difference-frequency source level obtained was 188 dB re 1 μPam, in a 3° beam produced by primaries at a source level of 225 dB re 1 μPam.

Recently, Konrad [30] has described one of the largest parametric systems to date, the so-called TOPS (Towed Parametric Sonar). The projector size is 0.5 x 2 m, f_o = 24 kHz, the primary source level is SL_o = 259 dB re 1 μPam and the half-power beamwidth of the primaries are 2 x 8°. The The difference-frequency range considered varies from 5 kHz down to 250 Hz having source levels SL_- from 230 dB re 1 μPam (at 5 kHz) to 186 dB re 1 μPam (at 250 Hz).

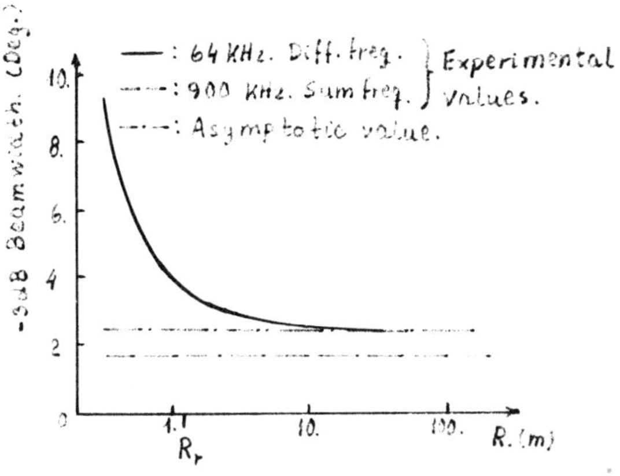

Figure 5. Beamwidth dependence on range. [20]

III. High-amplitude wave interactions in parametric transmitting arrays. Saturation effects.

Finite-amplitude effects, i.e. wave distortion and finite-amplitude absorption, in parametric transmitting arrays have been dealt with by Muir [19], Mellen & Moffett [18],[21], Merklinger [31], Bartram [34] and Fenlon [32],[33]. In all cases, quasi one-dimensional models were utilized among other things due to the fact that the treatment hitherto of finite-amplitude effects in parametric arrays is generally based on one-dimensional models and frequently on theories originally formulated for a monochromatic wave source. General effects of finite-amplitude distortion and absorption in a parametric array is an effective shortening of the array length leading to a broadening of the difference-frequency beam and a reduction in the difference-frequency source level relative to the source level values calculated through the use of expressions given in section II.

If a parametric source is saturation limited, almost certainly the saturation will occur within the nearfield, rather than the farfield, of the primary beam. For $\chi > 1$ in (32) the parametric array will be saturation limited in the nearfield and (32) will reduce to:

$$|Rp_s/(R_r p_o)| \simeq \chi/2(k_s/k_o)^2 \int_0^\infty dr/R_r \quad X$$
$$(1 + (k_s/k_o)^2 (r/R_r)^2)^{-1} = \pi/2(k_s/k_o)^2 \quad (44)$$

which shows a linear relation between p_s and p_o at an observation point outside the interaction region.

The original Mellen & Moffett equation for the parametric source level efficiency (32) involving the finite-amplitude effect taper function:

$$T_f = (1 + (\sigma/2)^2)^{-\frac{1}{2}} \quad (45)$$

where $\sigma = \chi \sinh^{-1}(r/R_r)$, together with the small amplitude taper $T_s = \exp(-\alpha_o r)$, can be numerically integrated for particular values of the down-shift-ratio $(k_o/k_s$ and of $\alpha_o R_r$.

Curves for the difference-frequency source level efficiency as a function of the scaled mean primary source level SL_o^* for the down-shift-ratio $k_o/k_s = 10$ with $\alpha_o R_r$ as a parameter is given in Figure 6. The primary source level SL_o is defined as the mean level of the two components: $SL_o = \frac{1}{2}(SL_1+SL_2)$, leading to the 1 kHz scaled value SL_o^* through:

$$SL_o^* = SL_o + 20 \cdot \log(f_o) \quad \text{dB re 1 µPam} \quad (45)$$

where $f_o = \frac{1}{2}(f_1+f_2)$ (kHz).

Using SL_o^* from (45) in Figure 6 leads for an appropriate $\alpha_o R_r$-value (and for $k_o/k_s = 10$) to the parametric source level efficiency (conversion efficiency), which added to SL_o gives the difference-frequency source level SL_- by:

$$SL_- = SL_o + 20 \cdot \log|Rp_s/(R_r p_o)| \quad \text{dB re 1 µPam} \quad (46)$$

Curves similar to the ones in Figure 6 for down-shift-ratios 5, 20 and 50 are given in [18] thus facilitating the evaluation of the parametric source level efficiency expressed by (32) for a broad variety of parametric array parameters. The asymptotic cases (39) and (44) are reflected in the shape of the curves in Figure 6. The absorption effect terminates the growth of the difference-frequency wave amplitude, leading to the square law dependence of p_s on p_o given by (39) for input levels below a certain value.

No higher order spectral interactions are included in the derivation of (32), but these interactions, which are taken into account in expressions derived in [33] for very high primary source levels, will in general give rise to enhanced finite-amplitude effects, thus causing a more rapid decrease in the parametric conversion efficiency. This rapid decrease can be seen by the dotted line curve in Figure 6.

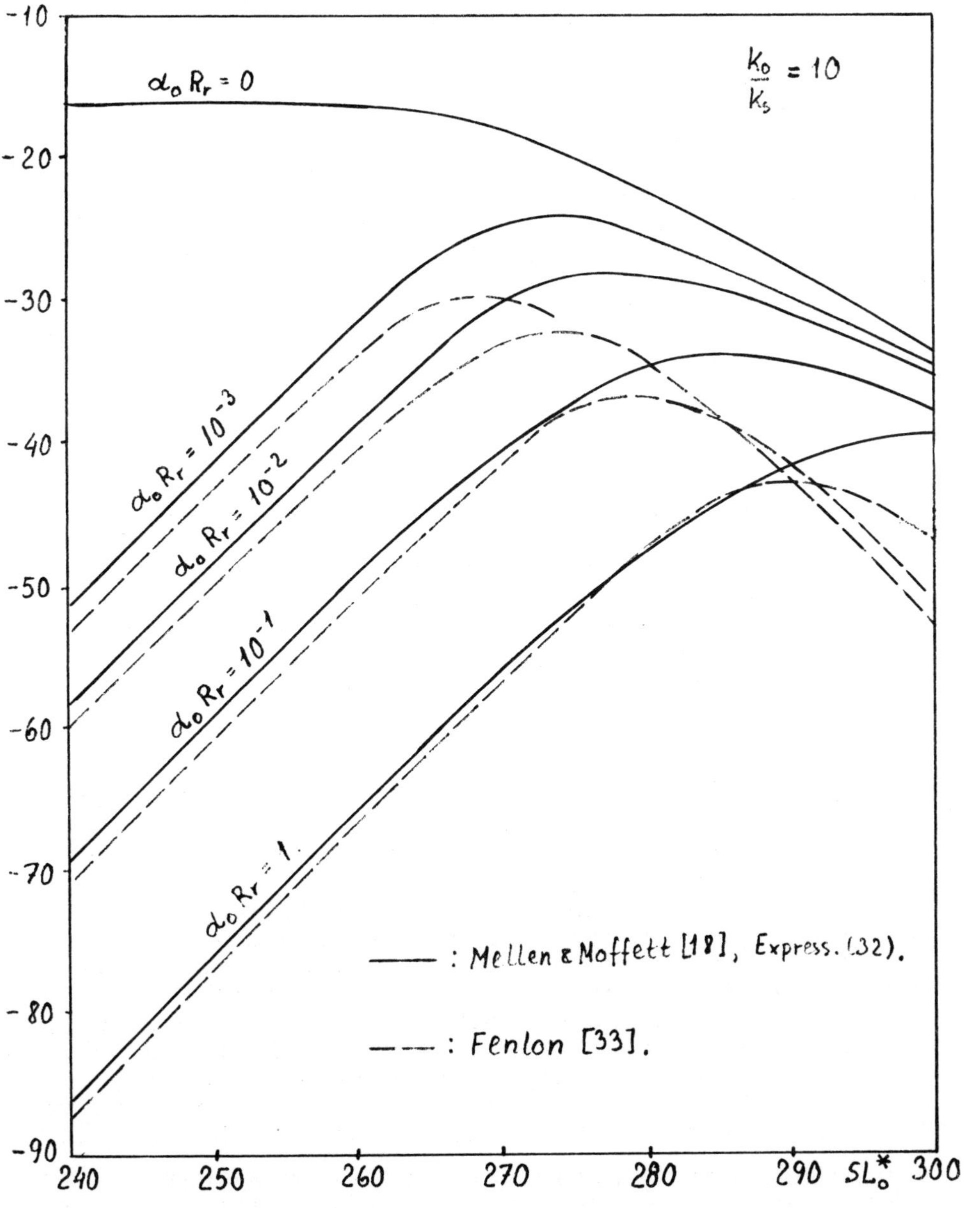

Figure 6. Abscissa: Scaled mean primary wave source level SL_o^* (dB re 1 μPam kHz, RMS)
Ordinate: Parametric conversion efficiency. $20\log|Rp_s/(R_r p_o)|$. (dB)

Finite-amplitude effect taper functions for parametric arrays have been worked out by various authors [18], [21], [31], [33] and [35]. Lockwood [35] used a modified form of the weak-shock-theory [8] for a derivation of an intensity taper function based upon the time-domain theory for difference-frequency wave generation given in [31]. This intensity taper function has then been introduced into a modified form of Muir & Willette's volume integral expression (43), which integrated numerically opens up the possibility of a determination of p_s for $\chi > 1$ and for the field point situated in the interaction region (here the nearfield), which cannot be performed using (32), due to the fact that (32) is derived assuming the field point well-outside the interaction region. Parametric conversion efficiencies calculated by the use of Lockwood's volume integral method [35] show a good agreement with those based upon (32) in their common field of validity.

The shortening of the parametric array length due to finite-amplitude effects at high primary source levels leads to a blundering of the difference-frequency beam patterns around the array axis with an increase in the sidelobe levels relative to the level of the major lobe. These phenomena are due to the finite-amplitude absorption dependence on the primary wave amplitudes, which in the farfield of a directive source are greatest on the acoustic axis.

IV. Parametric receiving arrays.

The possibility of developing a parametric receiving array was first mentioned by Westervelt [7] and in spite of the fact that most efforts in developing parametric arrays have been laid down into the transmitting arrays in order to obtain highly directional, low frequency sources using relatively small transducers, the parametric receiving array has recently received considerable interest. In a parametric receiver, the nonlinear interaction process takes place between a low-frequency, plane signal wave of low intensity and a high-frequency pump wave of higher intensity being generated locally. The sum- or difference-frequency signal are then received by a transducer on the acoustic axis of the pump wave.

The discussion of parametric receiving arrays may appropriately be devided into (a) low-amplitude and (b) high-amplitude parametric receiving arrays. The difference between the two approaches lay in the inclusion or not of finite-amplitude effects in the pump wave. Farfield reception was considered by Barnard et al [36] who studied a first-order sound field consisting of a "low-amplitude" spherical, harmonic pump wave of frequency f_1 and a plane, harmonic signal wave of frequency f_2, $(f_1 > f_2)$. No nearfield effects of the pump were assumed to be involved and only the pump wave was assumed undergoing absorption (viscous).

In polar coordinates $((r,\psi,\phi)$, where ϕ is the polar angle and ψ is the azimuthal angle) their first-order sound field can be written as:

$$p_f = p_1(r_o/r)((2J_1(k_1 a\sin\phi))(k_1 a\sin\phi)^{-1} \times (\exp(-\alpha_1 r))(\cos(\omega_1 t - k_1 r) + p_2\cos(\omega_2 t - k_2 z) \quad (47)$$

where p_1 is the peak sound pressure level of the pump wave at a distance r_o and p_2 is the sound pressure level of the signal wave at input to the parametric receiving array. a is the radius of the piston pump and $k_2 z$ is defined as the plane-wave phase at r, where: $z = r(\sin\nu\cos\phi - \sin\phi\cos\nu\cos\psi)$. The angle ν in the horizontal plane is the acute angle between the acoustic axis of the array and the plane wave front. Insertion of (47) into (6) and introduction of the viscous absorption in an "ad hoc" way, lead to the following volume integral expression for the pressure amplitude of the sum- or difference-frequency signal:

$$p_s(R_o,\nu) = \beta\omega_s^2 p_1 p_2 r_o(2\pi\rho_o c_o^4 a k_1)^{-1} \times \int_0^R \int_0^{\phi_{eff}} \int_0^{\pi} (J_1(k_1 a\sin\phi))\exp(i(k_1 r \pm k_2 z + i\alpha_1 r)) \times (\exp(i(k_s + i\alpha_s)r'))(r')^{-1} r\,d\psi\,d\phi\,dr \quad (48)$$

where $\omega_s = \omega_1 \pm \omega_2$, and $r' = (r^2 + R_o^2 - 2rR_o\cos\phi)^{\frac{1}{2}}$, while ϕ_{eff} is the angle between the acoustic axis and the first null of the pump farfield radiation pattern. R_o is the distance along the acoustic axis between the pump and the receiving transducer. (48) has been solved numerically and the results have been tested against experiments for the sum-frequency wave [36] showing a good agreement, both concerning beam patterns and range (R_o) for various signal frequencies. Later works along the line introduced by (48) have been published in [37] and nearfield reception has been treated theoretically by Rogers et al [38].
The half-power beamwidth of the parametric receiver is completely determined by:

$$\theta_h = 1.9(\lambda_2/R_o)^{\frac{1}{2}} \quad (49)$$

Only a small amount of work has been done in analyzing finite-amplitude effects in parametric receiving arrays. Experiments involving acoustic saturation in parametric receivers have been performed by Berktay & Al Temini[39],

and by Konrad et al [40]. A theoretical model involving plane pump and signal waves was worked out by Bartram [41] and recently Lockwood [35] has published a theoretical approach in which the axial response of the parametric receiving array has been calculated in two ways. (a) By the use of a one-dimensional model in which the secondary signal levels are predicted from the weak-shock-theory using a simple correction for nearfield geometry and (b) by numerical evaluation of the scattering integral solution, i.e. (43) for the transmitting array modified to be used by the parametric receiving array, in which the finite-amplitude effects are introduced through the use of nearfield and farfield taper functions derived from the one-dimensional solution.
An excellent agreement between the scattering integral solution and the experimental results in [40] is found, while the nearfield taper function used leads to results that sharply disagree with the experimental findings in [39]. Therefore, the parametric receiving array models developed in [35] should only be applied to farfield reception.
In general some serious problems still have to be overcome before a practical system can be realized. For instance, motion of the pump and the transducer during reception, water noise at the signal frequency or at the sum- or difference-frequencies, electronic noise in the equipment etc., can create serious problems for full-scale reception. But the promising results hitherto obtained point to solutions to the problems to be obtained in a not too distant future.

V. Improvements of the conversion efficiency of the parametric acoustic arrays.

The low conversion efficiency of the parametric acoustic arrays previously mentioned is its capital weakness. The second order effects on which it is based limit the energy transfer from the primary beams to the difference- and sum-frequency beams.
It was pointed out by Merklinger [26] that if the necessary bandwidth of the transmitter was available, a pulsed carrier type of transmission would give an improvement in conversion efficiency of between 2 and 6 dB depending on the system constraints. Experimental evidence for this prediction has later been created [25]. 100% amplitude modulation of a primary wave was shown in [25] and in [27] to lead to a 2.5 dB increase in the difference-frequency sound pressure level compared to a two-component primary of the same total power.

From the parameters involved in the expressions (7), (9), (33) and (35) for low-amplitude wave interactions in parametric transmitting arrays it may be concluded, that an increase in the virtual source strength and thus in p_s may be obtained primarily by the following procedures: a. An increase in the peak amplitudes of the carrier waves,and b. an increase in β and a decrease in ρ_o and in particular in c_o in the fluid. The influence of these parameters on the conversion efficiency of a parametric transmitting array has been studied in theory and through experiments in [25]. By putting the nearfield of the projector under pressure, cavitation effects could be avoided and the nearfield liquid was replaced by liquids showing more appropriate β and $\rho_o c_o$ values (methyl and ethyl alcohols), which led to an about 15 dB increase in the difference-frequency sound pressure level. Further, it was shown in [25] that some essential dB's at the difference-frequency sound pressure level could be gained through the use of an acoustic lens effect and through the use of a slow-waveguide antenna effect in the nearfield liquid cylinder, a subject which also recently has been studied by Ryder et al [42] for a silicone rubber cylinder in contact with the projector. The increase in β by using air-bubbles in the interaction region of the primaries has been attempted by Lockwood et al [43], but without considerably success.

The disappearance of β in the expression for p_s by high-amplitude wave interactions (44) points against an increase in the conversion efficiency for nearfield saturation limited arrays may be obtained through a decrease in the velocity of sound of the liquid in contact with the projector, a decrease in the primary mean frequency ω_o or an increase in the projector size (normally attempted to be kept small), which can be seen from the transformation of (44) into:

$$|p_s| \simeq \pi/2(\omega_s^2/\omega_o^2)(R_r/R)p_o = \pi/2(\omega_s^2/\omega_o^2)(k_oS/2\pi)(p_o/R) = \omega_s^2 S p_o (4Rc_o\omega_o)^{-1} \qquad (50)$$

using (27).
The last expression in (50) was also found by Bartram [34] through a consideration of the absorption in repeated shocks.

The increase in the conversion efficiency of the parametric acoustic arrays belongs to a research field into which more theoretical and experimental work must be invested before an optimum exploitation of the parametric array can be achieved .

Conclusions.

The parametric acoustic arrays have due to intensive studies went through a fast development, leading into a practical exploitation, since it was first suggested only 13 years ago. Some problems, certain nearfield effects, certain saturation effects, improvement of the conversion efficiency etc., still have to find their solutions, but the promising results hitherto obtained in theoretical studies and in laboratory and full-scale tests point to a still increasing activity in the field of the parametric acoustic arrays in the very close future.

List of generally used symbols.

ρ : fluid density.
c_o : sound velocity in the undisturbed fluid.
δ_{ij} : Kronecker delta.
x : cartesian coordinate
t : time
p : pressure
u_i : particle velocity vector.
q : source strength density function.
β : $1 + B/2A$
α : small signal absorption coefficient.
k : wave number.
η : shear viscosity of the fluid.
ζ : bulk viscosity of the fluid.
κ : coefficient of heat conductivity of the fluid.
ε : acoustic Mach number (u_o/c_o).
χ : acoustic saturation number.
Γ : Gol'dberg number.

References.

1. Helmholtz, H., Die Lehre von den Tonempfindungen als physiologische Grundlage für die Theorie der Musik. Brunswick, 1862.
2. Lamb, H., The dynamical theory of sound. Dover New York, 1960.
3. Lighthill, M.J., Proc. Roy. Soc.(london), 211A, 564, 1952.
4. Lighthill, M.J., Proc. Roy. Soc.(London), 222A, 1, 1954.
5. Westervelt, P.J., J. Acoust. Soc.Amer., 29, 199, 195
6. Westervelt, P.J., J. Acoust. Soc. Amer., 29, 934, 1957.
7. Westervelt, P.J., J. Acoust. Soc. Amer., 35, 535, 1963.

8. Bjørnø, L., Nonlinear Acoustics, in *Acoustics and Vibration Progress*, Vol. II, (Eds.) R.W.B. Stephens & H.G. Leventhall, Chapman & Hall, London 1976.
9. Bjørnø, L., *Ultrasonics International 1975, Conference Proceedings*, IPC Science and Technology Press Ltd., London 1975.
10. Bellin J.L.S. and Beyer, R.T., *J. Acoust. Soc. Amer.*, 34, 1051, 1962.
11. Naze, J. and Tjøtta, S., *J. Acoust. Soc. Amer.*, 37, 174, 1965.
12. Berktay, H.O., *J. Sound Vib.*, 2, 435, 1965.
13. Hobæk, H., *J. Sound Vib.* 6, 460, 1967.
14. Muir, T.G. anf Blue, J.E., *J. Acoust. Soc. Amer.*, 46, 227, 1969.
15. Gol'dberg, Z.A., *Soviet Phys.-Acoust.*, 2, 146, 1956.
16. Blackstock, D.T., *J. Acoust. Soc. Amer.*, 39, 1019 1966.
17. Shooter, J. A., Muir, T.G. and Blackstock, D.T., *J. Acoust. Soc. Amer.*, 55, 54, 1974
18. Mellen, R.H. and Moffett, M.B., Naval Underwater Systems Center, Technical Memo. 1971. PA4-229-71.
19. Muir, T.G., Ph.D. thesis, University of Texas at Austin, ARL-TR-71-1.
20. Muir, T.G. and Willette, J.G., *J. Acoust. Soc. Amer.*, 52, 1481, 1972.
21. Moffett, M.B., Naval Underwater Systems Center, Technical Memo. 1971. PA4-234-71.
22. Fenlon, F.H., *J. Acoust. Soc. Amer.*, 50, 1299, 1971.
23. Fenlon, F.H., *J. Acoust. Soc. Amer.*, 55, 35, 1974.
24. Berktay, H.O and Leahy, D.J., *J. Acoust. Soc. Amer.*, 55, 539, 1974.
25. Bjørnø, L., Christoffersen, B. and Schreiber, M.P., *ACUSTICA*, 35, 99, 1976.
26. Merklinger, H.M., *Proceedings of Symposium on Nonlinear Acoustics*, University of Birmingham, Apr. 1971.
27. Eller, I.A., *J. Acoust. Soc. Amer.*, 56, 1735, 1974
28. Lockwood, J.C. and Fransdal, L.A., AMETEK/STRAZA R&D Rept. 11-730004-00000E-11, 1973.
29. Walsh, G., *Proceedings of Symposium on Nonlinear Acoustics*, University of Birmingham, April 1971.
30. Konrad, W.L., *Finite-Amplitude Wave Effects in Fluids*, (Ed.) L. Bjørnø, IPC Science and TEchnology Press Ltd. London 1974.
31. Merklinger, H.M., Ph.D. thesis, University of Birmingham, 1971.
32. Fenlon, F.H., ibid Ref. 30.
33. Fenlon, F.H., *Parametric scaling laws*, Westinghouse Res. Labs. Nooo14-74-c-0214, 1974
34. Bartram, J.F., *J. Acoust. Soc. Amer.*, 52, 1042, 1972.

35. Lockwood, J.C., AMETEK/STRAZA, Tech. Rept. 11-1554E-76-6, 1976.
36. Barnard, G.R., Willette, J.G., Truchard, J.J. and Shooter, J.A., J. Acoust. Soc. Amer., 52, 1437, 1972.
37. Truchard, J.J., ibid Ref. 30.
38. Rogers, P.H., Van Buren, A.L., Williams Jr., A.O. and Barber, J.M., ibid Ref. 30.
39. Berktay, H.O. and Al Temini, C.A., J. Sound Vib. 9, 295, 1969.
40. Konrad, W.L., Mellen, R.H. and Moffett, M.B., Naval Underwater System Center, Technical Memo.,PA4-304-71 1971.
41. Bartram, J.F., J. Acoust. Soc. Amer., 53, 383, 1973.
42. Ryder, J.D., Rogers, P.H. and Jarzynski, J., J. Acoust. Soc. Amer., 59, 1077, 1976.
43. Lockwood, J.C. and Smith, D.P., AMETEK/STRAZA, Tech. Rept., 11-1354E-74-1, 1974.

DISCUSSION

Comment : P. DE HEERING

What happens to the beam pattern when you try to increase conversion efficiency by using slow propagation waveguides?

Reply : L. BJØRNØ

For the small half-power beamwidth of about 2° of the primary-wave beam patterns used in our experiments, no serious interaction between the primary waves and the cylindrical shell of the waveguide takes place, and the distortion influence of the shell on the beam patterns is very small. Broader beamwidth, primary waves will be exposed to a beamwidth distortion depending upon the length of the slow waveguide antenna, the critical angle, etc., which only in the case of a spreading-loss-limited array may be assumed to have an influence upon the formation of the difference frequency beam. This subject has not yet been fully investigated.

Comment : P. TOURNOIS

For high-amplitude sound waves in fluids, two physical effects have to occur: (i) the pressure increase with amplitude, which is a scalar effect and (ii) the displacement of the fluid, which can be similar to a doppler effect, and is a vectorial effect. Have you taken into account both effects and which is the leading one in water? I suppose it depends on the acoustic impedance of the fluid involved.

Reply : L. BJØRNØ

Two sources of non-linearity contribute to the cumulative distortion of a finite-amplitude wave during its propagation. The first is a kinematic one, being due to the fact that the phase velocity of the finite-amplitude wave is the sum of the local velocity of sound and the local particle velocity, thus leading to a "convection" of the sound velocity along with the particle velocity. The second source is a thermodynamic one, being due to the non-linear relationship between the pressure and the density of a fluid, being expressed by the B/A and the C/A ratios. In liquids the last source is the strongest, but in gases the first source will be dominating, see [8].

ACOUSTIC AND ELECTROMAGNETIC WAVES PROPAGATING IN A TENUOUS RANDOM MEDIUM

R.H. Clarke

Department of Electrical Engineering,
Imperial College, London, S.W.7.

ABSTRACT. Unification is sought of the various theories of random wave propagation which occur in connection with optical and radio wave propagation in the atmosphere and in acoustic propagation in the sea. Common ground may be found in the parabolic equation approximation combined with partial coherence theory.

1. PARABOLIC EQUATION APPROXIMATION [1]

Since our objective is to study the behaviour of fields propagating through tenuous (i.e., weak) random fluctuations in refractive index to possibly large distances, attention can be confined to the phenomenon of forward scatter due to irregularities in refractive index whose typical scale size is large compared to the propagating wavelength. Under these conditions both electromagnetic 2 and acoustic waves obey the scalar equation

$$\nabla^2 f + k^2 (1 + 2\mu) f = 0 \qquad (1)$$

in which $f(\underset{\sim}{r}, t)$ at position $\underset{\sim}{r}$ and time t is any electric-field component in the electromagnetic case, and the pressure in the acoustic case. The mean wavenumber k corresponding to the mean propagating wavelength λ is $k = 2\pi/\lambda$, and $\mu(\underset{\sim}{r}, t)$ is the small departure of the refractive index from its mean value of unity. In equation (1) the scalar field $f(\underset{\sim}{r}, t)$ is assumed to be quasi-monochromatic or narrow-band, since the refractive index fluctuates relatively slowly. The basic time dependence $\exp(j\omega t)$ is suppressed.

G. Tacconi (ed.), Aspects of Signal Processing, Part 1, 61-67.

In an analogous manner, since it is to be assumed that the field propagates essentially in one direction, take the main direction of propagation to be the z-axis and write

$$f(\underset{\sim}{r},t) = u(\underset{\sim}{r},t)\exp(-jkz) \tag{2}$$

where u now represents the propagating field and is a relatively slowly varying function of both position and time. Substituting (2) into (1) and assuming that

$$\frac{\partial^2 u}{\partial z^2} \ll 2k\frac{\partial u}{\partial z} \tag{3}$$

as a consequence of the assumption that the scale size of the irregularities is much larger than the wavelength, in Cartesian coordinates (1) becomes

$$\frac{\partial^2 u}{\partial x^2} + \frac{\partial^2 u}{\partial y^2} - j\,2k\frac{\partial u}{\partial z} + 2k^2\mu u = 0 \tag{4}$$

which is the parabolic equation (P.E.) approximation. For the sake of brevity in the ensuing discussion this equation will be used in its two-dimensional form by assuming uniformity in y. However, there is no difference in principle for the development in three dimensions.

2. A P.E. SOLUTION FOR A RANDOM MEDIUM

Following the spirit of Tappert and Hardin's [3] successful application of the parabolic equation to non-random problems in under water sound propagation by means of the "split-step Fourier algorithm", it is instructive to consider two corresponding steps towards a solution in the case of random propagation.

First, in the absence of refractive index irregularities ($\mu \equiv 0$) the scalar field in the now uniform half-plane $z \geqslant 0$ is given exactly by the Weyl representation [4]

$$f(x,z) = \int_{-\infty}^{\infty} F(S)\exp\left\{-jk(xS + zC)\right\} dS \tag{5}$$

where the plane-wave spectrum function F(S) is expressed in terms of the angle variable $S = \sin\theta$, where θ is the angle that each plane wave direction makes with the z axis, and $C = \cos\theta$. The representation (5) includes inhomogeneous (when $|S| > 1$) as well as homogeneous plane waves, although the former are not important for the present purpose but must be included for completeness.

The scalar field over the plane z = 0 is, from (5),

$$f(x,0) = \int_{-\infty}^{\infty} F(S)\ \exp(-jkxS)\ dS \tag{6}$$

whose Fourier inverse

$$F(S) = \frac{1}{\lambda}\int_{-\infty}^{\infty} f(x,0)\ \exp(+jkxS)\ dx \tag{7}$$

or symbolically $F(S) = \mathcal{F}\left[f(x,0)\right]$, indicating that the angular plane-wave spectrum F(S) is the Fourier transform of the field over the plane z = 0. It can also be shown that F(S) is the far-field (Fraunhofer) diffraction pattern of the aperture-field distribution f(x,0). It should be noted that complete specification of electromagnetic fields requires two angular spectrum functions, corresponding to the two tangential components $E_x(x,0)$ and $E_y(x,0)$ of the electric field in the plane z = 0, whereas acoustic fields require only one. The second spectrum function is easily included, and its omission merely restricts the discussion to linearly polarized aperture fields.

Continuing the discussion for fields diffracted in a uniform medium, if F(S) is narrowly confined to the direction of the z-axis (say F(S) is negligible outside the range $|S| \leqslant 0.2$, or $|\theta| \leqslant 12^{\circ}$) then the cosine can be approximated by $C \simeq 1 - \frac{1}{2}S^2$, and using (5) the field over the plane $z = \Delta z$ can be written approximately as

$$f(x;\Delta z) = \exp(-jk\Delta z)\int_{-\infty}^{\infty} F(S)\ \exp(\tfrac{1}{2}jk\Delta z\ S^2)\ \exp(-jkxS)\,dS \tag{8}$$

in which the integral expression, say $u_0(x;\Delta z)$, can be seen to be the solution of the parabolic equation in a uniform medium. This confirms that the parabolic equation is valid for propagating fields that are narrowly confined in angle.

Equation (8) can be written in alternative forms. In the manner of Tappert and Hardin [3] it is

$$f(x;\Delta z) = \exp(-jk\Delta z)\,\mathcal{F}^{-1}\left\{\mathcal{F}\left[f(x,0)\right]\exp(\tfrac{1}{2}jk\Delta z\ S^2)\right\} \tag{9}$$

or [5], by applying the convolution theorem, as

$$f(x;\Delta z) = \frac{\exp(-jk\Delta z)}{\sqrt{j\lambda\Delta z}}\int_{-\infty}^{\infty} f(x^1,0)\exp\left\{-(jk/2\Delta z)(x-x^1)^2\right\}dx^1 \tag{10}$$

which is Fresnel's diffraction formula.

The second step towards a solution of the parabolic equation in a weakly irregular medium is to start again with (4) this time omitting the one remaining second derivative. This is not too unreasonable in view of inequality (3), particularly for isotropic irregularities. The solution of the first-order differential equation that remains is

$$u(x,t;\Delta z) = u(x,0)\ \exp\left\{-jk\int_0^{\Delta z}\mu(x,z,t)\ dz\right\} \tag{11}$$

assuming that the initial field distribution over the plane z = 0 is strictly monochromatic Equation (11) is merely the initial field distribution over the plane z = 0 projected forward onto the plane z = Δz, with its phase modified by the integrated path length along rays between the two planes parallel to the z-axis.

The two artificially separated parts of the procedure, the first allowing for diffraction but ignoring the irregularities and the second suppressing diffraction but allowing for the effect of the random irregularities, can now be combined. Multiplying (8 by the phase integral part of (11) yields the field over the plane z = Δz as

$$f(x,t;\Delta z) = f_0(x;\Delta z)\ \exp\left\{\ j\ \phi(x,t;\Delta z)\right\} \tag{12}$$

where $f_0(x;\Delta z)$ is the field that would have existed over the plane in the absence of irregularities and

$$\phi(x,t;\Delta z) \ = \ -\ k\int_0^{\Delta z}\mu\ (x,z,t)\ dz \tag{13}$$

Equation (12) combined with (13) should not be taken literally as the field that exists at the point (x,Δz) at time t, but rather as the first step in a marching solution which will then be applied to the planar region between Δz and 2 Δz, and so on.

In the Tappert and Hardin method (12) and (13) are used with the step size Δz small enough for the integral of (13) to be replaced by the integrand. This could also be done if a Monte Carlo type solution of the random propagation problem were required. But for the purposes of a more conventional statistical analysis of the problem, giving estimates of fluctuation amplitudes, spectra, densities, and the like, it is necessary (see below) to choose Δz to be about ten times the horizontal scale size of the refractive index irregularities. The small magnitude of μ and the fact that, μ being zero-mean, there will be a certain amount of cancellation in the process of integration, which may favour the validity of the representation (12) with (13). Certainly the variance σ_ϕ^2 of the phase in (13) must be small.

Summarizing the conditions imposed so far, if (ξ_0, η_0, ζ_0) are the scale sizes of the irregularities in the (x,y,z)directions

$$\xi_0\ ,\ \eta_0\ ,\ \zeta_0 \gg \lambda \tag{14}$$

$$\Delta z \gg \zeta_0 \tag{15}$$

$$\sigma_\phi^2 \sim k^2 \sigma_\mu^2 \zeta_0 \Delta z \ll 1 \tag{16}$$

where σ_μ^2 is the variance of the refractive index irregularities. Conditions such as (14) - (16) are commonly imposed, and enable (12) with (13) to be applied repeatedly to subsequent planar regions.

To illustrate briefly the value of the representation (12) with (13), the mean field at the plane z = Δz is

$$\langle f(x,t;\Delta z)\rangle = f_0(x;\Delta z) \left\langle \exp\left\{ j\phi(x,t;\Delta z)\right\}\right\rangle \tag{17}$$

where the sharp brackets indicate taking the statistical expectation. (A similar expression holds for the mean field at the nth plane, i.e., at z = nΔz, with Δz in (17) replaced throughout by nΔz.) The expectation on the right of (17) is the characteristic function of the random process, which is particularly simple to evaluate if ϕ is Gaussian. This will occur if μ is Gaussian, or if Δz contains a sufficient number of horizontal irregularities for some form of the central limit theorem to apply when μ is non-Gaussian. The variance of the phase can be calculated from (13) if the autocovariance function of the refractive index irregularities is known. (See [2] for a similar calculation in terms of the structure function of the refractive index irregularities, but under the rather awkward assumption that μ is delta-function correlated in the direction of propagation.) It is interesting to use (12) to cast some light on Chernov's division of the propagating field into two parts, one for a uniform medium and the other a perturbation term to first order in σ_μ, as described in [6].

Representation (12) with (13) also gives information about the first-order density of the propagating field [7]: in a stationary random medium the real and imaginary parts of the field are independently Gaussian and the complete signal has the form of a Hoyt distribution [8], and not a Nakagami-Rice as is often supposed. Other distributions, such as the lognormal, may arise from the inherent nonstationarity of the atmosphere or ocean [9], particularly when large distances are involved.

3. PARTIAL COHERENCE DESCRIPTION OF THE PROPAGATING FIELD

The second-order statistics of the propagating field can be obtained from the lateral coherence function, if it is appropriately defined [10]. Since, for a statistically stationary random medium, the first-order statistics of the field are complex Gaussian, this together with the second-order statistics completely describes the field.

In a uniform medium the lateral coherence function propagates without change. For the non-random field f(x,0) over the plane z = 0, the lateral coherence function for points separated in the x direction by ξ is

$$\Gamma_o(\xi) = \int_{-\infty}^{\infty} f^*(x,0)\ f\ (x + \xi\ ,\ 0)\ dx$$

where the asterisk denotes the complex conjugate. According to the van Cittert-Zernike theorem $\Gamma_o(\xi)$ is the Fourier transform of $|F(S)|^2$. But since the magnitude of the angular spectrum is unchanged in a uniform medium, then $\Gamma_o(\xi)$ must also be unchanged.

However, in the presence of tenuous random irregularities in refractive index the original coherence of the field will be deteriorated by the progressive randomisation of the phase, as indicated by (12) and (13). Thus over the plane z = Δz the lateral coherence function in the presence of refractive index irregularities, is [10]

$$\Gamma(\xi, \tau; \Delta z) = \Gamma_o(\xi) \left\langle \exp\left\{ j \left[\phi(x + \xi,\ t + \tau;\ \Delta z) - \phi(x, t; \Delta z) \right] \right\} \right\rangle \qquad (18)$$

assuming that the intervening region is statistically stationary, and where τ is a time displacement.

If the region to the right of the plane z = Δz were uniform, the coherence function (18) would be identical over all subsequent planes. But if the random irregularities continue with the same statistical form then the lateral coherence function will change in a way which will exhibit the continuing degradation in the coherence of the propagating field. Over the nth plane, i.e., at z = n Δz, the lateral coherence function will be given by (18) with Δz replaced by nΔz throughout.

It will not be possible in the space available here to work out the implications of (18) in detail. But it should be noted that at the nth plane the spatial correlation is $\Gamma(\xi, 0; n\Delta z)$ and the temporal autocorrelation is $\Gamma(0, \tau; n\Delta z)$, from which the spectrum of the fluctuations can be derived.

It is worth emphasizing finally that the approximate analysis presented here relies heavily on the initial assumption that the propagating field is essentially forward scattered. This in turn requires that the fluctuations in refractive index are very small compared to unity, that the spatial scale sizes of the irregularities, though not necessarily the same in all directions, should all be very large compared to the wavelength, and that the angular spread of the initial field should be narrow.

REFERENCES

1. V.I. Tatarskii, The Effects of the Turbulent Atmosphere on Wave Propagation, Israel Program for Scientific Translations, 1971.
2. R.L. Fante, Electromagnetic Beam Propagation in Turbulent Media, Proc. IEEE, 63(12), 1975, pp.1669-1692.
3. F.D.Tappert and R. Hardin in C.W.Spofford, A Synopsis of the AESD Workshop on Acoustic-Propagation Modelling by non Ray-Tracing Techniques, AESD, Arlington, Va., May, 1973.
4. A.J.Devaney and G.C. Sherman, Plane-Wave Representations for Scalar Wave Fields, SIAM Review, 15 (4), 1973, pp.765-786.
5. J.A. Ratcliffe, Some Aspects of Diffraction Theory and their Application to the Ionosphere, Rep.Progr.Phys., 19, 1956, pp.188-267.
6. A.M.Prokhorov, et al., Laser Irradiance Propagation in Turbulent Media, Proc. IEEE, 63 (5), 1975, pp.790-811.
7. R.H.Clarke, Transmission of Underwater Sound through Internal Waves and Turbulence, Satellite Symposium on Underwater Acoustics, Eighth International Congress of Acoustics, University of Birmingham, 1-2 August, 1974.
8. P. Beckmann and A. Spizzichino, The Scattering of Electromagnetic Waves from Rough Surfaces, Pergamon, 1963.
9. M. Nakagami, The m-Distribution - A General Formula of Intensity Distribution of Rapid Fading, in W.C.Hoffman, ed., Statistical Methods in Radio Wave Propagation, Pergamon, 1960, p.25.
10. R.H.Clarke, Sound Propagation in a Variable Ocean, Jour. Sound Vibn., 34 (4), 1974, pp.457-477.

TIME-FREQUENCY-SPACE GENERALIZED COHERENCE AND SCATTERING FUNCTIONS

Robert LAVAL

Société d'Etudes et Conseils AERO
3 avenue de l'Opéra - 75001 PARIS, France

1. INTRODUCTION

The notion of scattering function is commonly used to describe the time-frequency spreading which characterizes the signals distorsion during the propagation in a random medium. For the underwater acoustic channel, the random character is introduced by the roughness of the boundaries (surface and bottom), the inhomogeneities in the volume of water affecting the sound velocity, and the interferences between a large number of rays or modes (particularly in shallow waters).

The spatial and angular aspect of the scattering is generally described independently of the time-frequency aspect by a spatial coherence function.

It has been proposed in a previous presentation /1, 2/ to unify these two concepts into a single multidimensional description of the scattering process, which has led to the definition of a generalized scattering function characterizing the time-frequency-angular spread introduced by the propagation.

The interest of this unification is multiple :

a. The generalized scattering function represents the ambiguity function in the most general space including all the variables.

Part of this work has been done during the summer 1976 at the SACLANT ASW RESEARCH CENTRE where the author was working as a temporary consultant.

G. Tacconi (ed.), Aspects of Signal Processing, Part 1, 69-87.

Its knowledge will be required to define the optimum processing of the signals received by a multi-element array. It is absolutely essential in particular for the design of a long range acoustic communication system where the coding of the transmitted signal, the transmitting and the receiving array and the receiver signal processing have to be optimized together.

b. The generalized scattering function has quite often a characteristic structure in its multidimensional form which does not appear clearly from the separate observation of the time-frequenc scattering and the spatial coherence functions.

c. In all the applications where the source and/or the receiver (or the target producing an echo) are mounted on platforms in motion, the space time and frequency effects are mixed together in a way which looks extremely complex at first glance. The generalized scattering function leads to elegant solutions of this type of problems, by simple rotations of its coordinate systems.

3. GENERALIZED COHERENCE AND SCATTERING FUNCTION

Let recall briefly the definitions of the generalized coherence and scattering function as they have been defined in ref. /1, 2/ :

By using the formalism of linear fifter theory, the channel can be characterized in the most general way by its transfer (or weighting) function :

$$H(f, t, \vec{s}, \vec{r})$$

H being a complex function giving the amplitude and phase of the quasi monochromatic signal which would be received at the time t by a point receiver at a location defined by a vector $\vec{r}$ if a monochromatic wave of unit level and frequency f was transmitted by a point source at a location $\vec{s}$.

If the propagation medium has random characteristics, H is a random function which can only be described by its statistical properties.

For the sake of simplicity, the description may be limited to the first and second order moments.

The first order moment will be the mean value H_o :

$$H_o(f,t,x,y,z) = \langle H(f,t,x,y,z)\rangle$$

where the sign $<\ldots>$ means ensemble average.

(We now assume the source $\vec{s}$ to be fixed at the origin, x, y, z being the coordinates of $\vec{r}$).

H will then decompose into a sum of two terms :

$$H = H_0 + \tilde{H}$$

where $\tilde{H}$ is a random function of zero mean value.

H_0 will be called the "coherent term" and $\tilde{H}$ the "incoherent term".

Theratio :

$$\gamma = \frac{H_0^2}{<H^2>} = \frac{H_0^2}{H_0^2 + <\tilde{H}^2>} \qquad \text{with} \qquad 0 \leqslant \gamma \leqslant 1$$

being the "coherence factor".

When $0 < \gamma < 1$ the process is said to be "partially coherent".

When $\gamma = 0$, $H_0 = 0$ the process is said to be "totally incoherent".

The incoherent terme $\tilde{H}$ can be described at the second order by its covariance function :

$$\Gamma(f_1, f_2, t_1, t_2, x_1, x_2, y_1, y_2, z_1, z_2) = \\ <\tilde{H}(f_1, t_1, x_1, y_1, z_1)\, \tilde{H}^*(f_2, t_2, x_2, y_2, z_2)>$$

which can also be written, if we call $\delta f = f_2 - f_1$, $\delta t = t_2 - t_1$, etc.

$$\Gamma(f, t, x, y, z, \delta f, \delta t, \delta x, \delta y, \delta z) = \\ <\tilde{H}(f, t, x, y, z)\, \tilde{H}^*(f + \delta f, t + \delta t, x + \delta x, y + \delta y, z + \delta z)>$$

By a generalization of what is done for the time-frequency scattering function, we may introduce the concept of a time-frequency-space WSSUS process (Wide Sense Stationary Uncorelated Scattering) for which the covariance Γ will not depend any more of the f, t, x, y and z terms, but will be a function of the difference terms δf, δt , δx, δy, δz only /3, 4/.

The five dimensions function :

$$\Gamma(\delta f, \delta t, \delta x, \delta y, \delta z)$$

will then be called the generalized, or time-frequency-space coherence function.

By Fourier transforming this function with respect to the five differential variables a new function is obtained :

$$R(\tau, \Phi, u, v, w)$$

which is the generalized scattering function.

Where :

. τ is a time spread variable (corresponding to δf by Fourier transformation).

. Φ is a frequency spread variable (corresponding to δt).

. u, v and w are the spatial frequency spread variables (corresponding to δx, δy and δz).

If the sound field produced by the source at the origin may be decomposed into elementary plane waves in the vicinity of the point r , the variables u, v, w can be expressed by :

$$u = \frac{f}{c}\alpha_x \qquad v = \frac{f}{c}\alpha_y \qquad w = \frac{f}{c}\alpha_z$$

with :

$$\alpha_x = \sin\theta_x \qquad \alpha_y = \sin\theta_y \qquad \alpha_z = \sin\theta_z$$

where θ_x is the angle of the elementary plane wave with the x axis, θ_y with the y axis, and θ_z with the z axis.

α_x, α_y , and α_z are also the direction cosines of the vector defining the direction along which the wave is propagating.

Changing to the variables u , v , w the generalized scattering function can then be expressed as a time-spread frequency-spread angular-spread function.

Within the hypothesis that the sound field can be decomposed into plane waves, the variables u, v and w, or the variables θ_x, θ_y , θ_z are not independent, but they are related by the formula :

$$\alpha_x^2 + \alpha_y^2 + \alpha_z^2 = 1$$

The coherence function of the scattering function will then be fully defined by the knowledge of a 4 (instead of 5) coherence function or scattering function, such as :

$\Gamma(\delta f, \delta t, 0, \delta y, \delta z)$ in which δx is equal to 0

or $R(\tau, \Phi, \theta_y, \theta_z)$

y being for instance the variable corresponding to the horizontal transverse axis, z to the vertical, and x to the horizontal axis in the direction source-receiver.

The relation of R as a function of θ_x will just be found by :

$$\theta_x = \arcsin\left[\frac{2\pi f}{c}\sqrt{1 - \sin^2\theta_y - \sin^2\theta_z}\right]$$

The 5 dimensions coherence function can be computed from the knowledge of a 4 dimensions one by the relation :

$$\Gamma(\delta f, \delta t, \delta x, \delta y, \delta z) = \iint \Gamma(\delta f, \delta t, \delta y, \delta z)\, e^{i\left(v\delta y + w\delta z + \frac{2\pi f}{c}\sqrt{1-v^2-w^2}\,\delta x\right)}\, dv\, dw$$

The generalized scattering function R may be considered as the ambiguity function of the medium : if a point source at the origin of coordinates transmits a signal of ideal ambiguity function (in time-frequency), a receiver composed of a highly directive array followed by a high resolution time-frequency processor which will try to locate the source will see a diffused "cloud" instead of a point in a range, doppler, angles space. In case of partial coherence the receiver will see a bright point (or a more complicated "stable part" of the picture) surrounded by a diffused halo.

3. COHERENCE FUNCTION AND SCATTERING FUNCTION OF A SINGLE VARIABLE

In classical theories, the coherence is generally expressed as a function of a single variable.

$\Gamma(\delta f) \equiv \Gamma(\delta f, 0, 0, 0, 0)$ is the "frequency coherence" or the "bifrequency correlation function". It is the section of the multi-dimension coherence function along the axis δf, where δt, δx, δy and δz are equal to zero.

The Fourier transform of $\Gamma(\delta f)$ is the "time-spread function" $R_s(\tau)$.

$R_s(\tau)$ is the projection of the 4 dimensions generalized scattering function on the τ axis :

$$R_s(\tau) = \iiint R(\tau, \Phi, \theta_y, \theta_z)\, d\Phi, d\theta_y\, d\theta_z$$

The time spread function $R_s(\tau)$ is the mean squared value of the impulse response h (τ) received at the point $\vec{r}$ if a Dirac pulse is transmitted at the origin :

$$R_s(\tau) = \langle h^2(\tau) \rangle$$

$R_s(\tau)$ should not be confused with the section of the scattering function along the τ axis, which would be :

$$R(\tau, 0, 0, 0)$$

where Φ , θ_y and θ_z are equal to 0 *; the Fourier transform of which would be the projection (and not the section) of the coherence function on the δf axis in other words :

$$R(\tau, 0, 0, 0) \Longleftrightarrow \iiint \Gamma(\delta f, \delta t, \delta y, \delta z)\, d(\delta t)\, d(\delta y)\, d(\delta z)$$

The direct measurement of R (τ , o, o, o) (or of its Fourier transform) cannot be isolated from the estimation of the multi-dimensional coherence or scattering function, as it does involve the coherence law in the full space of the variables.

In a similar way :

$\Gamma(\delta t)$ is the time coherence function. Fourier transform of $R_s(\Phi)$ which is the "frequency-spread function", resulting from the projection of R (τ , Φ , θ_y , θ_z)on the Φ axis.

$\Gamma(\delta x)$, $\Gamma(\delta y)$ and $\Gamma(\delta z)$ are the spatial coherence functions. With the conventions choosen for the axis they correspond respectively to the longitudinal, transversal and vertical coherence functions. Their Fourier transforms are respectively R_s (u), R_s(v) and R_s(w) which are the longitudinal, transversal and vertical "spatial frequency spread functions", projections of R on the axis u, v and w. By expressing these functions as functions of the angles θ_x, θ_y and θ_z , one gets the angular spread functions :

* As a consequence θ_x would be equal to $\pi/2$ and the section of R along τ expressed in 5 dimensions would have the dissimetrical form : $R(\tau, 0, \pi/2, 0, 0)$

$$R_s(\theta_x) \cdot R_s(\theta_y) \cdot R_s(\theta_z)$$

each one representing the angular ambiguity functions of the medium as it would be seen by a linear array in the x, y or z direction.

The one dimensional coherence or spread function have not necessarily to be taken along one of the 5 reference axis but they can also be evaluated along any oblique direction. Let us imagine that the receiving (or the transmitting) point moves in the x, y horizontal plane with a constant speed V, in a direction β with respect to the x axis. The signal received from a point source at the origin transmitting a signal of constant frequency f will be:

$$\mathcal{H}(t) = H(f, t, x_0 + V\cos\beta t, y_0 + V\sin\beta t, z_0)$$

and the time coherence function of this received signal will result from the following oblique section on the multidimensional coherence function $\Gamma(\delta f, \delta t, \delta x, \delta y, \delta z)$:

$$\Gamma[\mathcal{H}(t)] = \Gamma(0, \delta t, V\cos\theta\delta t, V\sin\theta\delta t, 0)$$

The Fourier transform of $\Gamma[\mathcal{H}(t)]$ will be the "frequency-spread scattering function applied to the received signal. It will be the projection of $R[\tau, \Phi, u, v]$ on the corresponding oblique axis in the τ, Φ, u, v space).

If a linear FM pulse had been transmitted instead of a pure tone, the section axis of the coherence function would also have been oblique with respect to the axis f, and the projection axis of the scattering function would have been oblique with respect to the τ axis.

4. RESTRICTION CONCERNING THE VALIDITY OF THE WSSUS CONDITIONS

The WSSUS conditions applied to the generalized time-frequency-space random channel can be expressed in two equivalent ways :

a. The random transfer function $\tilde{H}$ (f, t, x, y, z) should be wide sense stationary with respect to any of its variables, which means that its autocovariance function would only depend of the differential terms δf, δt, δx, δy, δz. It is then possible to define an autocovariance function $\Gamma(\delta f, \delta t, \delta x, \delta y, \delta z)$ which does not depend of f, t, x, y, z and will be called the coherence function.

b. The scattered energy which is received from different directions, from different frequency bands, or at different time delays is uncorrelated. The cross correlation function of this scattered energy is then :

$$\Delta(\delta t, \delta \Phi, \delta \theta_y, \delta \theta_z)\, R(\tau, \Phi, \theta_y, \theta_z)$$

where Δ is the symbol for the Dirac pulse and R $(\tau, \Phi, \theta_y, \theta_z)$ is the scattering function.

It is clear that these conditions are never fulfilled in a strict sense and this for several reasons :

. The random character which may characterize the physical properties of the medium (bottom roughness, surface waves, thermal microstructure and turbulences in the volum) may not be perfectly homogeneous in space and stationary in time.

. The scattering phenomena are very sensitive to small changes in the geometry. Even if the medium is homogeneous and stationary, the function H will not be stationary as a function of the range x and the depth z, as the structure of the multipath and the grazing angles of each ray on the surface and the bottom will change.

. The scattering phenomena are always frequency dependent. Most of the time they are governed by diffraction effects which makes the propagation laws to be dispersive. Even simple considerations of dimensions indicate the scattering to be frequency dependent . (We have seen that the relations between the coherence functions along the different axis are frequency dependent).

When the process is not stationary the second order statistics is a covariance function which depends of both the coordinates f, t, x, y, z and the displacements terms δf , δt , δx, δy, δz :

$$\Gamma(f, t, x, y, z, \delta f, \delta t, \delta x, \delta y, \delta z)$$

The concept of stationary can be saved, however, as long as the statistical properties of the process are varying slowly enough in time frequency and space, that the covariance function varies much more slowly in f than in δf, in t than in δt , in x than in δx, etc.

It is then possible to devide the f, t, x, y, z space into some elementary domains of local stationary within which the process can be considered as stationary, and both a coherence function and a scattering function can be defined.

It is necessary for this that the dimensions $\Delta f, \Delta t, \Delta x, \Delta y, \Delta z$ of the "local stationarity domain" be much larger than the corresponding dimensions $(\delta f)_{eff}$, $(\delta t)_{eff}$, $(\delta x)_{eff}$ etc. of the "elementary domain of effective coherence" within which the values of the transfer function H are correlated.

This condition will rarely be fulfilled for the variable f. Experimental results as well as theoritical modelling indicates that the stationarity of the frequency dependence of H rarely extends beyond a 1/3 octave, sometimes even less.

In order for the process to be locally stationary in f, the effective frequency coherence separation $(\delta f)_{eff}$ has to be much smaller, say no more than 1/100 octave. This means that the time spread function $R_s(\tau)$ will extend on more than a hundred of periods. In other terms, the time frequency product of the 1/3 octave impulse response has to be large. This condition may be fulfilled in shallow water, although it is generally quite marginal. It is practically never verified in deep water.

If the process is stationary for all the variables except f, the scattering function takes the form :

$$R(\tau, \delta\tau, \Phi, u, v, w)$$

the process being no more uncorrelated in the τ dimension.

The non stationarity may also occur along the depth axis z, particularly in shallow water where the number of wave length as a function of depth may not be very large. In this case a term (δw) will appear in the scattering functions, and the principle of decomposition of the sound field into elementary plane waves will have to be applied with much care, as the angles θ_x, θ_y and θ_z may become complex and not satisfy anymore the conditions $\alpha_x^2 + \alpha_y^2 + \alpha_z^2 = 1$.

5. DIMENSIONLESS COHERENCE AND SCATTERING FUNCTIONS

As already said, all the scattering effects are frequency dependent :

. the frequency spread $R(\Phi)$ is generally proportional to f as it is associated with the doppler of the scatterers [5, 6];

. the time spread $R(\tau)$ generally decreases with frequency as a consequence of the medium dispersion;

. the angular spread $R(\theta_x)$ etc. tends to be independent of frequency which makes $u = \frac{f}{c} \sin \theta_x$, v and w proportional to f.

It is then natural to use the dimensionless variables $\frac{\delta f}{f}$, $2\pi f \delta t$, $\frac{2\pi f}{c}\delta x$, $\frac{2\pi f}{c}\delta y$, $\frac{2\pi f}{c}\delta z$.

The coherence function expressed as a function of these dimensionless variables will then be :

$$\Gamma\left(\frac{\delta f}{f}, 2\pi f \delta t, \frac{2\pi f}{c}\delta x, \frac{2\pi f}{c}\delta y, \frac{2\pi f}{c}\delta z\right)$$

The Fourier transform of the function Γ with respect to these dimensionless variables will be :

$$R(\eta, \xi, \alpha_x, \alpha_y, \alpha_z)$$

where $\alpha_x = \sin\theta_x$, $\alpha_y = \sin\theta_y$, $\alpha_z = \sin\theta_z$

the variable η being a "time spread operator" expressing the time delays in periods of the transmitted wave, and the variable $\xi = \frac{\Phi}{f}$ being a "doppler-spread variable" (frequency spread divided by the central frequency).

If Γ and R are only depending of these dimensionless variables, we shall tell that the process is WSSUS in the dimensionless space. Such a process will be dispersive, as the time spread is inversely proportional to the frequency, and the ambiguity in the doppler estimation will be independent of frequency as $\xi = \frac{\Phi}{f}$ /5, 6/.

6. COHERENCE FACTOR ALONG THE VARIOUS AXIS

It may happen that the transfer function may be decomposed into a sum of functions some of them not being dependent of all the variables. As an example, a model for a shallow water channel could take the form :

$$H = H_0 + H_1(f,x,z) + H_2(f,x,y,z) + H_3(f,t,x,y,z)$$

where :

H_0 represents the coherent part of the process.

H_1 would result from the multipath (independent of t and y).

H_2 is related to the bottom scattering (independent of t).

H_3 dependent of the 5 variables would be associated with the scattering on the surface waves.

The mean value of H, which will be computed by averaging it along a given reference axis will not be the same for all the variable.

$$\frac{1}{\Delta f}\int_{f_0}^{f_0+\Delta f} H\,df = H_0$$

$$\frac{1}{\Delta t}\int H\,dt = H_0 + H_1(f,x,z) + H_2(f,x,y,z)$$

$$\frac{1}{\Delta x}\int H\,dx = H_0$$

$$\frac{1}{\Delta y}\int H\,dy = H_0 + H_1(f,x,z)$$

$$\frac{1}{\Delta z}\int H\,dz = H_0$$

The coherence factor will then be different for the various reference axis.

$$\gamma_f = \gamma = \gamma_z = \gamma = \frac{H_0^2}{\langle H^2\rangle}$$

$$\gamma_t = \frac{H_0^2 + \langle H_1^2\rangle + \langle H_2^2\rangle}{\langle H^2\rangle}$$

$$\gamma_y = \frac{H_0^2 + \langle H_1^2\rangle}{\langle H^2\rangle}$$

The coherence factor will also be equal to γ for all the oblique axis.

The axis t and y are then presenting the singular property with respect to all the other directions, of a higher coherence factor. We shall call them the "directions of partial invariance".

Much care has to be taken using the corresponding variables to carry out some integrations.

7. METHODS FOR THE DIRECT ESTIMATION OF THE GENERALIZED COHERENCE FUNCTION

The transfer function H (f, t, x, y, z) can be considered as a particular realization of a stochastic process. By associating the concept of ergodism with the one of stationarity the generalized multidimension coherence function can be estimated by the finite integral :

$$\Gamma(\delta f,\delta t,\delta x,\delta y,\delta z) \approx \frac{1}{\mathcal{A}}\iiiint\!\!\int_{\mathcal{A}} H(f,t,x,y,z)\,H^*(f+\delta f,t+\delta t,x+\delta x,y+\delta y,z+\delta z)\,df,dt,dx$$

The domain of integration $\mathcal{A}$ has to satisfy two conditions :

a. It has to be large enough to contain a sufficient number of statistically independent coherence domains (the dimensions of which are $(\delta f)_{eff}$, $(\delta t)_{eff}$, $(\delta x)_{eff}$ etc.) in order to give a statistically significant averaging.

b. It has to be contained within one elementary domain of local stationarity.

When this domain $\mathcal{A}$ has been defined, the sample of the function H contained within the domain $\mathcal{A} + \delta\mathcal{A}$ has to be measured. $\delta\mathcal{A}$ taking into account the maximum possible excursion of δf, δt etc. (We dont consider here the difficulty which may limit the precision in the measurement of H as both a function of f and t, as a consequence of the uncertainty principle. The ways to overcome this difficulty are well known in the classical theories of the scattering function).

It is not necessary however that the domain of integration extends on all the 5 variables. The integration can even be carried out along one single variable, the domain $\mathcal{A}$ being then reduced to a segment on a straight line. As an example, let us take f as the integration variable :

$$\Gamma(\delta f, \delta t, \delta x, \delta y, \delta z) \approx$$
$$\frac{1}{f_2 - f_1} \int_{f_1}^{f_2} H(f,t,x,y,z)\, H^*(f+\delta f, t+\delta t, x+\delta x, y+\delta y, z+\delta z)\, df$$

The estimation will be valid provided the following conditions are respected :

a. The variable should not correspond with an axis of partial invariance : if the process can be decomposed as in paragraphe 6 the variables t or y should not be used, but f, x, z, or any variable corresponding to an oblique axis in the multispace can be chosen.

b. The integration domain $f_2 - f_1$ has to be much larger than the frequency coherence distance $(\delta f)_{eff}$.

c. The process has to be stationary in f within the interval $f_2 - f_1$.

The conditions b and c can only be satisfied together if the process is locally stationary in f.

The practical experimental procedure to implement the above method could be the following :

A point source transmits a broadband signal (ideally a Dirac pulse) at some discrete time intervals. The signals are received by a number of point hydrophones distributed on a 2 or 3 dimensions array (2 are sufficient with the plane wave decomposition hypothesis).

The Fourier transforms of the signals received by each hydrophone from each transmission represent the transfer function H_{pm} (f, t_p, x_m, y_m, z_m) as a continuous function of f for the particular transmission time t_p and for the particular coordinates associated with the hydrophone m.

Computing the integral :

$$\frac{1}{f_2 - f_1} \int_{f_1}^{f_2} H_{pm}(f, t_p, x_m, y_m, z_m)\, H^*_{qn}(f+\delta f, t_q, x_n, y_n, z_n)\, df$$

for an arbitrary couple of signals corresponding to different transmission times and different hydrophone locations would give an estimate of $\Gamma(\delta f, \delta t, \delta x, \delta y, \delta z,)$ for the particular value of $(\delta t)_{pq} = t_q - t_p$, $(\delta x)_{mn} = x_n - x_m$ etc. and the value of δf which has been choosen.

Taking all the signals 2 by 2 and repeating the computation for all the pairs which can be formed, and for various values of δf would permit an estimation of the function Γ to be constructed point by point. The definition and the regularity of the distribution of the points in the δt, δx, δy, δz space depends on the number and geometrical distribution of the hydrophones on the array and the number and distribution of the transmitted pulses on the time axis. Geometrical spacing in space and in time tends to give more homogeneous distribution of interval for a minimum number of hydrophones and impulses.

An interesting aspect of this method is that the array dimension along x,y and z and the total duration of the series of transmitted pulses have to be just superior to the effective coherence distance $(\delta x)_{eff}$, $(\delta y)_{eff}$, $(\delta z)_{eff}$, $(\delta t)_{eff}$ along the corresponding axis.

The sample along the variable which has been choosen to carry out the integration (in this case f) is the only one which has to extend over a range of variation which has to be much larger than the effective coherence distance along the same axis :

If another variable is choosen to carry out the integration, such as the range x or the depth z, the transmitted signal can be much narrower in frequency (just superior to $(\delta f)_{eff}$) but the array has to be much larger along x or along z. Such a large array has not necessarily to be physically built, but can be synthetised by towing a smaller array (or, which is the same, by towing the sound source) along the range x or the depth z. In this case, the integration is carried out along an oblique axis in the t, x (or in the t, z) plane.

The generalized scattering function R $(\tau, \Phi, \theta_y, \theta_z)$ can be computed by Fourier transformation of the coherence function.

The direct estimation of the coherence function or the scattering in its multidimensional form is then possible but it requires an extremely heavy process which can only be carried out within fixed experimental installations involving large receiving arrays, stabilized transmission platforms and powerful signal processing facilities.

8. ESTIMATION OF THE ONE-DIMENSION COHERENCE AND SCATTERING FUNCTIONS

Most of the time, the experimental approaches will limit themselves to the separate estimation of the single variable coherence functions : $\Gamma(\delta f, o, o, o, o)$, $\Gamma(o, \delta t, o, o, o)$ etc.

The corresponding experiments are much simplified, but the pinciple remains the same. As an example lets consider the estimation of the transverse coherence function $\Gamma(o, o, o, \delta y, o) \equiv \Gamma(\delta y)$:

$$\Gamma(\delta y) \approx \int_{\mathcal{A}} H(f,t,x,y,z)\, H^*(f,t,x,y+\delta y,z)\, df, dt\, dx\, dy\, dz$$

The integration can be carried out over a volume $\mathcal{A}$ which should include a large number of statistically independent coherence domains, but should be contained within the limits of the local stationarity domain.

Here again, it is not necessary that the domain of integration should extend on the 5 variables. The integration can be performed along a single variable. In this cases two families of cases have to be considered, depending of the variable which has been choosen for the integration :

a. The integration can be performed on the same variable along which the coherence is being measured (in our example y) :

$$\Gamma(\delta y) \approx \frac{1}{y_2 - y_1} \int_{y_1}^{y_2} H(f,t,x,y,z)\, H^*(f,t,x,y+\delta y,z)\, dy$$

The experimental method based on this principle requires a linear array in the y direction. This array has to be much longer than the effective coherence distance $(\delta y)_{eff}$.

A point source at the origin transmits a pure tone signal of frequency f and the element number k of the array receives the signal $H_k e^{2\pi i f t}$. If δy_0 is the spacing between two successive elements of the array, the estimation of $\Gamma(\delta y)$ will be :

$$\Gamma(m\,\delta y_0) = \frac{1}{N-m} \sum_{k=1}^{N-m} H_k \cdot H^*_{k+m}$$

The spacing δy_0 between successive elements has to be smaller than the effective coherence length δy in order to describe $\Gamma(\delta y)$ with a sufficient definition in δy.

It is clear that the array has to be straight line with a precision corresponding to a small fraction of the wave length.

The method then requires a long straight line array with a large number of independent elements.

The processing of such an array can be treated in another way: forming the beams of the array in various directions θ_y with respect to the perpendicular to the array and measuring the energy D (θ_y) received as a function of θ_y, one will measure the angular scattering function which is the Fourier transform of $\Gamma\left(\frac{2\pi f\,\delta y}{c}\right)$.

b. The variable along which the averaging process is carried out can be different of the one along which the coherence is to be measured. As an example, the estimation of $\Gamma(\delta y)$ can be performed by averaging along the variable f :

$$\Gamma(\delta y) \approx \frac{1}{f_2 - f_1} \int_{f_1}^{f_2} H(f,t,x,y,z) \cdot H^*(f,t,x,y+\delta y,z)\, df$$

In this case the point source at the origin has to transmit a broad-band signal, ideally a Dirac pulse, which can be approximated by an explosive charge.

The resulting impulse responses are received by the hydrophones of a linear transversal array, which has just to be slightly longer than the effective correlation distance $(\delta y)_{eff}$, the spacings being arranged in order to cover a suitable range of intervals for a correct description of $\Gamma(\delta y)$.

Let call $h(\tau, y_m)$ the impulse response received by the hydrophones n° m in the array, its coordinate along the y axis being y_m and $h(\tau, y_n)$ the impulse response received by the hydrophone n° n. (Droping the variables t, x, z which are left constant in this application). The functions h being Fourier transform of the functions H, we can write $\Gamma(\delta y)$ in the form :

$$\Gamma(y_m - y_n) \approx \frac{1}{f_2 - f_1} \iint_{-\infty}^{+\infty} \int_{f_1}^{f_2} h(\tau, y_m)\, h(\tau + \delta\tau, y_n)\, e^{2\pi i f \delta\tau}\, d\tau\, d(\delta\tau)\, df$$

Integrating successively, first in τ, then in $\delta\tau$, finally in f corresponds to the following sequence of signal processing operations

1. The impulse responses received by two individual hydrophones m and n are cross-correlated, which gives a function of $\delta\tau$:

$$\rho_{mn}(\delta\tau)$$

2. $\rho_{mn}(\delta\tau)$ is Fourier transformed, which gives the cross-spectrum $g_{mn}(f)$ which is a complex function of f.

3. The cross-spectrum $g_{mn}(f)$ is averaged by integrating over f within the domain $f_2 - f_1$. The result of this integration is a complex value $C_{mn} e^{i\varphi_{mn}}$ which represents an estimation of $\Gamma(\delta y_{mn})$ along the axis joining the two hydrophones, and for the value of δy corresponding to their separation distance.

4. The phase term φ_{mn} will be discarded as it can be admitted, for reason of symmetry, that the coherence function along an axis perpendicular to the propagation is purely real. Only the amplitude term C_{mn} will then be retained. In fact, the phase term, if it does exist, is due to the fact that the line joining the two hydrophones does not coincide exactly with the y axis, so is not exactly perpendicular to the propagation. As long as the angle is relatively small, this deviation does not really

affect the amplitude term C_{mn}.

The total coherence function $\Gamma(\delta y)$ will be constructed point by point by taking the values of C corresponding to all the spacings $(\delta y)_{mn}$ of the hydrophones taken 2 by 2.

It is clear that the methods suppose that the process is locally stationary in both the y and f directions.

With the method b, the stationarity in y can only be checked by repeating the experiment for different positions of the array along the y axis.

The stationarity in f can be checked by applying some tests on the received impulse responses :

. the time-frequency product of the band-pass filtered impulse response should be large,

. the autocorrelation function of the filtered impulse response should not be larger than the impulse response of the band pass filter. (If not, the uncorrelated scattering condition is not verified on the variable τ).

The processing will be repeated for all the elementary frequency bands (in practice for all the 1/3 octave bands) which are available in the received signal, which will give a frequency dependent coherence function of δy :

$$\Gamma(f, \delta y)$$

As already mentionned in part 4, it will happen most of the time that the conditions of stationarity along the variable f are not fulfilled. In this case, the application of the above method will conduct to an estimation of the function $\Gamma(f, \delta y)$ which will be based on the averaging of an insufficient number of statistically independent coherence domains. (This means in practice that the time-frequency product of the impulse response filtered by 1/3 octave will not be large enough).

The estimated value of $\Gamma(f, \delta y)$ will then be affected by important fluctuations. If the measurement is carried out in several octave bands, these fluctuations can be reduced by expressing Γ as a function of the dimensionless variable $\frac{2\pi f}{c}\delta y$:

$$\Gamma\left(f, \frac{2\pi f}{c}\delta y\right)$$

the Fourier transform of which being the angular spread function :

$$R(f, \theta_y)$$

and by smoothing one or the other of these expressions along the variable f.

As it has been mentionned in part 5, the stationarity of the process tends to follow the dimensionless variables (such as $\frac{2\pi f}{c}\delta y$) better than the original ones (such as δy). If the process can be considered as being WSSUS in the dimensionless space, the expressions of Γ and R as a function of $\frac{2\pi f}{c}\delta y$ or θ_y will become independent of f.

The type of reasoning which has been used for describing a possible method for the estimation of the coherence function along the y axis by an averaging process which can be conducted along y axis itself, or along another axis such as the f axis can be easily transposed for finding the methods to estimate the coherence function along other axis (including oblique axis) in the space time-frequency space.

9. CONCLUSIONS

It is believed that the generalized coherence function and the scattering function, in their multidimensional space-frequency-time form represent the tools which are necessary for the solution of all the signal processing problems in a medium where the sound field generated by a distant source presents a random small scale structure.

An experimental estimation of these multidimensional functions is possible in principle, but it requires some very sophisticated equipment and processing facilities, which can only be done as a kind of exercice in very particular and limited number of conditions.

On the opposite, experimental estimations of the one-dimension spreading or scattering functions which are the sections or the projections of the above functions on the various axis can be done much more easily.

The reconstruction of the multidimension functions from these section or projection may present some real difficulty, as the structure of these functions in the multivariable space can be rather peculiar and is generally such that these functions are not "separable". The only chance to obtain a correct representation of these multidimension functions is to associate a reasonable amount of experimental investigations with theoritical studies which would describe the general shape of these functions as a function of the statistical properties of the medium.

The stationarity hypothesis which are generally assumed a priory when carrying these estimations have to be checked with much care, and possibly replaced by more realistic hypothesis when they are not verified in the real medium.

ACKNOWLEDGEMENTS

Many thanks are due to J. C. VETTORI for fruitfull discussions on this subject.

REFERENCES

1. LAVAL (R). Sound Propagation Effects on Signal Processing. Proceedings of the "NATO Advanced Study Institute on Signal Processing" held at Loughborough - UK, Academic Press (1973).
2. HOVEM (J. M). Resolution Limitations of a Random Inhomogeneous Medium. Saclantcen TR 214 (1973).
3. ELLINTHORPE (A. W), NUTTAL (A. H). Theoritical and empirical Results on the Characterisation of Undersea Acoustic Channels. 1st Annual Communication, IEEE Convention, Boulder, June 1965.
4. SOSTRAND (K. A). Measurements of Coherence and Stability of Underwater Acoustic Transmission. Proceeding of Nato Advanced Study Institute on Signal Processing. Enschede, 1968.
5. JOURDAIN (G). Espérance d'ambiguité dans le cas où l'écho subit une compression de temps vis-à-vis de l'émission. Annales Télécommunications, 28, N° 1-2, Jan. -Feb. 1971.
6. JOURDAIN (G). Caractérisation d'un milieu de transmission variable par un modèle de FAPV. Annales Télécommunications, Sept. -Oct. 1973.

SERIES EXPANSIONS OF VELOCITY PROFILE, HORIZONTAL RANGE AND TRAVEL TIME OF RAY THEORY IN OCEANS

Tamio Kuyama and Toshiaki Kikuchi

Department of Applied Physics, National Defense Academy, Yokosuka, Japan

ABSTRACT. By applications of the velocity profile with depth, which was proposed by M.A.Pedersen, the horizontal range and the travel time of the ray theory are represented by hypergeometric series. Especially, when there is the relative maximum of velocity profile, the properties of series expansions are investigated. The relations between the convergency of series expansions and the value of the exponent of the velocity profile equation are discussed.

1. INTRODUCTION

M.A.Pedersen[1] proposed the velocity profile which is indicated by Eq.(1). This velocity profile may be fitted easily with any observed form of velocity profile in the ocean when one increases the number of parameter C_k. Also, with this velocity profile, one can calculate series expansions of the horizontal range and the travel time of the ray in the ocean. M.A.Pedersen *et al*[2] presented the Taylor series expansions of horizontal range, travel time and range derivatives. However, it was limited to the case of expansions about the relative minimum point of velocity profile.

Present authors intended to investigate the case of expansions about the relative maximum point of velocity profile. In this case, one should separate the mode of ray propagation in the two distinct modes. One is the trapped mode and the other is the leaky mode. In both modes, to expand the horizontal range and the travel time in series expansions, the hypergeometric series expansion has been utilized. The series expansion method is compared with the conventional method. At the transition from the trapped to the leaky mode, there is a split-beam ray. Near the split-beam ray, the trouble of slow convergency of series expansions has been found. The trouble

G. Tacconi (ed.), Aspects of Signal Processing, Part 1, 89-94.

is not solved in our discussion. However, it may be necessary to apply the idea of the generalized WKB method which has been propose by E.L.Murphy[3] to solve this problem.

2. PROBLEM FORMULATION

The left schematic of Fig.1 represents the model of the velocity profile which has the point of relative maximum velocity c_0. The form is expressed by

$$z-z_0=\sum_{k=1} C_k |c-c_0|^{\sigma_k}, \tag{1}$$

where z_0 is the depth of the maximum velocity and vertical bars indicate to take the absolute value. The exponents σ_k may be any ordered set of positive real numbers. The right schematic of Fig.1 represents the ray of trapped, split-beam and leaky mode. The sound velocity c_m is the velocity where the ray becomes horizontal. The symbol S indicates a source and c_s is the velocity of sound at S. The symbol H_1, H_2 and H_3 indicate the receiving point. One ray which starts from S and goes to H_1 *via* c_m is the ray of trapped mode. In this case c_m is always smaller than c_0. When this ray of trapped mode touches horizontally on the axis of the maximum velocity at H_2, it becomes a split-beam ray. In this case, c_m is equal to c_0. The ray, which starts from S at the more larger grazing angle than the split-beam ray and penetrates into the lower layer through H_3, is the ray of leaky mode. In this mode c_m is greater than c_0. The range SH_1 is the horizontal cycle range of the ray of trapped mode. This range is indicated by $\tilde{R}$. The travel time corresponding to this cycle range is indicated by $\tilde{T}$.

3. HORIZONTAL RANGE AND TRAVEL TIME OF THE RAY OF TRAPPED MODE

The horizontal range $\tilde{R}$ in Fig.1 is calculated by the ray theory and the result reduces to

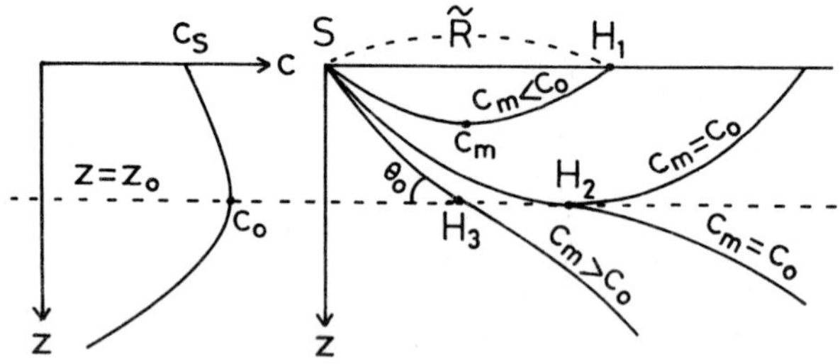

Fig.1. Profile model and the ray of trapped, split-beam and leaky mode.

$$\tilde{R}=c_s\delta^{\frac{1}{2}}\sum_{K=1}B_k(c_0-c_s)^{\sigma_{K-1}}\bar{\mathbb{E}}(\varepsilon)\mathbb{D}(\delta)[\beta(n+1,\tfrac{1}{2})F(n+1,1-\sigma_k;n+\tfrac{3}{2};\bar{z})]^T, \quad (2)$$

where n(=0,1,2,···) indicates the element of column vector, β() is beta function, F() is hypergeometric series,

$$B_k=2C_k\sigma_k, \quad (3)$$

$$\bar{\mathbb{E}}(\varepsilon)=[1,\varepsilon,0,0,\cdots], \quad (4)$$

$$\mathbb{D}(\delta)=\begin{vmatrix} 1 & -\frac{1}{2}\delta & \frac{1}{2}\frac{3}{2}\frac{1}{2}\delta^2 & -\frac{1}{2}\frac{3}{2}\frac{5}{2}\frac{1}{3}\delta^3 & \cdot \\ 0 & 1 & -\frac{1}{2}\delta & \frac{1}{2}\frac{3}{2}\frac{1}{2}\delta^2 & \cdot \\ 0 & 0 & 1 & -\frac{1}{2}\delta & \cdot \\ 0 & 0 & 0 & 1 & \cdot \\ \cdot & \cdot & \cdot & \cdot & \cdot \end{vmatrix}, \quad (5)$$

$$\bar{z}=(c_m-c_s)/(c_0-c_s), \quad (6)$$

$$\delta=(c_m-c_s)/(c_m+c_s), \quad (7)$$

$$\varepsilon=(c_m-c_s)/c_s . \quad (8)$$

The symbol $\bar{\mathbb{E}}(\varepsilon)$ is a row vector, $\mathbb{D}(\delta)$ is a matrix and $[\cdots]^T$ indicates a column vector. For the trapped mode, there is the relation $c_s<c_m<c_0$. Therefore, it is clear that $0<\bar{z}\leq 1$. The symbol δ and ε are very small quantity in the ocean. The hypergeometric series is very rapid convergent when $\bar{z}<1$ and $\sigma_1>0.5$. However, when $\bar{z}=1$ and $\sigma_1<0.5$, this series does not converge. When $\bar{z}=1$, the trapped mode becomes the split-beam ray. $\tilde{R}$ of the split-beam ray does not converge when $\sigma_1=0.5$. By the same method as above, one can calculate the travel time of trapped mode by

$$\tilde{T}=c_m c_s^{-1}\delta^{\frac{1}{2}}\sum_{K=1}B_k(c_0-c_s)^{\sigma_{K-1}}\mathbb{E}(\varepsilon)\mathbb{D}(\delta)[\beta(n+1,\tfrac{1}{2})F(n+1,1-\sigma_k;n+\tfrac{3}{2};\bar{z})]^T, \quad (9)$$

where

$$\mathbb{E}(\varepsilon)=[1,\ -\varepsilon,\ \varepsilon^2,\ -\varepsilon^3,\cdots\ ]. \quad (10)$$

Other symbols in Eq.(9) are the same as in Eq.(2).

4. HORIZONTAL RANGE AND TRAVEL TIME OF THE RAY OF LEAKY MODE

The horizontal range of the ray of leaky mode from S to H_3 in Fig.1 is given by

$$R=\frac{c_s}{2}(c_m^2-c_s^2)^{-\frac{1}{2}}\sum_{K=1}B_k(c_0-c_s)^{\sigma_k}\bar{\mathbb{E}}(\varepsilon)\mathbb{D}(\delta)[\beta(n+1,\sigma_k)F(n+1,\tfrac{1}{2};\sigma_k+n+1;\bar{z})]^T, \quad (11)$$

where

$$\bar{z}=(c_0-c_s)/(c_m-c_s), \quad (12)$$

$$\delta=(c_0-c_s)/(c_m+c_s), \tag{13}$$
$$\varepsilon=(c_0-c_s)/c_s . \tag{14}$$

For the ray of leaky mode, there is the relation $c_m>c_0>c_s$. Therefore, $\bar{z}$ defined by Eq.(12) is always smaller than one. But, $\bar{z}$ becomes one for the split-beam ray. The travel time of the leaky mode corresponding to Eq.(11) is obtained similarly

$$T=\frac{c_s^{-1}}{2}c_m(c_m^2-c_s^2)^{-\frac{1}{2}}\sum_{K=1}B_k(c_0-c_s)^{\sigma_k}\mathbb{E}(\varepsilon)\mathbb{D}(\delta)[\beta(n+1,\sigma_k)F(n+1,\tfrac{1}{2};\sigma_k+n+1;\bar{z}) \tag{15}$$

Eqs.(11) and (15) do not converge when $\bar{z}=1$ and $\sigma_1\leq 0.5$. When $\bar{z}<1$ and $\sigma_1>0.5$, the convergency of these equations is very rapid.

5. NUMERICAL EXAMPLES

It is assumed that a source is located at the surface of the surface sound channel and that $c_1=55.55$ sec, $\sigma_1=1$, $c_s=1525.004$, $c_0=1526.804$ m/sec, $z_0=100$ m, $c_2, c_3,\cdots; \sigma_2, \sigma_3,\cdots=0$ in Eqs.(2), (9),(11) and (15). The horizontal range and the travel time in the surface sound channel are calculated and the results are compared with the results which are calculated by the conventional method. The horizontal range of trapped mode *vs* c_m and that of leaky mode *vs* c_m/c_0 and θ_0 are indicated in Figs.2 and 3 respectively. In Fig.2 the coincidence is very good. In Fig.3 the coincidence becom bad when c_m approches to c_0. For the travel time, we have obtained the similar results as above.

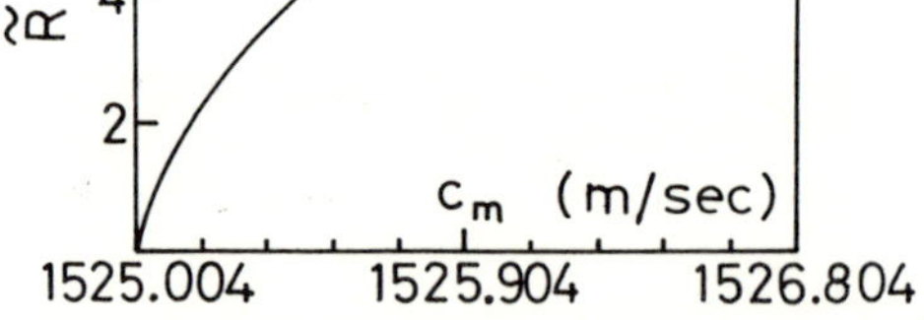

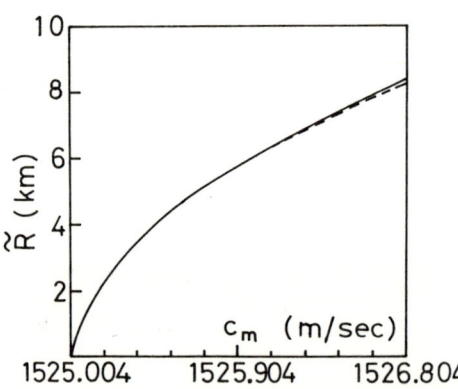

Fig.2. Horizontal range of trapped mode *vs* c_m.
(———)Conventional
(-----)Series expansion with $\bar{z}=0$

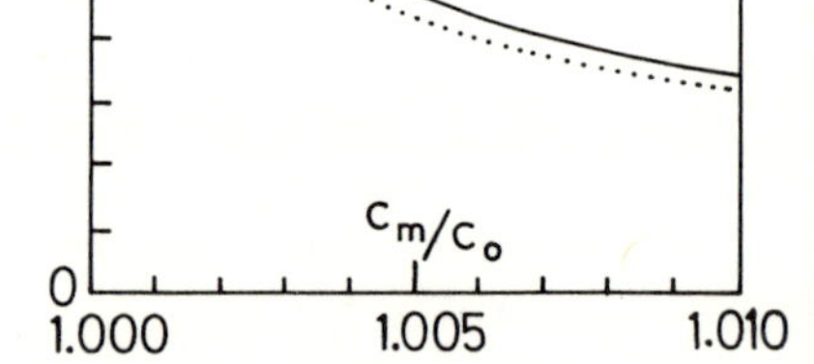

Fig.3. Horizontal range of leaky mode *vs* c_m/c_0.
(———)Conventional
(-----)Series expansion up to $\bar{z}^7$
(•••••)Series expansion with $\bar{z}=0$

6. DISCUSSIONS

When $\bar{z}=1$ and $\sigma_1<0.5$, the series expansion of the horizontal range and that of the travel time do not converge. When $\bar{z}=1$ and $\sigma_1>0.5$, these series expansions converge slowly. When $\bar{z}<1$ and $\sigma_1>0.5$, these series expansions converge very rapidly. The properties of convergency of the series expansion depend largely upon the value of $\bar{z}$ and σ_1.

It is assumed that the exponent σ_k are written in the set of $\sigma_1<\sigma_2<\sigma_3<\cdots$. Then the velocity profile form near the axis of the maximum velocity is chiefly characterized by the value of σ_1, because the value of $|c-c_0|$ becomes very small near the axis. The relations between the form of the velocity profile near the axis and σ_1 are indicated in Fig.4. Near the axis, the form of the velocity profile is indicated nearly by the equation,

$$z-z_0=C_1|c-c_0|^{\sigma_1} \quad . \tag{16}$$

In Fig.4, the ordinate represents the value of $(z-z_0)/c_1$ and the abscissa represents the value of $|c-c_0|$. The parameters of the each profile are σ_1. When $\sigma_1=1$, the curve is a linear profile. When $\sigma_1=0.5$, the curve indicates a parabolic profile. When the value of σ_1 is larger than 0.5, the profile maximum form becomes more sharper than the parabolic profile. For these profile form we can calculate the value of $\tilde{R}$, $\tilde{T}$, R and T when $\bar{z}=1$. When the value of σ_1 is smaller than 0.5, the profile maximum form becomes more dull than the parabolic profile. For these profile form we cannot calculate the value of $\tilde{R}$, $\tilde{T}$, R and T when $\bar{z}=1$. It is seen that near the maximum point of the velocity profile, the profile form is very important factor for the ray theory calculation and the classical ray theory is not applied to the case when $\bar{z}=1$ and $\sigma_1<0.5$.

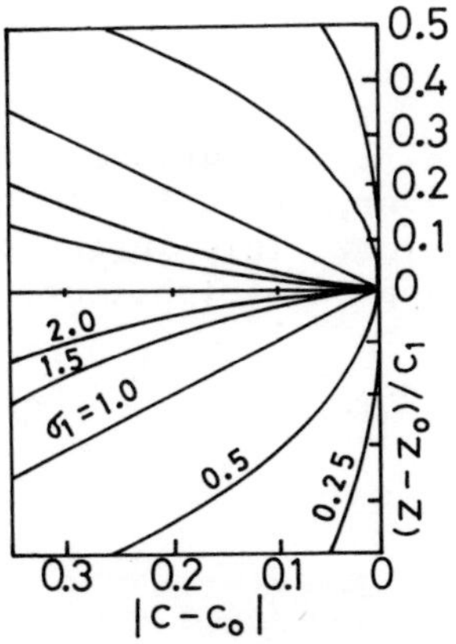

Fig.4. The relation between the velocity profile form and the value of σ_1.

7. CONCLUSIONS

The series expansion of the velocity profile is very convenient for the fitting with the observed velocity profile in the ocean. The analytical computations of the series expansion of the horizontal range and the travel time are not complicated and we have presented the closed form solutions. However, it is a future problem to improve the slow convergency of the series expansions near the sprit-beam ray when $\sigma_1 \leq 0.5$ and $\bar{z} \approx 1$ and the divergency of the series expansions when $\sigma_1 \leq 0.5$ and $\bar{z}=1$. To solve this problem it may be necessary to apply the idea of the generalized WKB method which has been proposed by E.L.Murphy.

REFERENCES

(1) M.A.Pedersen, Proc. 7th ICA,4,413-416(1971)
(2) M.A.Pedersen, D.White and D.W.Johnson, J.Acoust.Soc.Am.,58,78-96(1975)
(3) E.L.Murphy and J.D.Davis, J.Acoust.Soc.Am.,56,1747-1760(1974)
E.L.Murphy, J.Acoust.Soc.Am.,47,899-908(1970)

DISCUSSION

Comment : R. MACKINNON

Have you considered also the calculation of intensity using the series expansion methods? If so, would you comment on the intensity calculated at a turning point?

Reply : T. KUYAMA

I have calculated the intensity distribution in the surface sound channel by the series expansion. However, it is impossible to calculate the intensity at the turning point. This intensity should be calculated by the extended WKB approximation, for which E.L. Murphy has presented reports.

HORIZONTAL SPATIAL COHERENCE MEASUREMENTS WITH EXPLOSIVES AND CW-SOURCES IN SHALLOW WATER

B. Scholz

Forschungsanstalt der Bundeswehr für Wasserschall- und Geophysik, Kiel, Germany

ABSTRACT. Transverse and longitudinal horizontal spatial coherence measurements were conducted in the North Sea and the Baltic Sea with explosives and cw-sources. The data reveal a range, frequency and receiver depth dependence. A significant difference in the frequency dependence is observed between transverse and longitudinal coherence. Only a small influence of the rough sea surface on the spatial coherence is observed. A comparison of the data of the different areas and seasons indicate a much stronger influence of the bottom and the spatially varying sound speed field.

1. INTRODUCTION

Since 1968 FWG is investigating systematically shallow water sound propagation in the North Sea and the Baltic Sea. To know the propagation loss is not sufficient for the development of special SONAR-equipment and signal processing for shallow water. Therefore measurements were made to get knowledge about the time and space dependent characteristics of the transmitting channel. The space dependent properties are described by the spatial coherence of the sound field:

- the transverse horizontal spatial coherence
- the longitudinal horizontal spatial coherence
- the vertical spatial coherence

G. Tacconi (ed.), Aspects of Signal Processing, Part 1, 95-108.

Due to the available equipment vertical coherence measurements were not possible.

The spatial coherence is described by the degree of coherence of signals received at two locations spaced a distance apart as a function of the spacing. The degree of coherence is a measure of the similarity of two signals and is evaluated as the maximum of the normalized cross-correlation function. Generally the degree of coherence will decrease with increasing spacing. The spacing, multiplied by the wavenumber $2\pi/\lambda$, at which the spatial coherence function falls below 0.6 is called normalized coherence length. This coherence length is a measure of the angular uncertainty caused by the transmitting medium [1, 2, 3].

The measurements were conducted using line arrays of 10 non-equally spaced hydrophones which provide 45 different spacing combinations. One array of 74 m length was suspendend to a certain depth from the research vessel PLANET when using airdropped broadband explosive sound sources. In the other experiment a fixed range was built up with a rigid array of 50 m length moored to the ground and a bottom mounted transducer for frequencies from 300 Hz to 5 kHz. In both experiments the hydrophone signals were recorded and evaluated ashore.

Whereas for the explosive data the integration time is determined by the shot duration of the order of 100 milliseconds, the fixed set-up allowed measurements with cw-signals and investigations of the spatial coherence with different integration times and time fluctuations.

All presented data are evaluated for appointed center frequencies and a bandwidth of 200 Hz.

2. TRANSVERSE HORIZONTAL SPATIAL COHERENCE MEASUREMENTS WITH EXPLOSIVES

2.1 Range dependence

The first example presents the data for the simplest case, a series from the central North Sea with nearly constant water depth of about 65 m and winter conditions with isothermal water. Figures 1, 2 and 3 exhibit for all frequency bands a strong range dependence of the spatial coherence function. Data averaged over all frequency bands for each propagation distance illustrate that relation (figure 4). The coherence lengths increase from about 100 for 1 nautical mile propagation distance to much more than 1000 for 27 nautical miles. This pronounced dependence was found to be typical for the North Sea under winter conditions.

During the summer the strong range dependence gets smaller. An experiment under late summer conditions show nearly no range dependence.

For comparison of the North Sea data of the different seasons figure 5 depicts the coherence lengths as a function of the propagation distance. A reduced coherence length defined at a cross-correlation maximum of 0.8 is used instead of 0.6 due to the line array being too short. The range dependence is caused by smaller grazing angles for larger propagation distances. A possible explanation of the distinct difference between winter and summer/fall-conditions is the stronger influence of the bottom introduced by the surface layer and the marked thermocline during summer, which are downward refracting conditions.

Measurements from the Baltic Sea show a range dependence too, but there was found no environmental situation which revealed a strong dependence as it was found in the North Sea during winter conditions. Data of all seasons show results similar to the North Sea summer results. This is a consequence of the main difference to the North Sea, the salinity layer with increasing sound speed near the bottom in the Baltic Sea (figures 12, 14).

2.2 Frequency dependence

As it was expected there is also a frequency dependence of the spatial coherence. This effect is observed for data of the North Sea and Baltic Sea. Figure 6 presents one example of the Baltic Sea.

The coherence lengths mostly decrease with increasing frequency in the case of transverse coherence. Furthermore the frequency dependence is stronger for the shorter propagation distances. This relation is strongest for the North Sea winter data. For propagation distances of 5 nautical miles and more the frequency dependence vanishes (figures 1, 2, 3).

2.3 Receiver depth dependence

Caused by the sound speed profile the influence of the environment on the sound on the way from the source to the receiver is different, when the receiver depth is varied. A

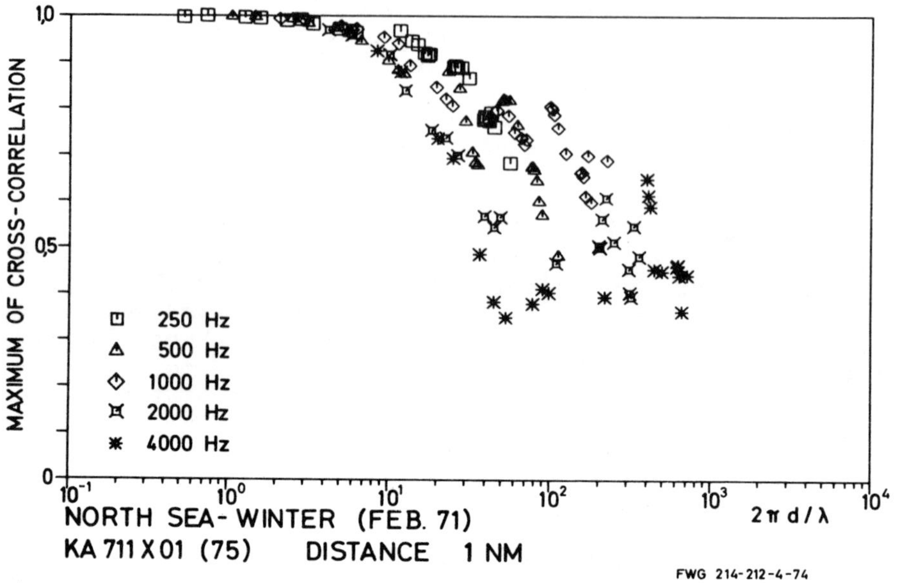

FIG. 1 Horizontal transverse spatial coherence
propagation distance 1 nautical mile
North Sea, water depth 65 m, winter conditions
wave height 0.5 m

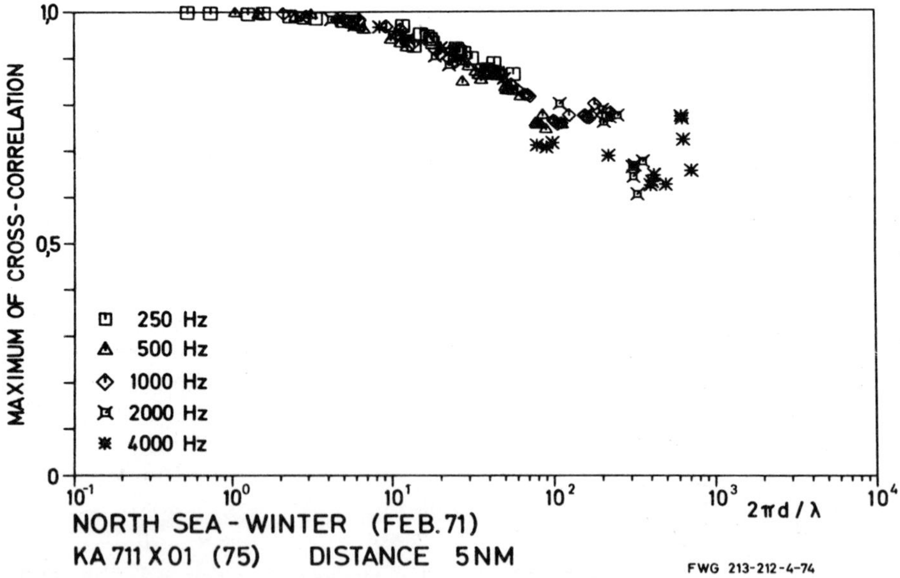

FIG. 2 Same as fig.1, but propagation distance
5 nautical miles

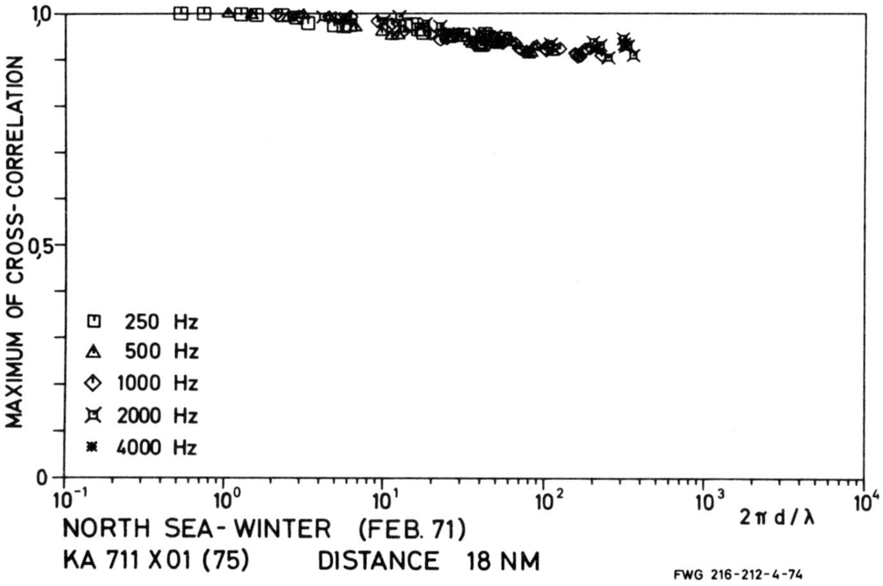

FIG. 3 Same as fig.1, but propagation distance 18 nautical miles

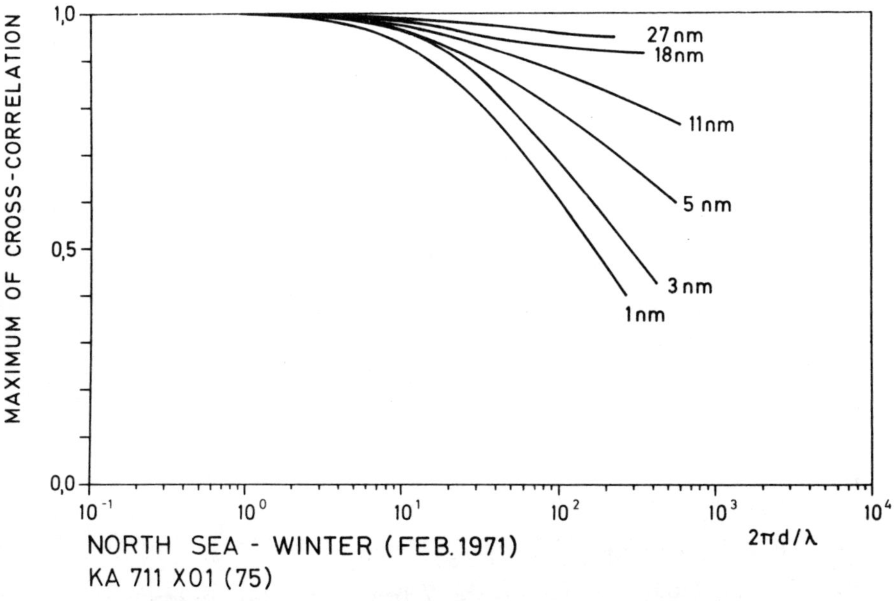

FIG. 4 Horizontal transverse spatial coherence data averaged over all frequencies for each propagation distance
same experiment as fig.1, 2 and 3

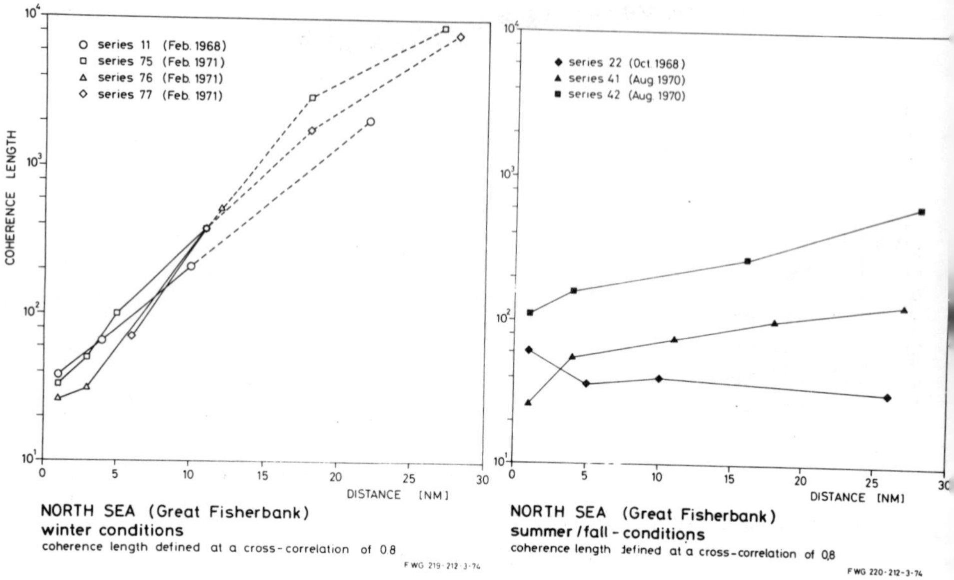

FIG. 5 North Sea spatial coherence lengths as a function of the propagation distance comparison between winter and summer data

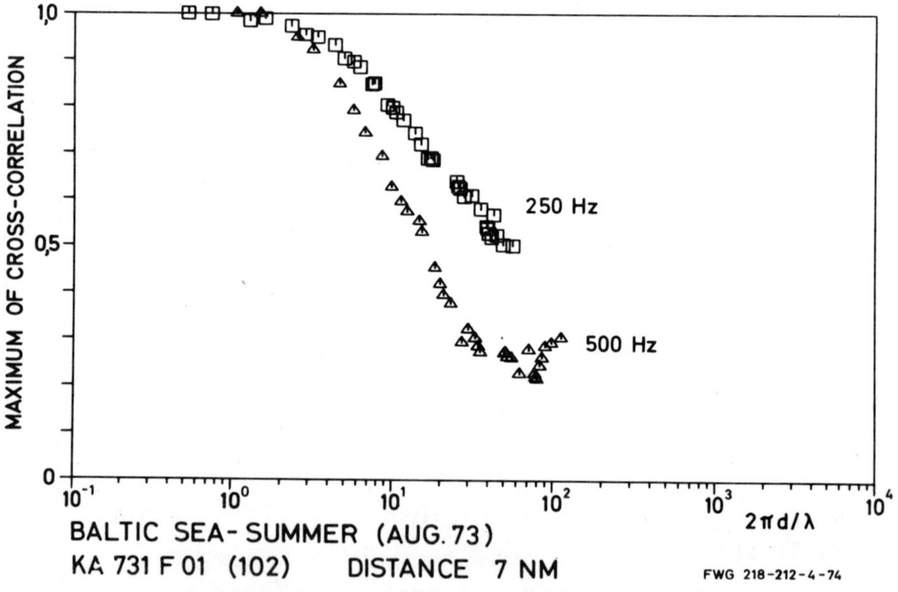

FIG. 6 Horizontal transverse spatial coherence propagation distance 7 nautical miles Baltic Sea, water depth 90 m, summer conditions very calm sea

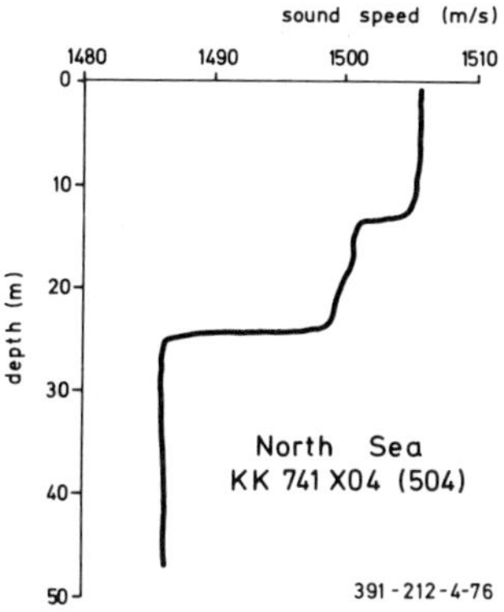

FIG. 7

Sound speed profile
North Sea, June 1974
water depth 49 m

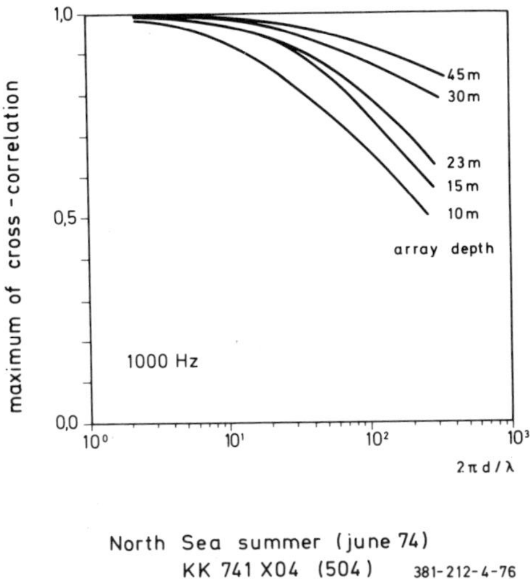

FIG. 8 Horizontal transverse spatial coherence dependence on depth of receiving array propagation distance 3 nautical miles

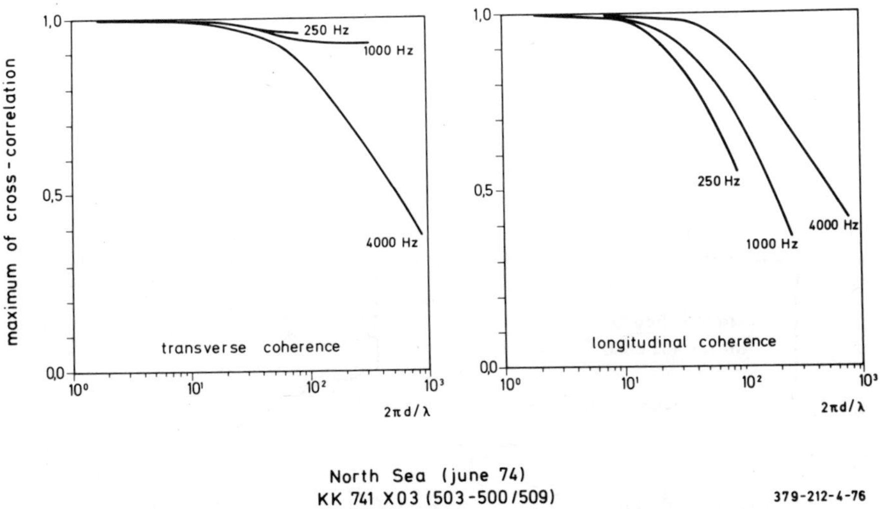

FIG. 9 Comparison of horizontal transverse and longitudinal spatial coherence, North Sea, June 1974, water depth 49 m, sound speed profile fig.10, wave height 0.8 m, explosive source depth 26 m, receiving array depth 30 m, propagation distance 5 nautical miles

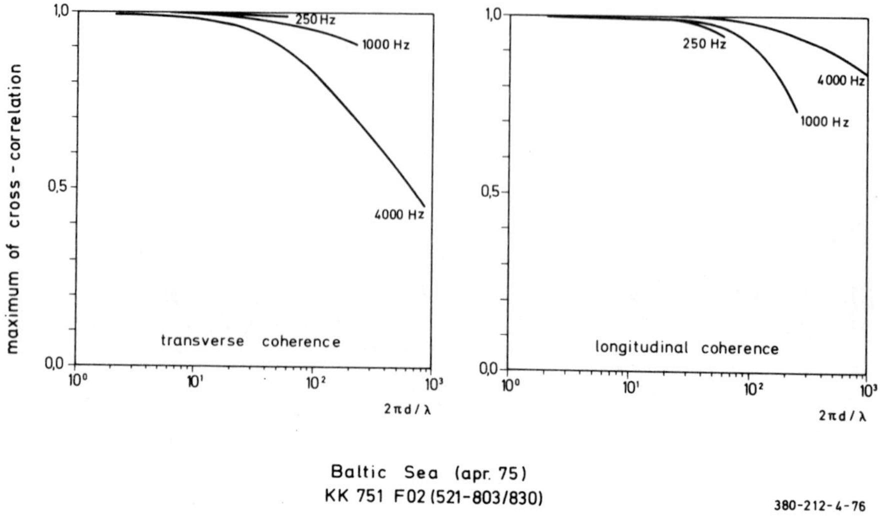

FIG. 11 Comparison of horizontal transverse and longitudinal spatial coherence, Baltic Sea, April 1975, water depth 93 m, sound speed profile fig.12, wave height 0.6 m, explosive source depth 26 m, receiving array depth 23 m, propagation distance 8 nautical miles

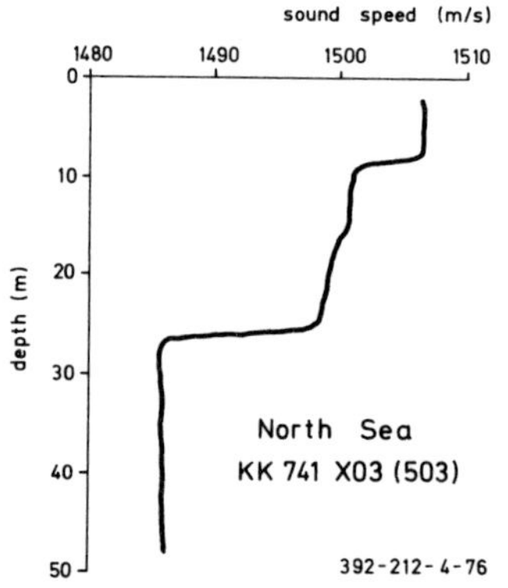

FIG. 10

Sound speed profile
North Sea, June 1974
water depth 49 m

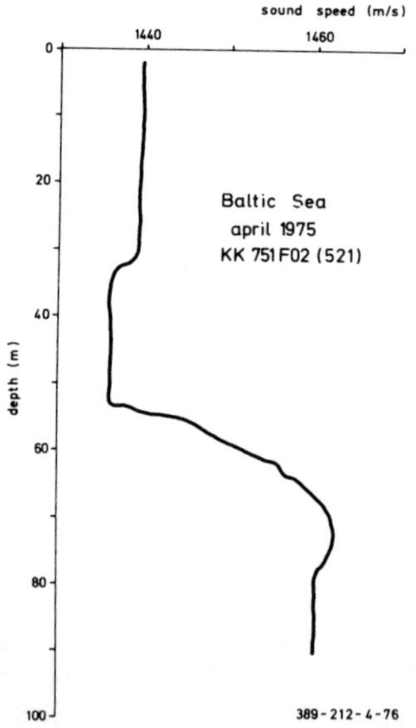

FIG. 12 Sound speed profile
Baltic Sea, April 1975
water depth 93 m

measurement from the North Sea with a typical temperature layer at the surface of June 1974 (figure 7) will clearly demonstrate this. Explosive sound sources were detonated below the thermocline in 30 m water depth. After a propagation of 3 nautical miles the receiving array was positioned in depths from 45 m to 10 m. The sound below the thermocline is less disturbed than above. Therefore the spatial coherence of the sound field is distinctly higher below the thermocline. Figure 8 presents the results for the 1000 Hz band. For lower frequencies this effect is not so strong.

3. LONGITUDINAL HORIZONTAL SPATIAL COHERENCE MEASUREMENTS WITH EXPLOSIVES

The longitudinal horizontal spatial coherence is also range and frequency dependent. But a remarkable different is revealed by the data. The transverse coherence is higher for the lower frequencies, whereas the longitudinal coherence shows the inverse relation. Data of the North Sea (figure 9) for 5 nautical miles propagation and of the Baltic Sea (figure 11) for 8 nautical miles propagation clearly illustrate this behavior. The coherence lengths for 250 Hz differ from $2\pi d/\lambda = 80$ (longitudinal) to more than $2\pi d/\lambda = 1000$ (transverse) whereas the lengths are about the same (about 350) in the case of 4000 Hz for the North Sea experiment.

As presented by two measurements of the Baltic Sea (figures 13 and 15) there was strong range dependence for the lower frequencies whereas nearly no dependence was observed for the higher frequencies.

4. TRANSVERSE HORIZONTAL SPATIAL COHERENCE MEASUREMENTS WITH CW-SOURCES

Investigations on the time fluctuations of spatial coherence were conducted using the fixed acoustic range which was installed in coastal water of 20 m depth in the Baltic Sea. The propagation distance was about 3 km. Figure 17 depicts the degree of coherence for two hydrophone spacings as a function of time. The center frequency was 3000 Hz and the integration time was 2.73 seconds. It is not surprising, that the variance of the data is small when there is high coherence (a) whereas the variance increases for increasing hydrophone spacing and lower coherence (b).

Theoretical considerations on horizontal transverse coherence state that there is a decay in the degree of

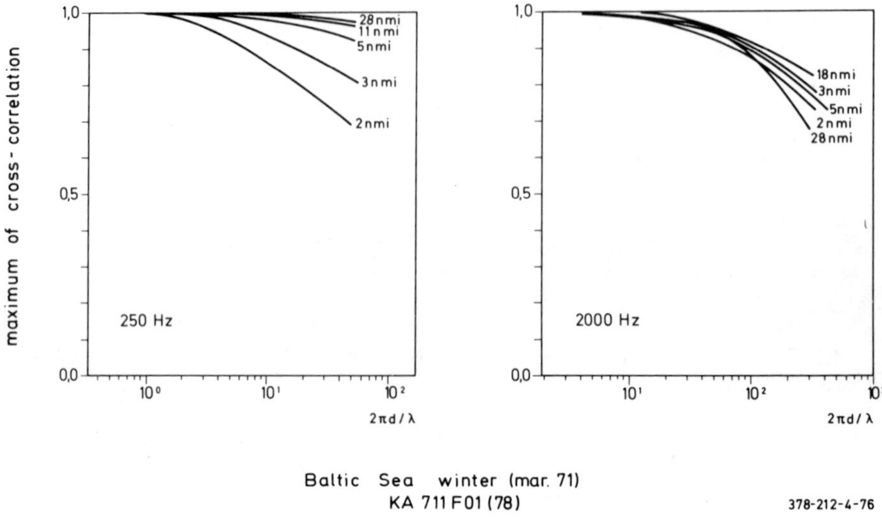

FIG. 13 Horizontal longitudinal spatial coherence, range and frequency dependence, Baltic Sea, March 1971, water depth 83 m, sound speed profile fig.14, wave height 2.0 m, explosive source depth 26 m, array depth 23 m

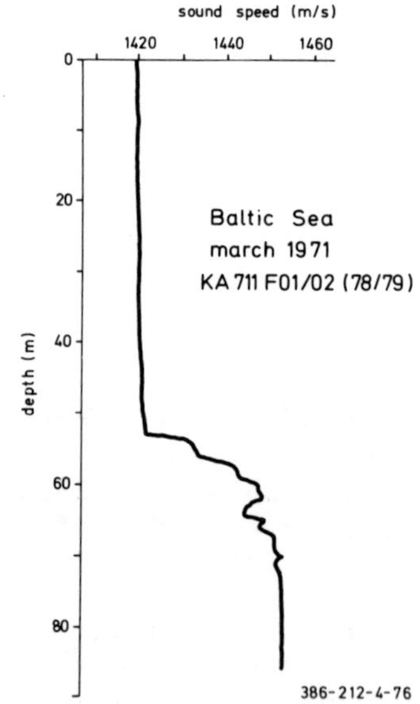

FIG. 14

Sound speed profile water depth 83 m

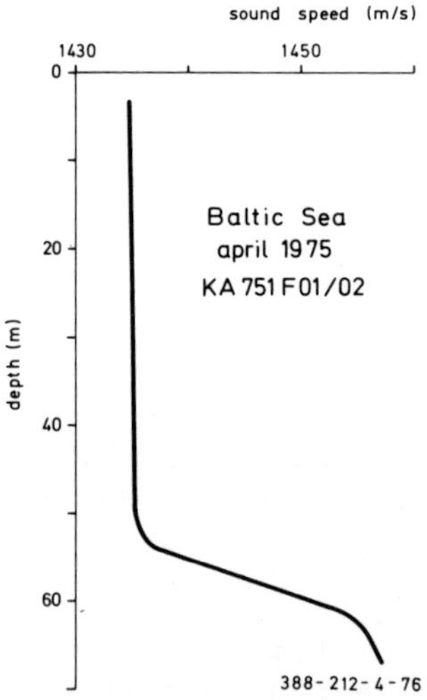

FIG. 16

Sound speed profile Baltic Sea, April 1975

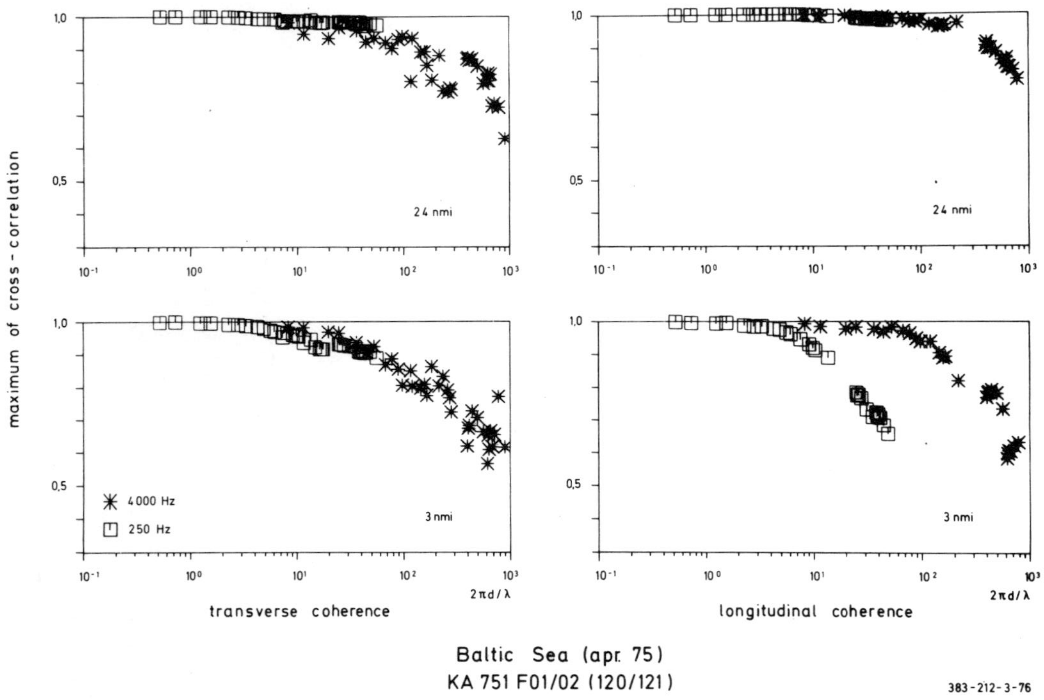

FIG. 15 Comparison of horizontal transverse and longitudinal spatial coherence, range and frequency dependence, water depth 70 m, sound speed profile fig.16, wave height 0.45 m, explosive source depth 26 m, receiving array depth 26 m, propagation distance 3 and 24 nautical miles

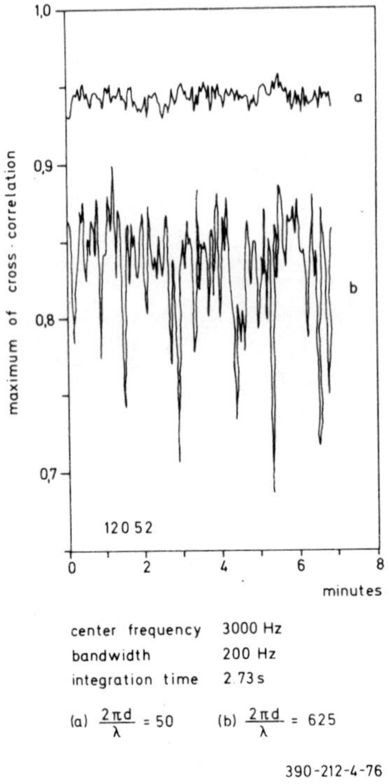

FIG. 17

Time fluctuations of spatial coherence measured with cw-signals propagation distance 3000 m

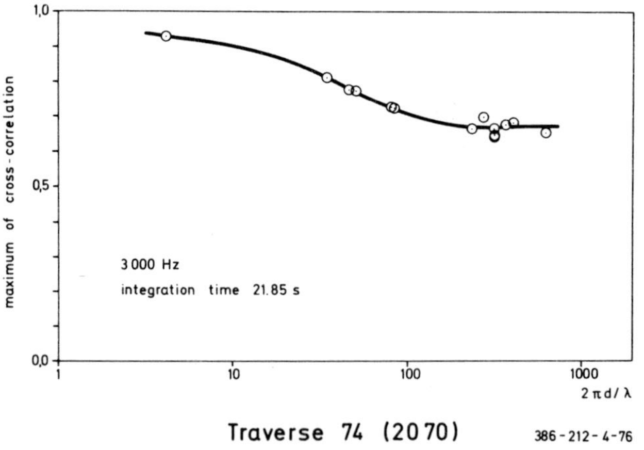

FIG. 18 Horizontal transverse spatial coherence measured with cw-signals, propagation distance 3000 m, water depth 20 m, wave height 0.33 m, main wave period 2.3 s

coherence introduced by the rough sea surface when the receiver separation in the sound field is in the order of the correlation distance of the surface roughness 2, 3. For increasing separations the degree of coherence will maintain a constant value. Measurements with explosives do not fit this functional form, the degree of coherence decreases for the whole possible range of separations up to 74 meters and do not reach a constant value [5]. On the other hand measurements with the fixed range and longer integration times seem to confirm the described influence of the sea surface. Figure 18 depicts the data of a 3000 Hz measurement which fit the functional form very well and show the decay in the order of the estimated correlation distance of the surface.

5. CONCLUSION

Measurements of the horizontal spatial coherence were conducted in the North Sea and the Baltic Sea under different seasonal environmental and experimental conditions. The data exhibit range, frequency and receiver depth dependence. Furthermore a significant difference in the frequency dependence is observed between transverse and longitudinal coherence. The relations described in this paper are found to be typical for the different situations. On the other hand some very few measurements result in an opposite behavior. But this may be due to the imperfect knowledge of the environmental conditions, e.g. the bottom is varying in roughness and its impedance over the propagation path.

One intention of these measurements was to learn about the influence of single environmental parameters on the spatial coherence. This problem is not yet solved and more experiments should be conducted systematically with a fixed acoustic range.

One basic fact can be stated, there is an influence of the sea surface on the spatial coherence, however a small one shich was observed from the data with the fixed range. The comparison of the winter and summer data of the North Sea, the theoretical considerations and the difference to the explosive data indicate a much stronger influence of the bottom and the spatially varying sound speed field.

REFERENCES

1. P. WILLE and R. THIELE, "Transverse horizontal coherence of explosive signals in shallow water", J. Acoust. Soc. Amer. 50, 348 - 353 (1971)

2. A. WASILJEFF, "Spatial horizontal coherence of acoustical signals in shallow water", SACLANTCEN SM-68, SACLANT ASW Research Centre, La Spezia, Italy, 1975

3. P.W. SMITH, Jr., "Statistics of fluctuations in measures of sound propagated in shallow water", Bolt Beranek and Newman Inc., Report 2498, 1973

4. R. ANCEY, "Coherence spatiale de signeaux acoustiques propagés par petit fonds" in: Groupe d'étude du traitement du signal, Nice, France, 1973

5. B. SCHOLZ, "Horizontal coherence measurements with explosive sources in shallow water", in: SACLANTCEN Conference Proceedings CP 13, 1974, Vol. I
also: FWG-Bericht 1975-2

6. R.J. URICK, "The coherence of signals, noise and reverberation in the sea", Naval Ordnance Laboratory, Report NOLTR 72-279, 1972

MEASUREMENT OF THE WEIGHTING FUNCTION OF THE TIME-VARIANT SHALLOW WATER CHANNEL

R. Thiele

Forschungsanstalt der Bundeswehr für Wasserschall- und Geophysik, Kiel, Germany

ABSTRACT. To describe the behaviour of randomly time-variant linear channels is comparably easy assuming a wide sense stationary uncorrelated scattering (WSSUS) channel. Its properties are formulated by the scattering function. Some measured scattering functions are presented. The scattering function is shown to comprehend the sea state influence only. However the observed variability is often due to changes in sound speed layers or water level. These slow fluctuations effect an unstationarity during the time intervals of local stationarity of the sea state.

1. INTRODUCTION

The properties of a linear transmission channel are described by the weighting function. The oceanic sound transmission channel is a linear, however a time-space-variant one with stochastic behaviour. That is why its weighting function is not deterministic and possibly not continuous, and therefore it cannot be used directly for parameterisation. To overcome this difficulty BELLO [1, 2] made the proposal to classify such a channel by the assumption of stationarity in the time and the frequency domain. This class of transmission channels he calls Wide Sense Stationary Uncorrelated Scattering (WSSUS)-Channels. Uncorrelated scattering demands a completely variant transmission and a weighting function without an invariant part. To meet these conditions, the weighting function has to be separated in an invariant part and in a "scattered part". This is admissible due to the linearity of the channel. The ratio

G. Tacconi (ed.), Aspects of Signal Processing, Part 1, 109-122.

of the invariant part to the complete weighting function is called degree of coherence.

Because of the linearity, the large number of scatterers causing the variability and the central limit theorem it is reasonable to classify the scattered part as Gaussian. The characteristics of a Gaussian process are completely described by a correlation function. The simplest type of a correlation function of a random time-variant Gaussian channel is the scattering function which simply relates to the ambiguity function of a signal transmitted in the channel [3]. The claim of stationarity may be weakened to a local stationarity [4, 5].

This rational method of describing a time variant channel stimulated ELLINTHORPE and NUTALL [6] to use this conception for the underwater sound channel.

The space variability may be classified in the same manner. Therefore LAVAL [5] extended the conception to the time and space variant channel by a five dimensional scattering function.

2. THE MEASURED SCATTERING FUNCTION

The weighting function (w.f.) of an invariant channel is measured as the answer of the channel to a transmitted pulse. For a slowly varying channel this is the momentary w.f. The channel fluctuations may be sampled by pulse sequences. A cw-pulse of e.g. 1 kHz bandwidth and a total measuring time of 40 minutes will result in about $5 \cdot 10^6$ samples. This number may be reduced if the time spread of the channel is less than the time from pulse to pulse.

If we use a time window for each pulse of 40 ms and a pulse repetition time of 200 ms, each momentary w.f. will have 80 numbers and the complete series will have $12 \cdot 10^3$ momentary w.f., thus a total of 960.000 numbers. These data may be handeled by a computer. Fig.1 shows the scattering function (s.f.) in the Baltic Sea of a 3 km transmission path at about 20 m water depth in coastal waters off the island Fehmarn. The time spread relates to the horizontal axis and the frequency spread to the vertical axis. The contour lines have a spacing of 3 dB. The maxima of the side lobes are found at a spread-frequency identical with the main sea state frequency and have a slightly increased delay compared to the coherent, invariant part at the spread-frequency $\mu = 0$. This is the "basic structure" of a s.f. Almost the same result are found in fig.2 with

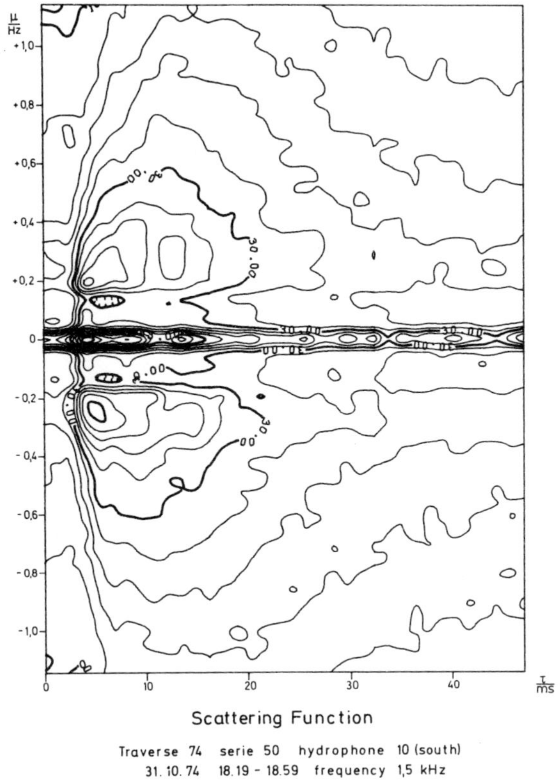

FIG. 1

Scattering function which shows the "basic structure". Measured in coastal water off the island Fehmarn/Baltic Sea. Water depth about 20 m, propagation distance 3 km

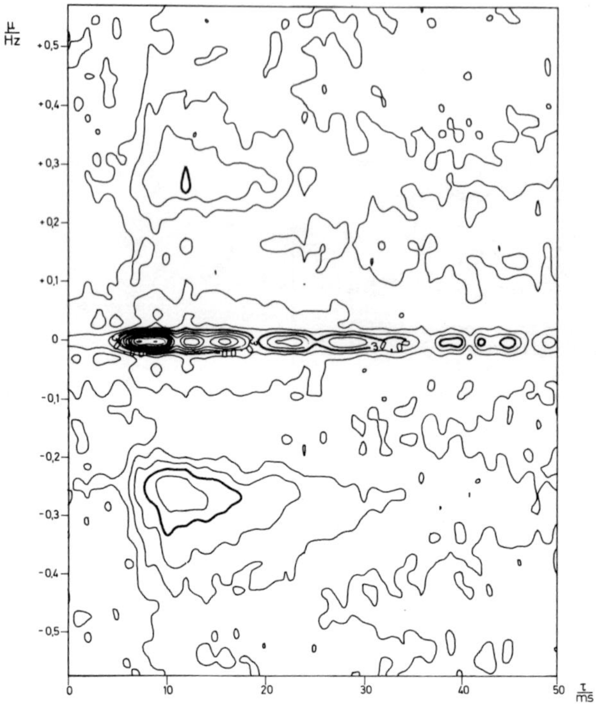

Scattering Function

Traverse 74 serie 79 hydrophone 10 (south)
6.11.74 18.05 - 18.45 frequency 1 kHz

FIG. 2

The same type of measurement as fig.1 with the same range but with another carrier frequency, measuring time and frequency resolution of the evaluation

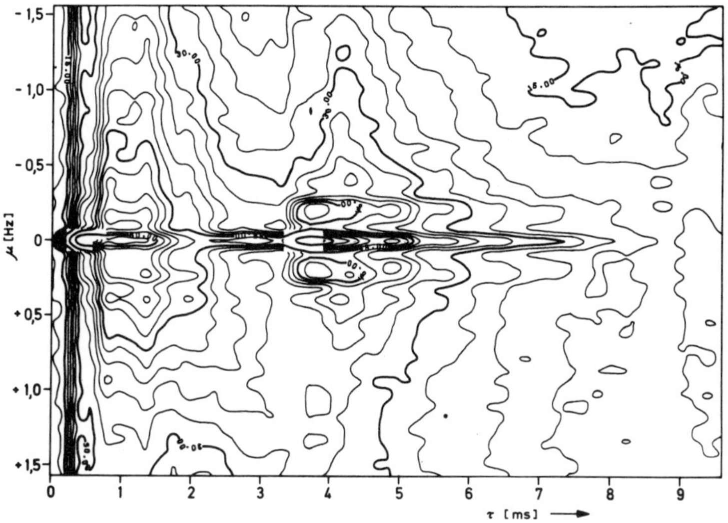

FIG. 3

Scattering function with 15 kHz carrier frequency measured in the Bornholm Basin/ Baltic Sea with 50 m water depth and 1300 m propagation distance [7]

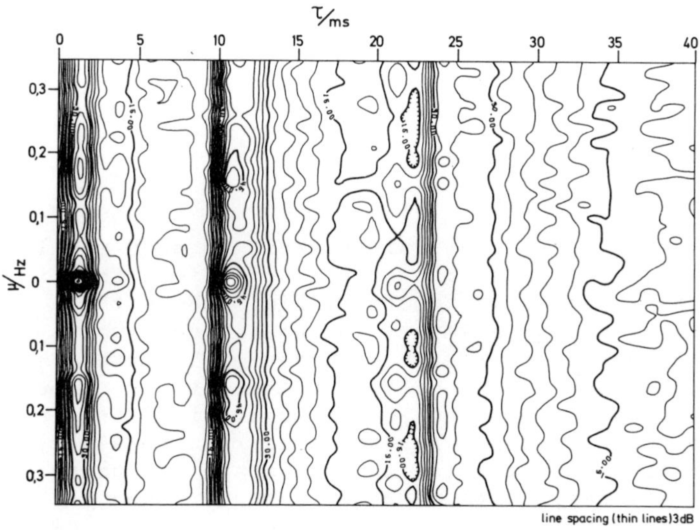

FIG. 4

Scattering function with 15 kHz carrier frequency measured at the Great Fisherbank/ North Sea with 70 m water depth and 1500 m propagation distance [7]

the same range but on another day and at a different center frequency. In fig.3 we see a s.f. of a 15 kHz measurement in the middle of the Baltic. Two groups in the time spread direction are not completely resolved [7]. Fig.4, a s.f. of a North Sea measurement with 15 kHz shows 3 groups of multi-paths with a very different degree of coherence. Much more complicated is the structure of an example (fig.5) from the AFAR (Azores Fixed Acoustic Range) field with 800 Hz center-frequency and a distance of 18 nmi.

The examples show the following common properties

1. Each distinctable path or mode has to be looked on separately. When the difference of traveltime of the individual paths is to small, we can possibly comprehend them to separated traveltime groups. Each of the groups has the basic structure of the scattering function.

2. There are maxima of the sidelobes at the significant seastate frequency with a steep slope on the small spread frequency side and a gentle slope on the large spread frequency side. With low coherence - as for the latest group of the North Sea measurement example - the frequency spread is larger and the maxima diminish. These results agree well with frequency spread measurements [8] as expected.

3. The scattered part versus the coherent part has a small time delay which is increasing with the spread frequency. The lower the degree of coherence the larger is the time spread.

3. AMPLITUDE FLUCTUATION

In general there is more practical interest in such simple phenomena as e.g. amplitude or energy fluctuations than in the very abstract scattering function. Fig.6 shows these fluctuations for the AFAR example. The amplitude fluctuations are dominated by interference. The visible interference has such a low frequency spread that it cannot be resolved by the scattering function. A scattering function which resolves this frequency spread would require a very long measuring time during which the sea state influence would not be stationary. With respect to the frequency spread of sea state influence we have to call these slow variations as nearly invariant or unstationary coherent. Looking to the upper curve of the energy transmission of the pulses of 200 Hz bandwidth at 800 Hz centerfrequency we see

with the hydrophone spacing d and sea wave correlation distance Λ [11, 12]. For a large hydrophone spacing the spatial coherence should be the same as the time coherence. The strongest decay should be at the sea wave correlation distance.

This disagrees with measuring results [13] about spatial coherence with explosives. There is not significant decay at the sea wave correlation distance (fig.9) and no observed constant coherence at larger hydrophone distances. One effect is given by the integration time of the measurement. In fig.10 a time variability experiment with 1,5 kHz frequency with two hydrophones 47 m apart shows the influence of the integration time. The shortest integration time over a single impulse agrees with the correlation time of explosive signals. The larger integration times are obtained by integrating over sequences of pulses. At the correlation time of the sea waves we find a decay of spatial coherence. This means that the momentary roughness gives not enough statistics of the mean roughness influence. On the other hand the short time spatial coherence of 0.68 is significantly lower than the time coherence of 0.78 for the one hydrophone and of 0.83 for the other. The governing effect for the spatial coherence is not the sea state but the bottom structure and the spatial distribution of the sound velocity layering. It is expected that these influences are not spatially stationary.

5. CONCLUSION

The conception of the WSSUS-channel is very useful for studies on the sea state influence on sound propagation. The scattering function is well measurable but with a high computational effort and sometimes confusing displays of the results. A simplification is possible by defining travel time groups of sound paths or modes with individual time coherence.

Looking at time or spacing variability this gives a second order effect. It gives only the shorttime respectively the small distance limitation of variability. The governing influences are caused by slow variations which break the WSSUS conception. A general conception for the description of these slow variations is still missing.

The conception of the WSSUS channel should be applied very carefully to practicle sonar problems. The strong variability of the sound transmission restricts both sonar range prediction and advanced signal processing methods. Therefore

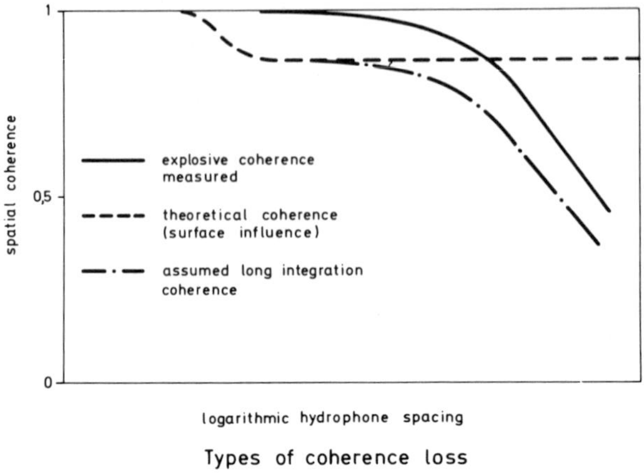

FIG. 9

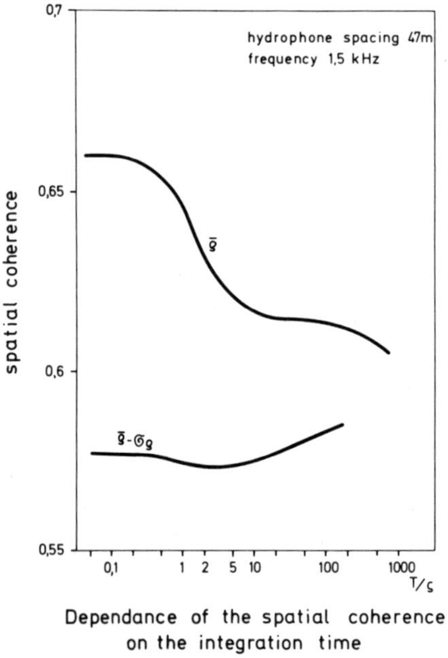

FIG. 10

Averaged spatial coherence between one pair of hydrophones with a spacing of 47 m. The left end of the curve gives the integration over one pulse. The integration time is increased by integrating over sequences of pulses.

more knowledge about the variability is necessary. The WSSUS-conception gives an important contribution to solve this task. But there are more practical observations necessary to find and overcome the limitations of this kind of channel description.

REFERENCES

1. BELLO, P.A. "Characterization of Randomly Time-Variant Linear Channels", IEEE Trans., CS-11:360, (1963)

2. BELLO, P.A. "Time-Frequency Duality", IEEE Trans., IT-10:18, (1964)

3. PRICE, R. and GREEN, P.E.Jr. "Signal Processing in Radar Astronomy", M.I.T. Lincoln Lab., T.R. 234, (1960)

4. THIELE, R. "Anwendung der Theorie zeitveränderlicher Filter zur Beschreibung des Flachwasser-Schallkanals", (Application of the Time-Variant Filter Theory to the Description of the Shallow Water Sound Channel), Thesis T.U. Hannover (1972)

5. LAVAL, R. "Sound Propagation Effects on Signal Processing", in "Signal Processing" (Proceedings of the NASI on Signal Processing) Academic Press, London (1973), P. 223

6. ELLINTHORPE. A.W. and NUTALL, A.H. "Theoretical and Empirical Results on the Characterization of Undersea Acoustic Channels", IEEE 1st Ann. Comm. Conv.: 585, (1967)

7. THIELE, R. "Shallow Water Channel Analysis by the Time Variant Weighting Function", SACLANTCEN Conference Proc. No. 14, Vol. II:173 (1974)

8. RODERICK, W.I., CRON, B.F. "Frequency Spectra of Forward Scattered Sound from the Ocean Surface", J. Acoust. Soc. Am., 48:759 (1970)

9. BECKMANN, P. and SPIZZICHINO, A. "The Scattering of Electromagnetic Waves from Rough Surfaces", Pergamon Press, Oxford (1963), p. 125

10. WILLE, P. and THIELE, R. "Horizontal Coherence of Explosive Signals in Shallow Water", J. Acoust. Soc. Am. 50:348 (1971)

11. SMITH, P.W. Jr. "Statistics of Fluctuations in Measures of Sound Propagated in Shallow Water", Bolt Beranek and Newman Inc., Report 2498 (1973)

12. WASSILJEFF, A. "Spatial Horizontal Coherence of Acoustical Signals in Shallow Water", SACLANTCEN SM-68 (1975)

13. SCHOLZ, B. "Horizontal Coherence Measurements with Explosive Sources in Shallow Water", SACLANTCEN, Conference Proc., No. 13, Vol. I (1974)

S U B J E C T 2

DETECTION ESTIMATION AND TRACKING TECHNIQUES

POWER-SPECTRUM ESTIMATION

O. L. FROST

ARGOSystems, Inc., Palo Alto, California, U.S.A.

1. INTRODUCTION

Spectral estimation often forms the basis for distinguishing and tracking signals of interest in the presence of noise and for extracting information from the received data. The application of Fourier techniques to the problem of estimating the properties of sinusoids in noise dates back as far as Shuster (1898). Fourier spectrum analysis is the basis for almost all spectral-estimation equipment,* including the common sweeping-filter spectrum analyzer, the parallel filter bank, the fast Fourier transform (FFT), the delay-line time compressor (Deltic), and the compressive spectrum analyzer (Microscan). A problem with Fourier spectrum analysis, however, is that it makes implicit assumptions concerning data outside the observation interval and, frequently, these physically unrealistic assumptions reduce the quality of the estimates. During the past decade, two radically different non-Fourier spectral-estimation techniques have emerged -- maximum-entropy spectrum analysis (Burg, 1967) and spectral decomposition (Pisarenko, 1973). These techniques offer alternative and often more realistic data models which, in many cases, lead to better estimation performance. This paper reviews the Fourier methods and compares them to the new techniques in terms of signal models assumed by the three basic methods and their ability to distinguish multiple sinusoids in noise.

---

*To be distinguished from frequency discriminators that measure the frequency of only one signal at a time. The performance of such discriminators degrades drastically when two or more signals of approximately equal power are present simultaneously in the input frequency band.

G. Tacconi (ed.), Aspects of Signal Processing, Part 1, 125-162.

Background

Until recently, the power spectrum of windowed time-series data has usually been computed by taking the Fourier transform of the data's known or estimated autocorrelation lags or by taking the modulus of the Fourier transform of the data values themselves. In their discrete form, each method implicitly assumes that the autocorrelation or the data is cyclic outside the observation window. In their continuous form, both approaches assume that the autocorrelation or the data outside the observation window is zero. As a result and predictable from the uncertainty principle in the continuous case (Claerbout, 1976), Fourier methods can distinguish two sinusoids in the actual data stream only when their frequency spacing is approximately equal to or greater than the reciprocal observation window width; in addition, estimation of their power can be quite difficult if they are closely spaced.

Recently, an increasing number of researchers have demonstrated the enhanced resolving power of maximum-entropy spectrum analysis. This technique produces an autoregressive (AR) model of the time series as though it were generated by an m-tap feedback filter excited by white noise. This model can be used to predict data outside the observation window with completely different values than the cyclic or zero-value predictions developed by Fourier methods. The enhanced resolution obtained by the AR techniques, therefore, may be the result of an autoregressive model that forms a least-squares match to the observed data and extends the data outside the observation window in a physically reasonable way; because of its longer duration, it is not surprising that application of Fourier analysis to the extended data often yields higher resolution. The consistency of such results depends on the appropriateness of the model and on the accuracy with which the data were observed. Measurements by Marple (1976) have indicated that, when the product of the input signal-to-noise ratio (SNR) and the number of taps m in the autoregressive model are sufficiently large, the AR method can produce significantly higher two-sinusoid resolution than will the Fourier-transform techniques.

A more recent high-resolution spectrum-analysis technique called "Pisarenko decomposition" (PD) develops a time-series model of sinusoids (not necessarily harmonically related) plus white noise. When the product of SNR and m is sufficiently large, PD obtains essentially the same excellent results as does the autoregressive method; however, at lower SNR, it enjoys substantially finer resolution than the autoregressive technique because its model is more appropriate. In addition, PD provides a convenient means for accurately estimating the power of the sinusoids.

2. POWER SPECTRAL ESTIMATION

The Fourier power spectral-estimation problem is to find the squared magnitude of complex multiplicative coefficients that enable harmonically related sinusoids to best represent observed data. From these coefficients, the presence of signals (such as sinusoids) in the time-series data is inferred, and such properties as their amplitude and frequency can be estimated.

Fourier Spectral-Estimation Methods

The Fourier power spectrum of windowed time-series data has usually been computed by the periodogram method, by averaging periodograms, or by the autocorrelation method. The periodogram technique (Schuster, 1898) computes the squared magnitude of the Fourier transform of the data to obtain an estimate of the power spectrum. The Fourier transform is

$$X(\omega) = \sum_{i=0}^{m} x(i) \exp(-j\omega i) \tag{2.1}$$

where x(i) are time-series data taken over m+1 points. The periodogram or power spectrum estimate taken from data starting at time 0 is then

$$\hat{G}_o(\omega) = \frac{1}{m+1} |X(\omega)|^2 \tag{2.2}$$

The averaged periodogram technique (Bartlett, 1950), reduces statistical fluctuations in noisy data by averaging power-spectra estimates. The averaged periodogram is obtained by summing periodograms taken m+1 samples apart on L successive data windows,

$$\hat{G}(\omega)_L = \frac{1}{L} \sum_{i=0}^{L-1} G_{i(m+1)}(\omega) \tag{2.3}$$

The autocorrelation approach (Blackman and Tukey, 1958) computes the Fourier transform on some fraction of the data's autocorrelation lag values. Blackman and Tukey recommended that these values be estimated by

$$\hat{c}_{BT}(k) = \frac{1}{N-|k|} \sum_{i=0}^{N-|k|-1} x(i)x(i+|k|) \qquad 0 \le |k| \le m \tag{2.4}$$

where x(1), x(2), ..., x(N) are the time-series data taken over N points and k is the time lag. This estimate is termed "unweighted" because it forms an unbiased estimate of the true autocorrelation.

The power spectrum of the data $G(\omega)$ is determined by a cosine transform of the autocorrelation function to take advantage of its symmetry about zero lag,

$$G(\omega) = c(0) + 2 \sum_{k=1}^{m} c(k) \cos(\omega k) \qquad (2.5)$$

where $\omega = 2\pi f$ is the radian frequency relative to the radian sampling frequency ($|\omega| \leq 2\pi$), and f is the frequency relative to the sampling frequency ($|f| \leq 1$). Assuming the maximum lag m to be a small fraction of the number of data points N reduces the statistical variability of the spectral estimate at the expense of reducing its resolution.

A problem with the unbiased estimator [Eq. (2.4)] is that it may not be a positive-definite function and occasionally produces spectral estimates having negative power. An autocorrelation estimate that solves this problem and always yields nonnegative power spectra is the Bartlett estimate.

$$\hat{c}_B(k) = \frac{1}{N} \sum_{i=0}^{N-|k|-1} x(i)x(i+|k|) \qquad 0 \leq |k| \leq m = N-1 \qquad (2.6)$$

in which the full number of m=N-1 nonzero lags is computed from the N data points. Bartlett weighting, however, broadens the single sinusoid response by a factor of 1.44 (Mohammed and Smith, 1975), thereby degrading the two-signal resolution.

When not all the lag points are used in the spectral estimate and $m < N-1$, the weighting [Eq. (2.6)] is called "triangular" and will be denoted by $\hat{c}_T(k)$. Unlike Bartlett weighting, the triangular weighting on the data points does not always produce nonnegative power spectral estimates.

Relationships between the Fourier Methods

The Fourier techniques are closely related; most of their differences are based on their implementation and on whether averaging is performed. For example, Oppenheim and Schafer (1975) observed that the power spectral estimate obtained by Fourier transforming via Eq. (2.5) the Bartlett-weighted autocorrelation estimate [Eq. (2.6)] is identical to the power spectrum obtained by the periodogram method. Averaging L distinct N-point periodograms therefore yields the same result as averaging L distinct N-point Bartlett-weighted autocorrelation power spectra.

The relationship between the triangular autocorrelation estimate [Eq. (2.6) with $m < N-1$] and the Blackman-Tukey estimate [Eq. (2.4)] is

$$\hat{c}_T(k) = \frac{N-|k|}{N} \hat{c}_{BT}(k) \tag{2.7}$$

If the largest lags considered are less than some finite number m, the triangular and Blackman-Tukey autocorrelation estimates become identical as the number of data points N goes to infinity. In this case, the estimates are also equal to the true autocorrelation c(k) if the process x(i) is ergodic, or

$$\hat{c}_T(k) \rightarrow \hat{c}_{BT}(k) \rightarrow c(k) \qquad \text{as } N \rightarrow \infty \quad |k| \leq m \tag{2.8}$$

These relationships will be useful later when comparing specific Fourier techniques to other types of spectral estimates.

Fourier Time-Series Model

The discrete Fourier transform of time-series data x(i), i=0, 1, ..., m is

$$X(K\delta\omega) = \sum_{i=0}^{m} x(i) \exp(-jK\delta\omega i) \tag{2.9}$$

and the discrete power-spectrum estimate is

$$G_D(K\delta\omega) = \frac{1}{m+1} |X(K\delta\omega)|^2$$

where K is an integer and $\delta\omega = 2\pi/(m+1)$ is the angular frequency spacing between evaluations.

The inverse transform recovering the time series is

$$x(i) = \frac{1}{m+1} \sum_{K=0}^{m} X(K\delta\omega) \exp(jK\delta\omega i) \tag{2.11}$$

from which the discrete Fourier transform can be considered a fit of a model to the data, where the model is the weighted sum of sinusoids uniformly spaced in frequency as illustrated in Fig. 1. The symmetry of the coefficients about half the sampling frequency for real data has been used to simplify the model.

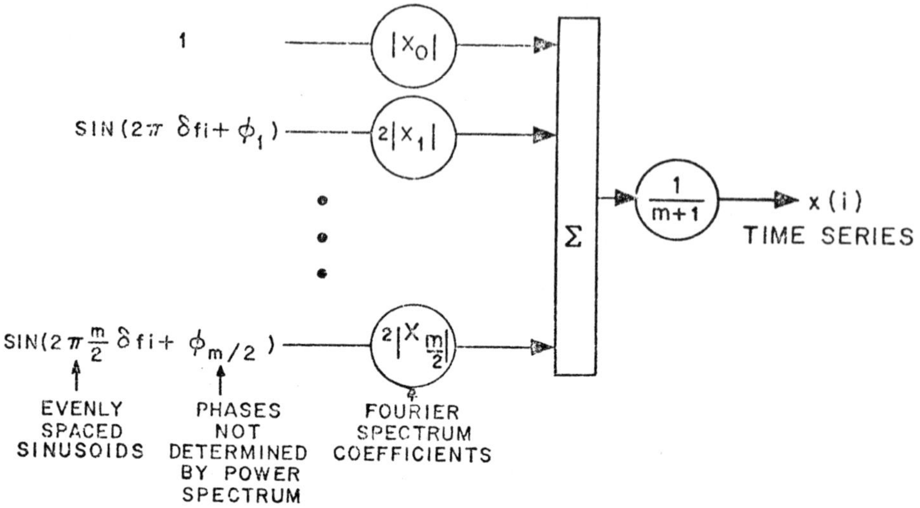

Fig. 1. FOURIER SPECTRAL ESTIMATION TIME-SERIES MODEL: Harmonically related sinusoids.

Examination of Eq. (2.11) for sample times outside the observation window reveals that $x[i \pm L(m+1)] = x(i)$ for $0 \leq i \leq m$ and L an integer (the model assumes that the data are cyclic with period m+1 outside the observation window). The effect of this implicit assumption is benign (Fig. 2a) when the data are cyclic or nearly cyclic; however, grave distortions (Fig. 2b) may occur when a fraction of a cycle is observed.

The continuous Fourier transform on the sampled data given in Eq. (2.1) has a different interpretation. The inverse transform to recover the data is

$$x(i) = \frac{1}{2\pi} \int_{-\pi}^{\pi} X(\omega) \exp(j \omega i) \, d\omega \tag{2.12}$$

which is the limiting case of a sinusoidal model where the frequency spacing between sinusoids has gone to zero and the repetition period is infinite. By direct substitution of Eq. (2.1), it can be shown that the inverse transform is zero outside the observation window. This implicit assumption of zero data outside the observation window results in a spreading of the power spectral density given in Eq. (2.2) and shown in Fig. 2c. This assumption could also produce severe distortion when part of a cycle (Fig. 2d) is observed.

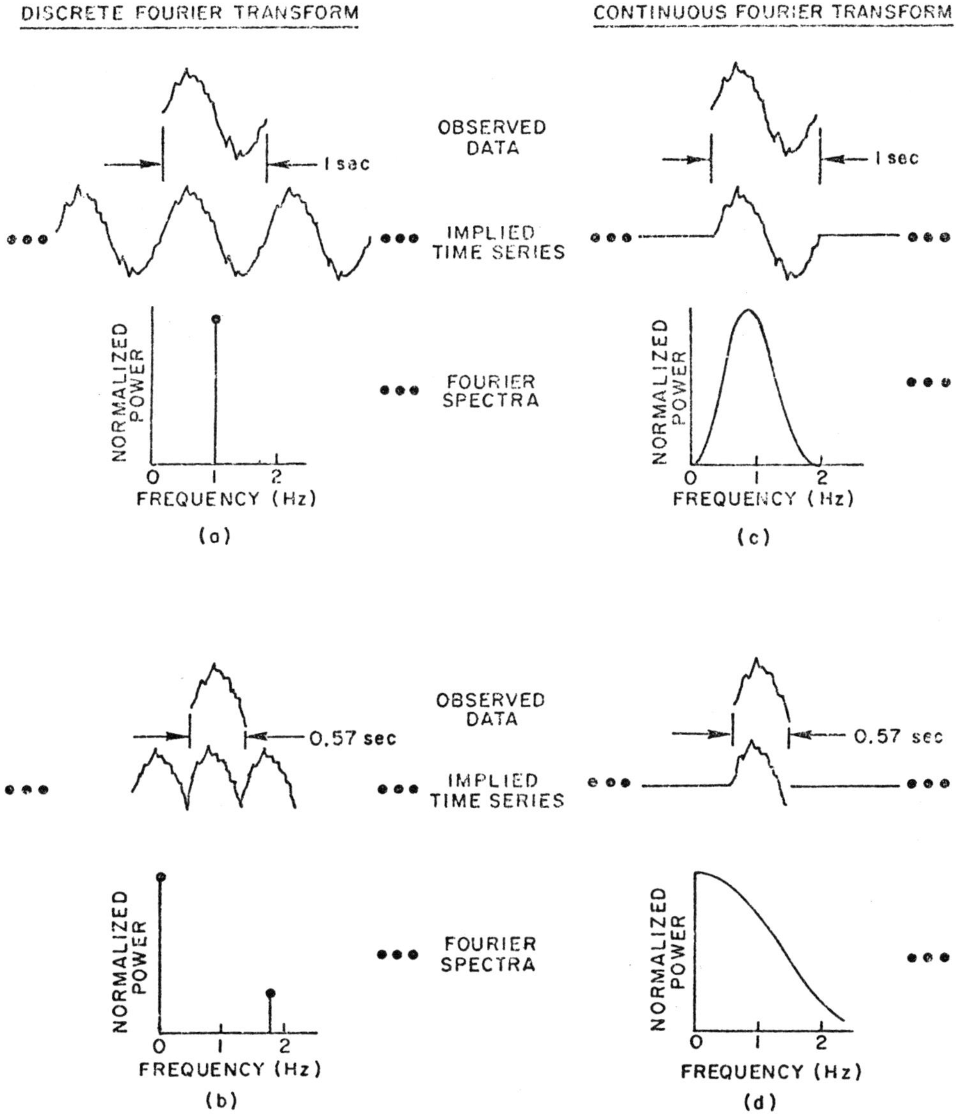

Fig. 2. FOURIER SPECTRA OF 1 HZ SINUSOID WITH 10% NOISE. The time series implied outside the observation window severely distorts the spectra in (b) and (d). After Ulrych (1972).

Cyclic time series composed entirely of sinusoids at the model frequencies are well-suited to spectral estimation by the discrete Fourier power spectrum. An example of subjectively "good" performance is illustrated in Fig. 3 (upper) where the discrete power spectrum of a sinusoid is at one of the harmonic model frequencies. The lower sections of the figure, however, show the "leakage" phenomenon that occurs when in the discrete Fourier spectrum of a pure sinusoid having a frequency not equal to a model frequency. In most applications, this leakage is undesirable because it can interfere with the detection of multiple sinusoids in noise. The model of evenly spaced sinusoids also makes it difficult to resolve two sinusoids (distinguish the presence of more than one) when they are closer in frequency than the model frequency spacing. Zero packing does not help

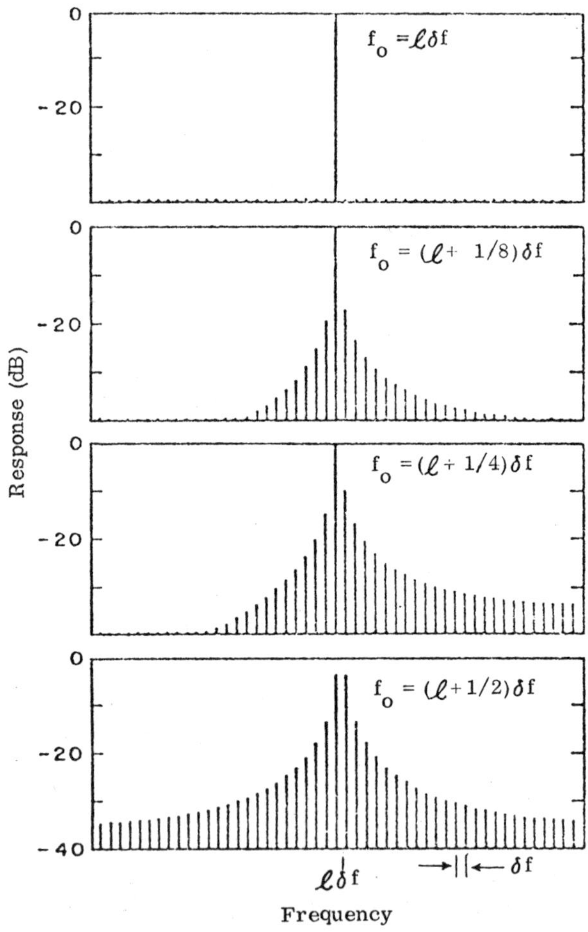

Fig. 3. LEAKAGE FOR SINUSOIDS OF DIFFERENT FREQUENCIES (from Mohammed and Smith).

to distinguish closely spaced sinusoids because it assumes that the data values are zero outside the observation window and does not improve two-sinusoid resolution.

3. MAXIMUM-ENTROPY (AUTOREGRESSIVE) SPECTRUM ANALYSIS

To alleviate problems caused by the zero-value or cyclic data extensions assumed by Fourier methods, Burg (1967) proposed an extension of the autocorrelation function outside the known interval. The spectrum computed from his extended autocorrelation tends to yield higher resolution because the time interval over which the transform is taken is longer. This new type of spectrum analysis, known as the maximum-entropy method is described in detail by Lacoss in another paper presented at this conference.

Van den Bos (1971) observed that the maximum entropy method is equivalent to least-squares fitting an m-coefficient autoregressive model to the available time-series data. Errors are modeled by white noise exciting the autoregressive (feedback) filter, as shown in Fig. 4. After

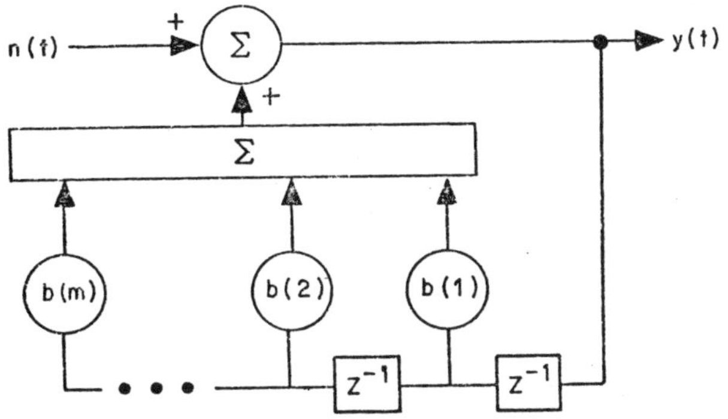

Fig. 4. AUTOREGRESSIVE FILTER. Time-series generation model for autoregressive spectrum analysis.

the feedback filter is initialized by the observed data, it generates a time series that extends beyond the observation window on the basis of the past values, thereby avoiding the Fourier method's a priori selection of uniformly spaced model frequencies. The maximum-entropy power spectrum is equal to the Fourier transform of the AR filter output autocorrelation when the filter is excited by white noise. Equivalently, the spectrum is taken to be the magnitude-squared transfer function of the AR filter. Historically, this spectrum has been called the autoregressive spectrum (Schaerf, 1963) and is so designated throughout the remainder of this paper. The advantages of this approach for obtaining spectral estimates over short observation windows are illustrated in Fig. 5 and may be compared to the Fourier spectra of Fig. 2; Chen and Stegen (1974) have shown, however, that the performance may not always be as good as indicated in the figure because the location of the spectral peak varies somewhat with the starting phase of the sinusoid. This variation is small when the number of samples per cycle is ten or more.

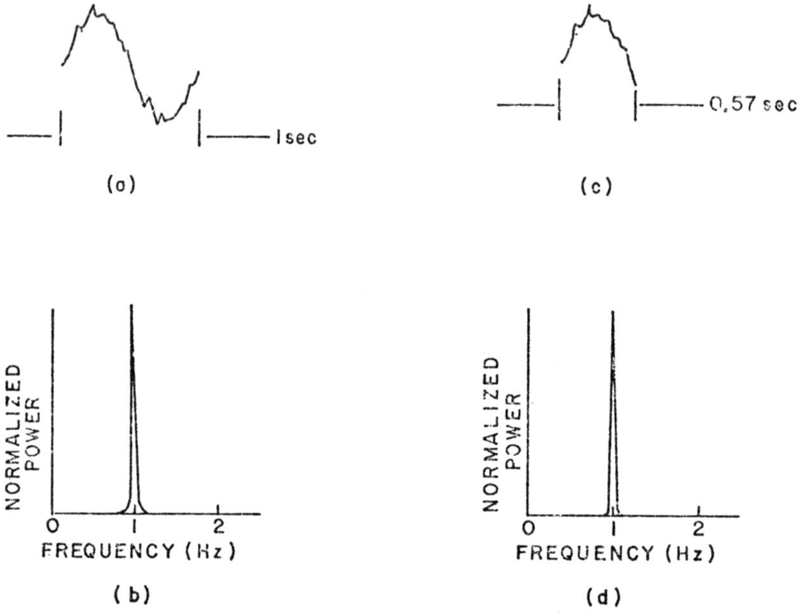

Fig. 5. AUTOREGRESSIVE SPECTRA OF 1 Hz SINUSOID WITH 10% NOISE. After Ulrych (1972).

Van den Bos also shows that the AR method is equivalent to making m+1 autocorrelation lag-value estimates from the data and constructing an AR filter whose first m+1 autocorrelation lag values exactly match these estimates. Subsequent autocorrelation lag value are those of the process generated by the autoregressive filter when excited by white noise and are generally different from the autocorrelation extensions assumed by the Fourier methods. Figure 6 is an example of better multiple-sinusoid resolution obtained by the autoregressive method applied to autocorrelation estimates.

Burg's technique often yields high resolution because of the AR filter model. Normally, an AR filter has an infinite impulse response and an infinite output autocorrelation function in response to white noise. The Fourier transform of the infinite-duration autocorrelation function

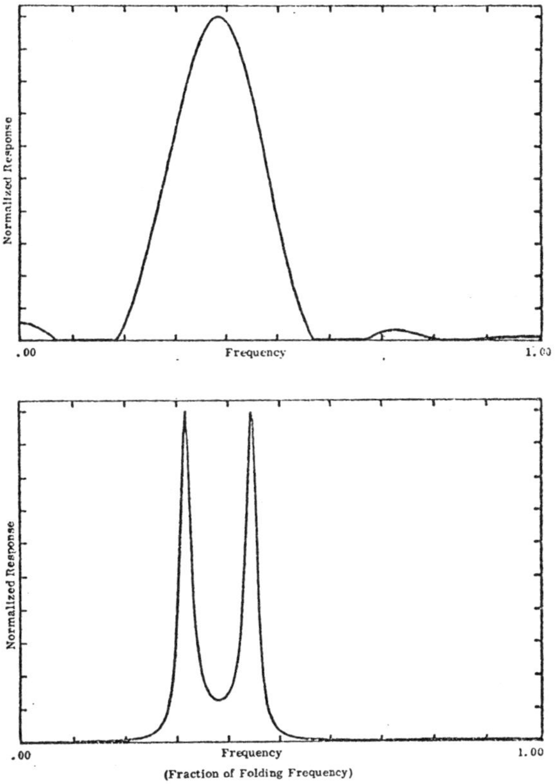

Fig. 6. COMPARISON OF RESOLUTION OBTAINED BY FOURIER (UPPER) AND AUTOREGRESSIVE (LOWER) TECHNIQUES WITH PERFECTLY KNOWN AUTOCORRELATION FUNCTION FOR SINUSOIDS WITH 10 dB SNR. Number of nonzero autocorrelation lag values is m = 9.

generally has higher resolution than does the Fourier transform of the autocorrelation measurements taken over an observation window. The physical relevance of the high-resolution spectrum is related to the appropriateness of the AR model and the accuracy of the measured autocorrelation lag values extended by the model.

The AR filter has m/2 pole pairs that can be placed on the Z-plane to form the modeling filter; spectral resolution depends on the proximity of the poles to the unit circle. By proper selection of feedback coefficients, the AR filter is in principle capable of synthesizing arbitrarily narrow passbands at m/2 different frequencies, and these frequencies are not required to be harmonically related. If the time series consists of two sinusoids at a high SNR, the poles can be placed close to the unit circle and high resolution is obtained as shown in Fig. 6. If the AR filter has too few degrees of freedom (too few coefficients), however, it cannot model the sinusoids well in the presence of strong noise and the poles must be moved back from the unit circle, thereby reducing the resolving power.

Marple (1976) has documented the two-sinusoid resolving power of the AR method and has compared it to conventional Fourier techniques as a function of SNR, number of data points and autocorrelation lags (or Fourier-transform size), and the method for estimating the autocorrelation function. His results reveal that the AR method usually produces better average two-signal resolution than do the Fourier techniques described in Section 2 although variance in the resolution may occasionally result in poorer AR resolution. Resolution improvement can be significant if the SNR or number of autocorrelation lags is sufficiently large; when they are small, the resolution degrades to that of the Fourier methods.

The Autoregressive Power Spectrum

Following Ulrych and Clayton (1975), the output of the mth order AR filter in Fig. 4 is y(t), where

$$y(t) = \sum_{k=1}^{m} y(t-k)b(k) + v(t) \tag{3.1}$$

and v(t) is a white-noise process with $E[v(t)v(t-k)] = \sigma_v^2 \delta(k)$. Here, $\delta(k)=1$ for k=0 and $\delta(k)=0$ for $k\neq 0$. Defining the m+1 vectors, B and Y(t), as

$$B^T = [1, -b(1), -b(2), \ldots, -b(m)] \tag{3.2}$$

and

$$Y^T(t) = [y(t), y(t-1), y(t-2), \ldots, y(t-m)] \tag{3.3}$$

Eq. (3.1) becomes

$$Y^T(t)\, B = v(t) \tag{3.4}$$

Premultiplying by Y(t) and taking the expected value obtains

$$E[Y(t)Y^T(t)]\, B = E[Y(t)v(t)] \tag{3.5}$$

or

$$C_{YY}B = C_{Yv} \tag{3.6}$$

where the (m+1)x(m+1) correlation matrix is defined by

$$C_{YY} \triangleq E[Y(t)Y^T(t)] \tag{3.7}$$

and the m+1 cross-correlation vector is

$$C_{Yv} \triangleq E[Y(t)v(t)] = [\sigma_v^2, 0, 0, \ldots, 0]^T \tag{3.8}$$

The last equality follows from Eq. (3.1) and the white-noise specification so that past values of the filter output are uncorrelated with the present driving noise v(t); that is,

$$E[y(t-k)v(t)] = \sigma_v^2\, \delta(k) \tag{3.9}$$

Given a set of estimated autocorrelations $\hat{c}(0), \ldots, \hat{c}(m)$, the estimated correlation matrix $\hat{C}_{YY}$ can be formed and Eq. (3.6) can be solved to obtain the unknown AR coefficients b(1), b(2), ..., b(m) and the noise power σ_v^2.

The autoregressive power spectrum is the power spectrum $G_{yy}(\omega)$ of the filter output y(t). The $G_{yy}(\omega)$ of the output of a filter with a magnitude-squared transfer function $|H(\omega)|^2$ and an input power spectrum $G_{vv}(\omega)$ is

$$G_{yy}(\omega) = |H(\omega)|^2\, G_{vv}(\omega) \tag{3.10}$$

Equation (2.5) shows that

$$G_{vv}(\omega) = \sigma_v^2 \tag{3.11}$$

and, from a simple calculation,

$$|H(\omega)|^2 = |1 - \sum_{k=1}^{m} b(k) \exp(-j\omega k)|^{-2} \tag{3.12}$$

therefore, the AR power spectrum is

$$G_{AR}(\omega) = G_{yy}(\omega) = \sigma_v^2 |1 - \sum_{k=1}^{m} b(k) \exp(-j\omega k)|^{-2} \tag{3.13}$$

Correlation Estimation

Given N data points, there are a number of ways to estimate the autocorrelation lags c(0), c(1), ..., c(m). The AR spectrum has proven to be quite sensitive to the method by which these estimates are made. Of the autocorrelation estimation procedures described in Section 2, the Blackman-Tukey technique usually obtains the best two-signal resolution but, occasionally, produces negative power spectra and an unstable AR filter. Bartlett weighting guarantees nonnegative power spectra but has lower resolution and also occasionally yields an unstable AR filter.

Burg (1968) derived a new approach for determining the autoregressive filter coefficients and driving noise power directly from the data samples, bypassing an explicit estimation of the autocorrelation function; however, from Eq. (3.6) this estimate is implicitly determined by the filter coefficients and driving noise power. Marple (1976) has shown that the implied Burg autocorrelation lag values and the Burg AR coefficients can be generated simultaneously. Generation of the Burg coefficients is not computationally simple, but the implied correlation function is nonnegative definite (yielding nonnegative power spectra) and the AR filter is not unstable. Experimentation by Ulrych and Bishop (1975) and measurements by Marple (1976) have revealed that the Burg technique usually yields an AR spectrum with higher resolution than that obtained by solving Eq. (3.6) with the Blackman-Tukey or Bartlett autocorrelation estimates. The Burg procedure has been described by Andersen (1974) and a computer program in FORTRAN is given by Ulrych and Bishop(1975).

Two-Sinusoid Resolution

One of the desirable properties of AR spectrum analysis is an enhanced two-sinusoid resolution relative to Fourier methods. Consider two sinusoids with frequencies f_1 and f_2. In this study, two-sinusoid resolution is defined as the frequency separation $\Delta f = |f_1 - f_2|$ at which the power spectrum evaluated at the midpoint between the signals $G[(f_1+f_2)/2]$ is equal to the average of the power spectra evaluated at the two signal frequencies; that is, two signals are termed "just resolved" when

$$G\left(\frac{f_1+f_2}{2}\right) = \frac{1}{2}\left[G(f_1) + G(f_2)\right] \tag{3.14}$$

Motivation for the definition was based, in part, on the nonlinearity of the new spectrum-analysis techniques in which such traditional concepts as "-3 dB response width" or "3 dB dip between peaks" have less meaning than with Fourier techniques. The criterion [Eq. (3.14)] is independent of such measures and is computationally simple to evaluate. Additional motivation stemmed from the fact that, for signals well-separated in frequency, the spectral-response peaks occur at f_1 and f_2 (Fig. 7a); therefore, before simulations were made, it was assumed that Eq. (3.14) would result in a flat-topped spectrum at the point of resolution (Fig. 7b).

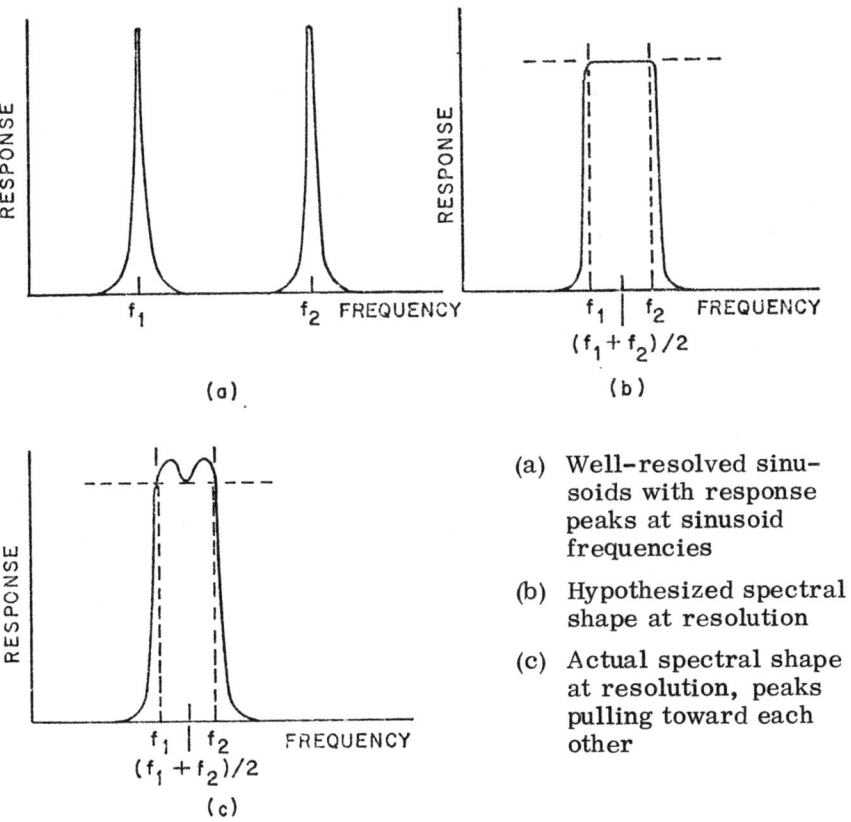

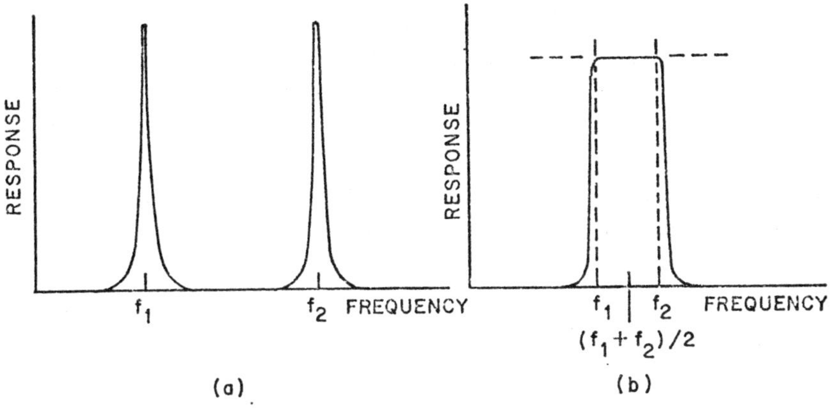

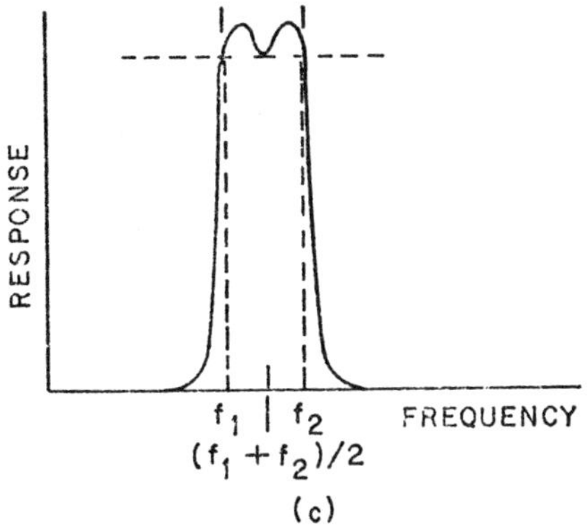

(a) Well-resolved sinusoids with response peaks at sinusoid frequencies

(b) Hypothesized spectral shape at resolution

(c) Actual spectral shape at resolution, peaks pulling toward each other

Fig. 7. FREQUENCY PULLING EFFECT. This effect near the resolution limit is common to all spectral-analysis methods.

For both the Fourier and AR spectral-estimation methods, however, simulation shows that, as the signals are brought closer together, the spectral-response peaks remain near the signal frequencies until just prior to merging into a single peak, at which point they move toward each other, as seen in Fig. 7c. The magnitude of this effect is approximately the same in both techniques.

Equipment-Limited Case. The two-sinusoid resolution of Fourier and AR techniques can be determined most easily when perfect knowledge of the autocorrelation function is available. This condition is approximated in the case of a stationary environment where the number of observable data points N is much larger than the number of data points that can be processed and stored simultaneously by the available equipment. This limitation could occur as a result of the number of points that can be processed and stored by the autocorrelator, FFT equipment, or integrator. In this section, therefore, it is assumed that the number of nonzero autocorrelation lags is some finite number m and that the number of observable data points is without bound ($N \rightarrow \infty$) thereby producing perfect knowledge of the m+1 autocorrelation lags c(0), c(1), ..., c(m).

Marple (1976) employed two methods to determine resolution when perfect knowledge of the time-series autocorrelation function can be assumed. The first uses the autocorrelation function for two sinusoids in noise:

$$c(k) = \sigma_n^2 \delta(k) + P_{S1} \cos(2\pi f_1 k) + P_{S2} \cos(2\pi f_2 k) \quad (3.15)$$

where σ_n^2 is the noise power and P_{S1} and P_{S2} are the signal powers. The SNR of the ith sinusoid is $10 \log (P_{Si}/\sigma_n^2)$. Defining $V_i^T = [1, \exp(j2\pi f_1), \ldots, \exp(j2\pi f_1 m)]$, the real autocorrelation matrix for two sinusoids in noise can be written as

$$C_r = \sigma_n^2 I + \frac{1}{2} P_{S1}(V_1{}^*V_1{}^T + V_1V_1{}^{*T}) + \frac{1}{2} P_{S2}(V_2{}^*V_2{}^T + V_2V_2{}^{*T}) \quad (3.16)$$

where I is the (m+1)x(m+1) identity matrix.

As noted by Lacoss (1971), the analytical work can be greatly simplified by using the complex form of the autocorrelation matrix,

$$C_c = \sigma_n^2 I + \alpha_1 V_1{}^*V_1{}^T + \alpha_2 V_1{}^*V_1{}^T \quad (3.17)$$

where α_1 and α_2 are real constants. This matrix is the basis for the second approach. Complex data can be generated by the real and imaginary components of a discrete Fourier-transform frequency cell or when a narrowband process is downconverted to baseband in quadrature and the two signals x(t) and y(t) thus obtained are treated as a single complex signal x(t) + jy(t). Using the complex matrix, Eq. (3.6) can be solved for the AR coefficients by twice applying a well-known matrix-inversion lemma (Bryson and Ho, 1969, p. 12). Substitution of the resulting coefficient values into Eq. (3.13) produces an extensive analytical expression for the AR spectrum of two sinusoids in noise that requires no matrix inversions to evaluate. Iterative computer evaluation of this expression to determine the frequency separation Δf resulted in the resolution curves in Fig. 8.

The resolution plotted is the dimensionless quantity of normalized resolution

$$R' = 2\pi m \Delta t \Delta f$$

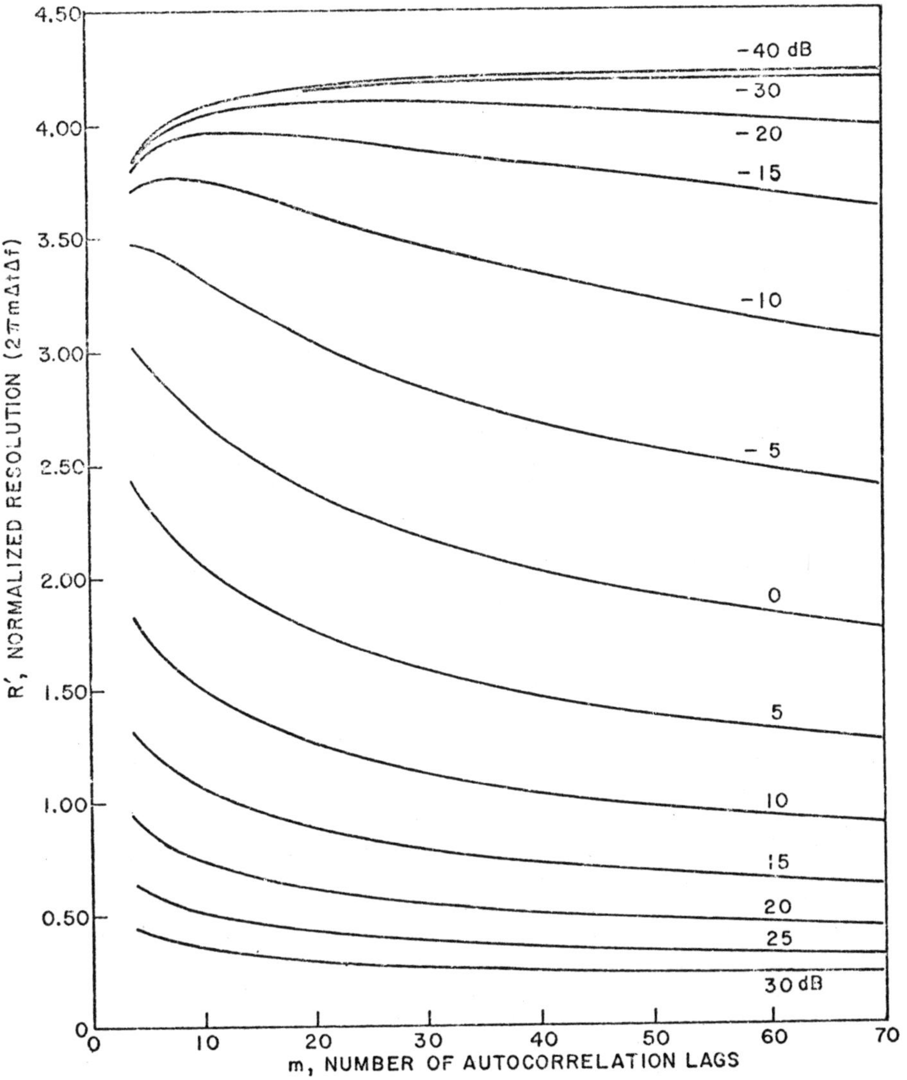

Fig. 8. NORMALIZED TWO-SINUSOID RESOLUTION OF COMPLEX AUTOREGRESSIVE SPECTRAL ESTIMATION FOR A PERFECTLY KNOWN AUTOCORRELATION FUNCTION. $N \to \infty$, m finite.

If the time between samples Δt is in seconds, the resolution in hertz is

$$\Delta f = \frac{R'}{2\pi m \Delta t}$$

Unless otherwise noted, the average frequency of the two sinusoids is half the folding frequency $[\frac{1}{2}(f_1+f_2) = 1/4\,\Delta t]$.

The AR resolution for a known real autocorrelation was evaluated by Marple (1976) through the numerical solution of Eq. (3.6) with the real autocorrelation matrix (3.16). This is a time-consuming approach but is more accurate when real signals are involved. The resolution thus obtained is labeled "real autoregressive" in Fig. 9. The jagged line of the normalized real AR resolution may lead one to believe that, at some points, the resolution worsens if the number of lags is increased by 1. This appearance is the result of the normalization as shown in Fig. 10 which plots an unnormalized real AR resolution

$$R = 2\pi \Delta t \Delta f$$

vs the number of nonzero lags m. It can be seen that, when m is even, increasing the number of lags does not improve the resolution because, when the two sinusoids are nearly half the folding frequency (near 1/4 the sampling frequency), the odd coefficients $b(i)$ are zero; therefore, adding a zero feedback coefficient to the AR filter has no effect. At other frequencies in the band, however, adding more coefficients usually improves resolution and never degrades it.

The resolution obtained by Fourier transforming the perfectly known autocorrelation values to m nonzero lags is also plotted in Figs. 9 and 10. This resolution is independent of the SNR because from Eq. (3.15), noise in the perfectly known autocorrelation function only affects the zero lag value. From Eq. (2.5), this value adds a constant to the spectrum and does not affect the resolution. Because the number of points N is infinite for m finite, the weighting is the same (unity) for both the Blackman-Tukey and triangular approaches and they produce identical results. Bartlett weighting over the m nonzero autocorrelation lags degrades the resolution by an additional amount. Transforming this perfectly known Bartlett autocorrelation function is equivalent to averaging an infinite number of periodograms, each taken on $m+1$ data points. The periodogram resolution data plotted here are the numerical averages of periodogram resolutions measured for successive 10° steps in the starting phase offset between the two sinusoids.

For perfect knowledge of the autocorrelation, the two-signal resolution of the AR spectrum analysis is always equal to or better than the Blackman-Tukey or Bartlett approaches. For large numbers of autocorrelation lags or high SNR, the two-signal resolution of the AR analysis is significantly better than that of conventional Fourier approaches. For example, for 60 nonzero lags and 20 dB SNR, AR spectrum analysis produces more than ten times better resolution than does the Blackman-Tukey technique; for the same number of lags and a SNR of unity (0 dB), the AR method obtains more than a factor of 2 better resolution.

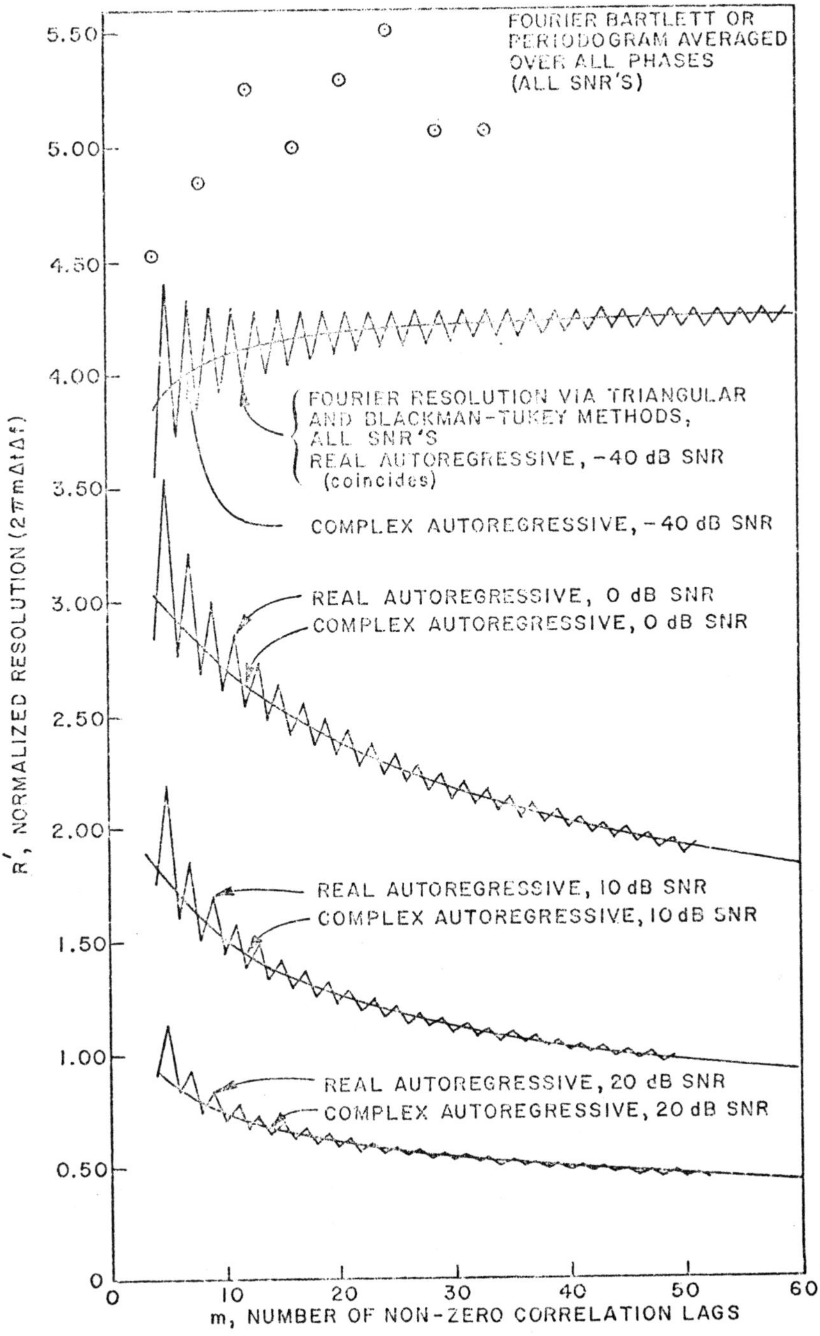

Fig. 9. NORMALIZED TWO-SINUSOID RESOLUTION WITH PERFECTLY KNOWN AUTOCORRELATIONS. $N \rightarrow \infty$

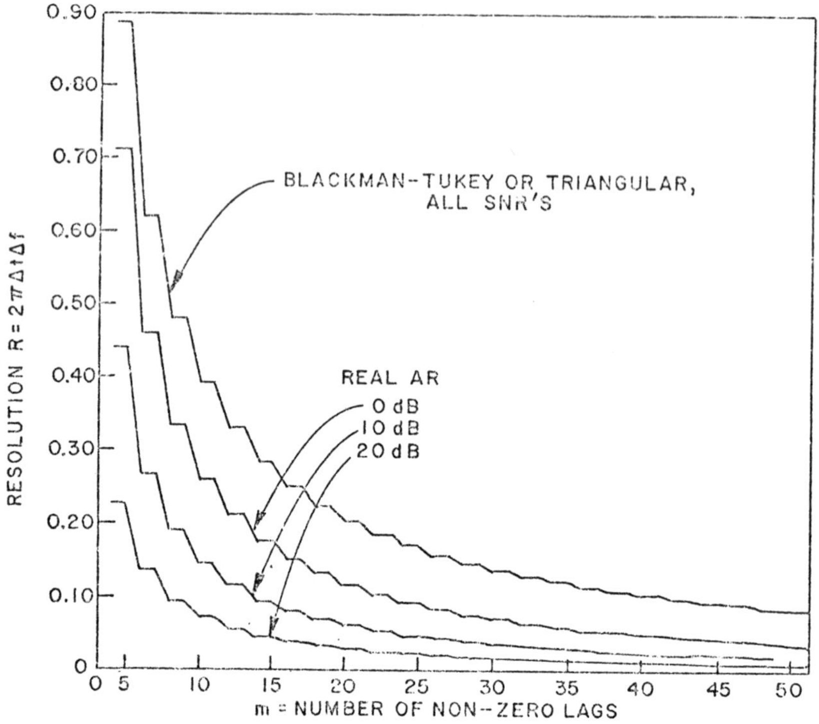

Fig. 10. TWO-SINUSOID RESOLUTION FOR PERFECTLY KNOWN AUTOCORRELATION FUNCTIONS. $N \to \infty$, m finite.

Both Fourier and AR techniques have somewhat different two-sinusoid resolutions, depending on the location of the signals in the band (Fig. 11). The magnitude of the variations is smaller for large numbers of coefficients.

Observation-Time-Limited Case. The case discussed in this section occurs when the observation time is limited and it is desirable to make the very best use of all available data. Here, the number of non-zero autocorrelation lag estimates that can be formed is one less than the number of data points (m=N-1).

Marple (1976) measured the two-signal resolution for this case by computer simulations in which time-series data were formed by adding sinusoids to an epoch of white gaussian noise. During each resolution measurement, the noise was held constant while the frequencies of the sinusoids were brought together gradually until the resolution criterion was met. For each series of measurements, this procedure was repeated over independent noise epochs, and the resolution measurements for each series were averaged at least ten times and until a Chebychev inequality

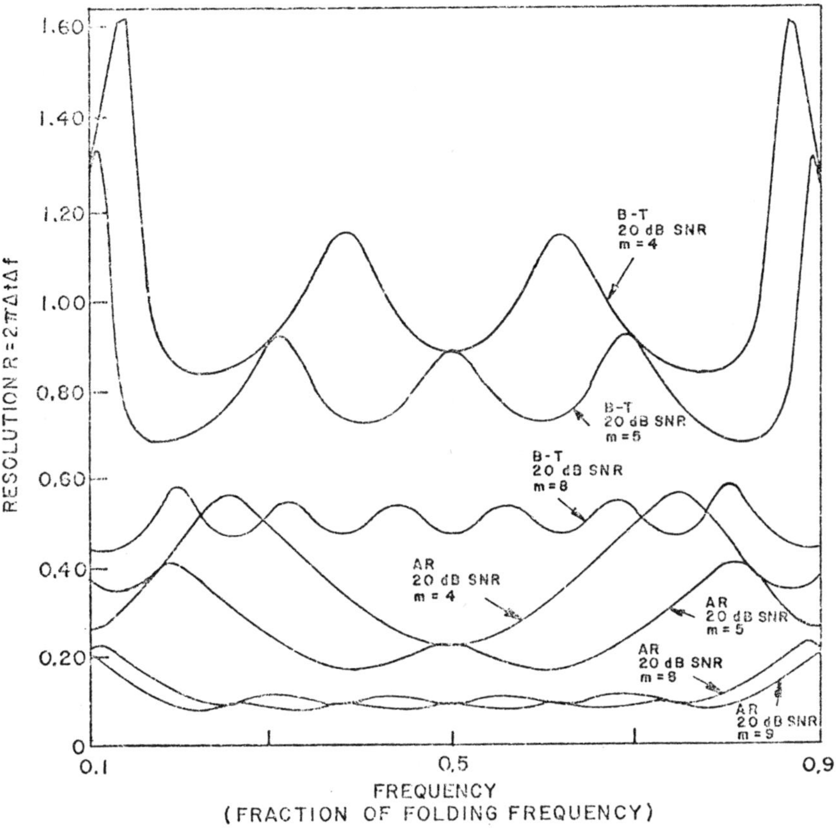

Fig. 11. VARIATION OF RESOLUTION AS A FUNCTION OF AVERAGE FREQUENCY. Evaluated for the perfectly known autocorrelation function. B–T = Blackman–Tukey; AR = Autoregressive.

on the measurement variance indicated that the average resolution had a 90 percent confidence of being within 10 percent of its true value.

Measured resolutions for the observation-time-limited case (Fig. 12) are usually not as good as the resolutions obtained when an infinite amount of data is available; however, they converge rapidly to the asymptotic resolution as the number of data points N increases, as illustrated in Fig. 13. With finite SNR, the statistical variation in the resolution is caused by the noise; the standard deviations are indicated by vertical bars about the data points in the figure.

Resolution also has a deterministic variation as a result of the relative starting phase of the two sinusoids; this variation was not found in the case of perfectly known autocorrelations because autocorrelation is not a function of the relative starting phase. Although the magnitude

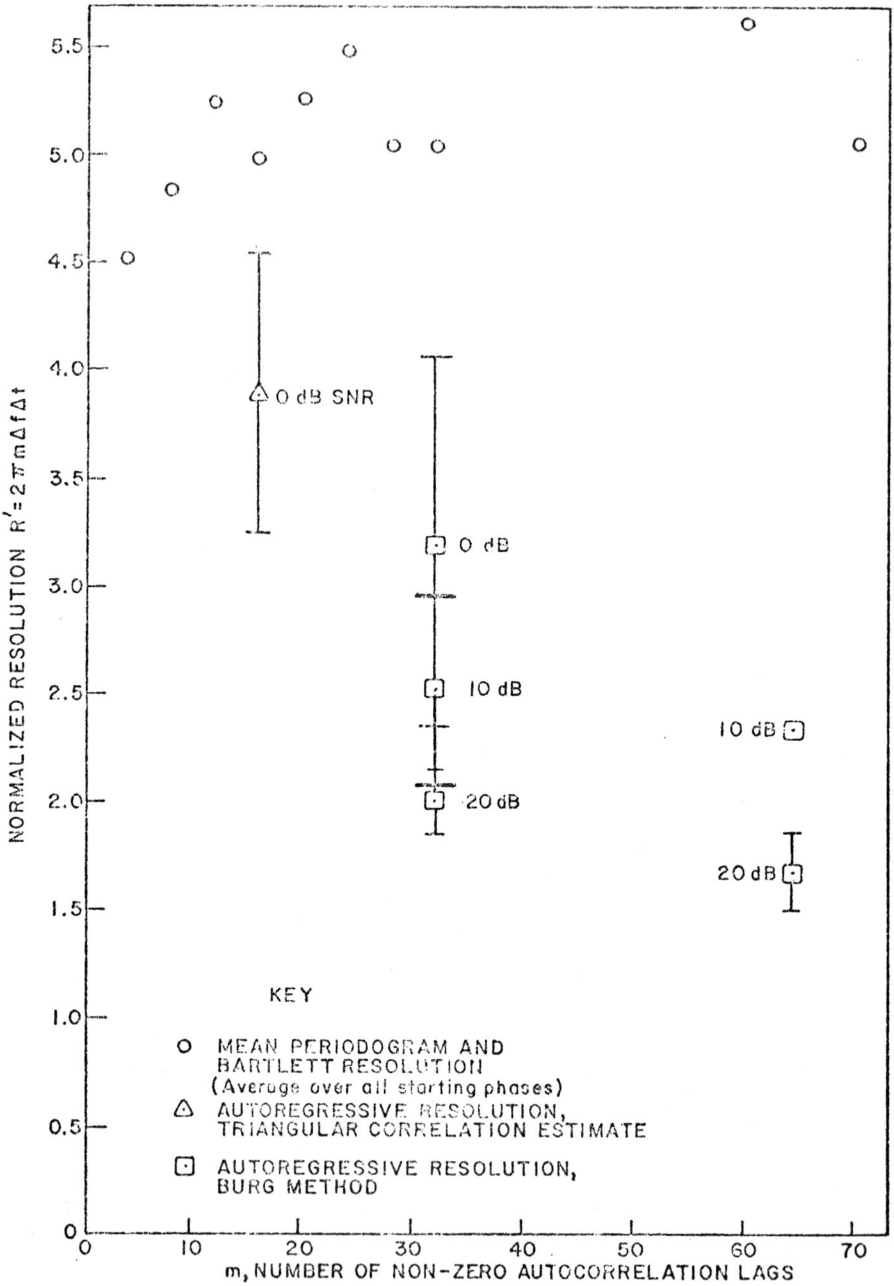

Fig. 12. NORMALIZED OBSERVATION TIME-LIMITED TWO-SINUSOID RESOLUTION, m=N-1. Vertical bars indicate standard deviation.

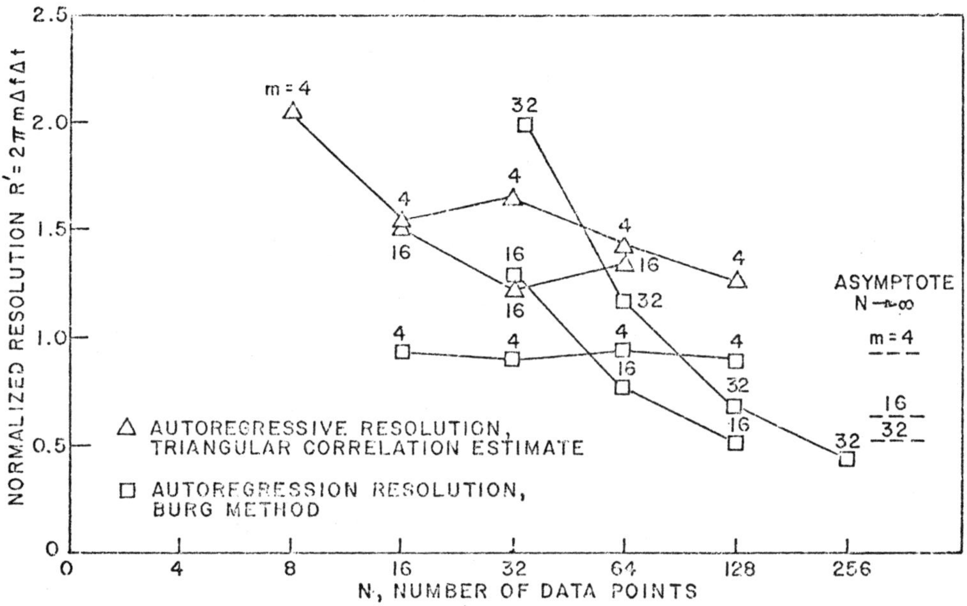

Fig. 13. NORMALIZED RESOLUTION IMPROVEMENT WITH ADDITIONAL DATA. Standard deviations not shown for simplicity.

of change in resolution R with the starting phase decreases as the number of data points N increases, the variation in normalized resolution R' is virtually independent of N because the starting phase is an "end effect" whose magnitude is inversely proportional to the number of data points. Multiplying by the number of data points to normalize the resolution makes the large standard deviation essentially constant ($\cong 2.1$). In the Fourier methods, the starting phase was found to cause variations, ranging typically from a finest resolution of 1.1 to a coarsest of 8.6 for nonzero lags between 4 and 32. No attempt was made to assess the effect of phase variation on the AR spectra because of the amount of computation required to obtain Chebychev measurement confidence of the AR simulations with noise.

Although the AR technique has higher two-signal resolution than does the Fourier technique, it is difficult to measure the power of the sinusoids. Lacoss (1971) observed that the amplitude of the AR spectral response is not a good indicator of sinusoid power but that the area under the response peak is proportional to the power. This area is difficult to measure if two sinusoids are close to each other in frequency, near the limits of resolvability. In the next section, a method for accurately measuring signal power is described and an even higher resolution spectral-estimation method is presented.

4. PISARENKO SPECTRAL DECOMPOSITION

In some applications (speech encoding and vocal-tract simulation, for example) an autoregressive filter excited by white noise appears to be a good model of the waveform generator (Atal and Hanauer, 1971). This filter is particularly attractive for modeling strong resonances such as those that can occur in the vocal tract.

Sinusoids may also be considered as strong resonances of an extremely narrowband filter. As a result at high SNR, they are well modeled by an autoregressive filter excited by white noise; however, at lower SNR, particularly when the filter has too few degrees of freedom (coefficients), the AR filter model of the total signal consisting of sinusoids with additive noise has poorer resolution, as shown in Fig. 8.

Pisarenko (1973) suggested an alternate spectral-estimation technique based on a time-series model of sinusoids plus additive white noise (Fig. 14). Known as "Pisarenko decomposition" (PD), it greatly alleviates the problem of degraded sinusoid resolution at low SNR. When the model is accurate and the autocorrelation function of the time-series data is perfectly known (the "equipment limited" case of Fig. 9), the PD two-sinusoid resolution is infinitely fine, regardless of the SNR; for this case, no finer two-sinusoid resolution can be obtained by any other method.

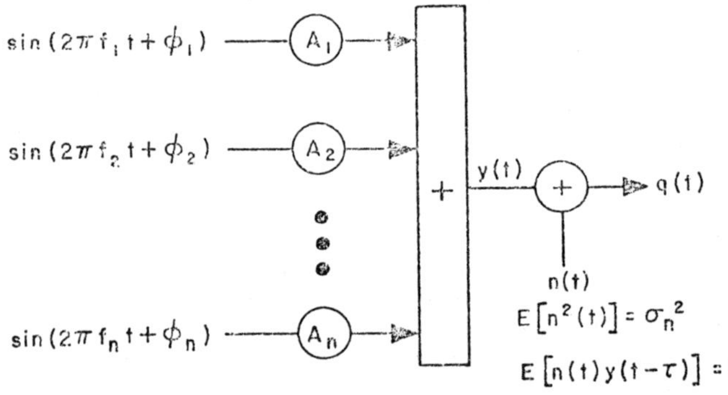

Fig. 14. PISARENKO TIME-SERIES MODEL OF SINUSOIDS WITH ADDITIVE WHITE NOISE. Frequencies are not necessarily harmonically related.

Concept

Pisarenko's spectrum-analysis method is based on the fact that the autocorrelation function of n sinusoids in white noise [see Eq. (3.15)] is

$$c(k) = \sigma_n^2\,\delta(k) + \sum_{i=1}^{n} \frac{A_i^2}{2} \cos(2\pi f_i k) \qquad 0 \leq k \leq m \tag{4.1}$$

Note that the white noise affects only the zero lag autocorrelation value. If the number of known nonzero lag values m is greater than or equal to twice the number of sinusoids ($m \geq 2n$), the noise power can be exactly determined and subtracted from the autocorrelation zero lag value, leaving the signal autocorrelation function

$$c_S(k) = c(k) - \sigma_n^2\,\delta(k) \tag{4.2a}$$

$$= \sum_{i=1}^{n} \frac{A_i^2}{2} \cos(2\pi f_i k) \qquad 0 \leq k \leq m \tag{4.2b}$$

Application of AR spectrum analysis to this noisefree (infinite SNR) autocorrelation function yields infinitely fine two-sinusoid resolution, as may be expected from the trend illustrated in Fig. 9.

Examples

Two examples are presented to illustrate the effectiveness of Pisarenko decomposition. In the first example, PD spectra are compared to the AR spectrum of three sinusoids at 10 dB SNR (Fig. 6 and repeated in Fig. 15a). A perfectly known autocorrelation function to nine nonzero lags is assumed. Fig. 15b plots the result of subtracting 90 percent of the noise power [computing the AR spectrum of $c(k) - 0.9\,\sigma_n^2\,\delta(k)$]; for the perfectly known autocorrelation function, this has precisely the same effect as increasing the SNR by 10 dB. Figure 15c plots the result of subtracting 99 percent of the noise power from the zero lag value, which again increases the SNR by 10 dB. At this point, a third signal is clearly observed between the first two; this signal was always present, but the Fourier and AR techniques failed to detect it (see Fig. 6). Comparison of the peak locations in Fig. 16 to the actual signal frequencies (indicated by arrows) reveals that subtraction of noise power from the autocorrelation function results in a more accurate determination of the signal frequencies.

A second example (Fig. 16) illustrates the effect of imperfect knowledge of the autocorrelation function. In this example, only mild performance degradation was obtained when the ten autocorrelation-function values were estimated by the Blackman-Tukey method via 300 samples of time-series data simulating three 10 dB SNR sinusoids in noise. Statistical fluctuations in the noise caused small differences in the spectral-response peaks from those obtained in Fig. 15 and the spectrum was positive even when 102 percent of the ensemble mean noise power was removed.

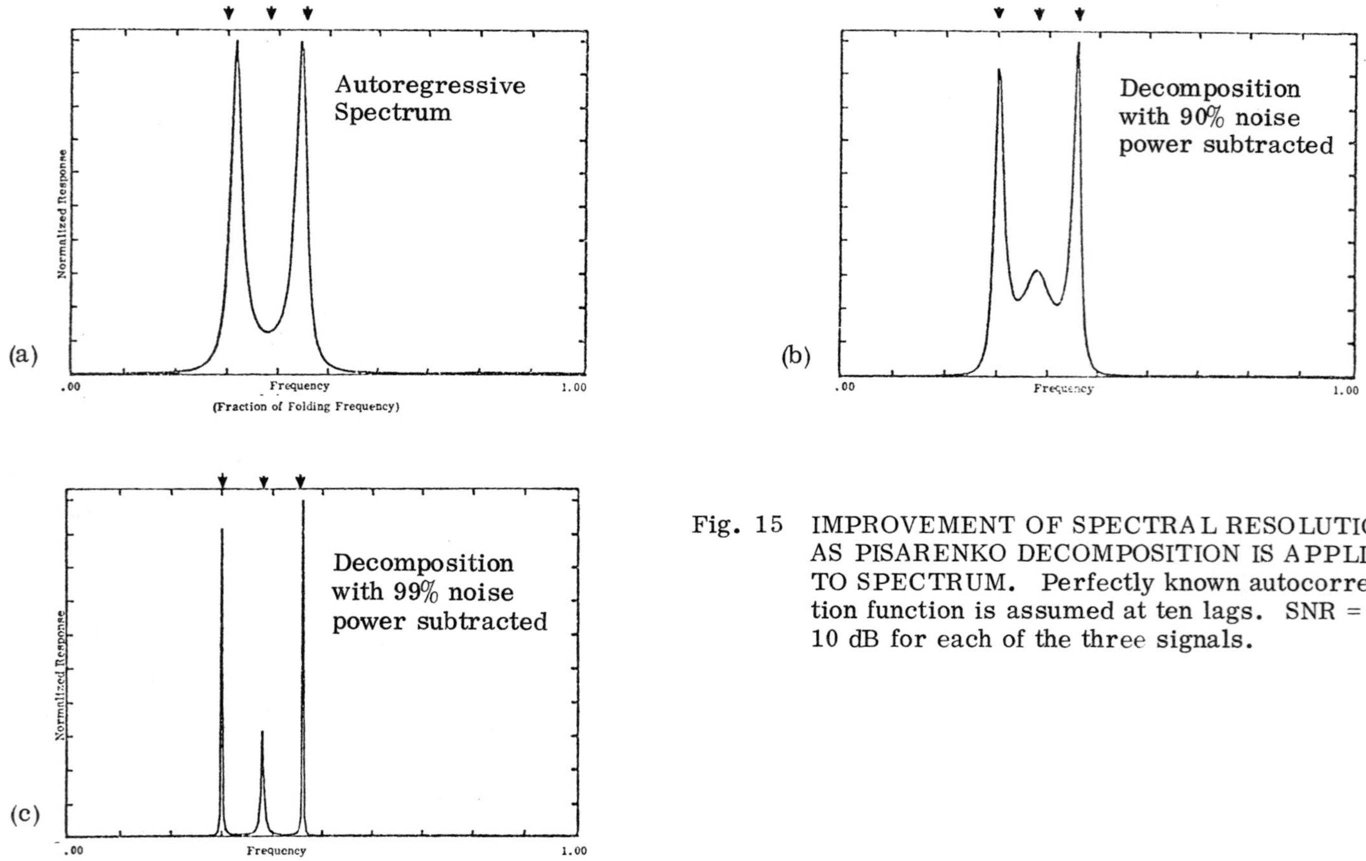

Fig. 15 IMPROVEMENT OF SPECTRAL RESOLUTION AS PISARENKO DECOMPOSITION IS APPLIED TO SPECTRUM. Perfectly known autocorrelation function is assumed at ten lags. SNR = 10 dB for each of the three signals.

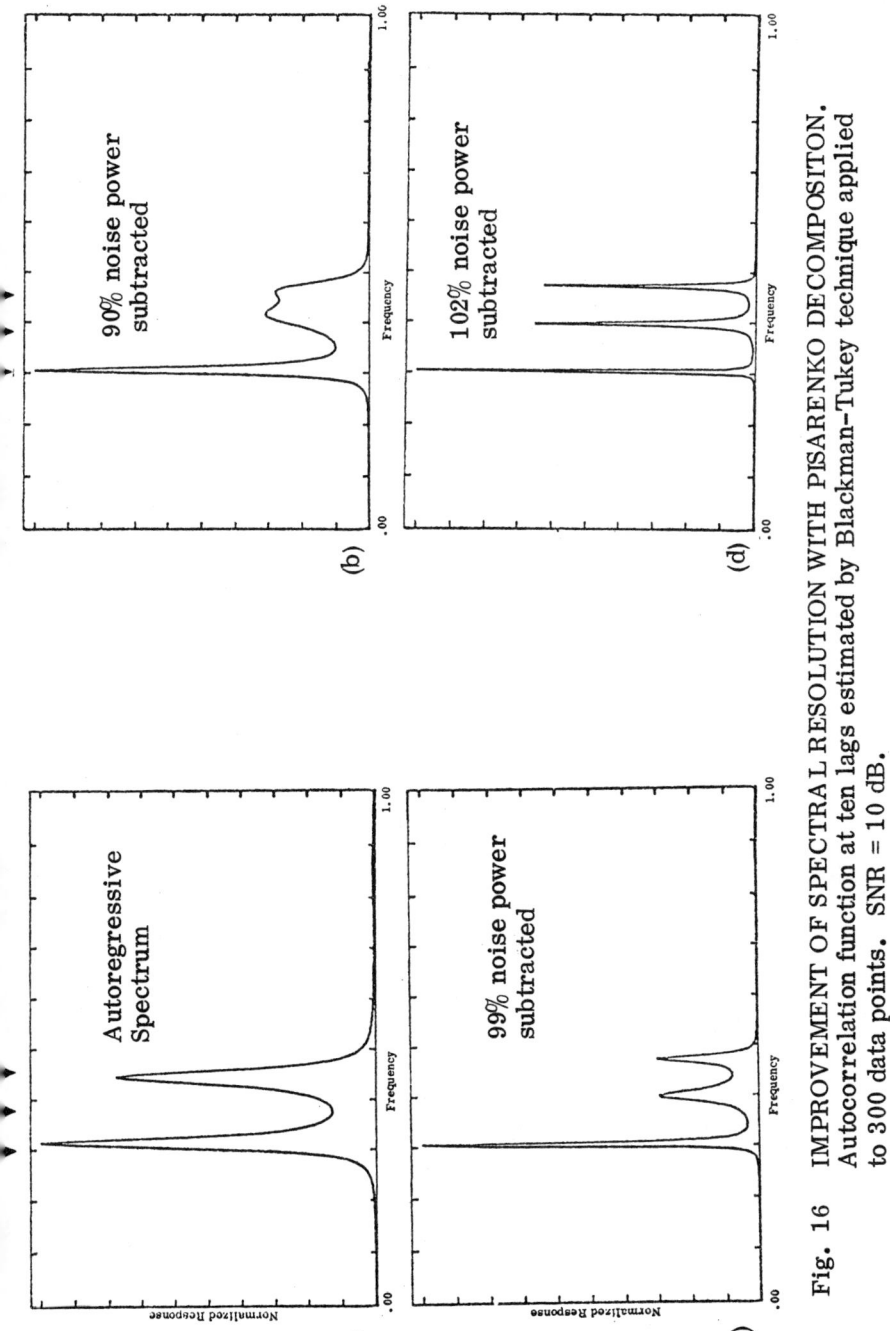

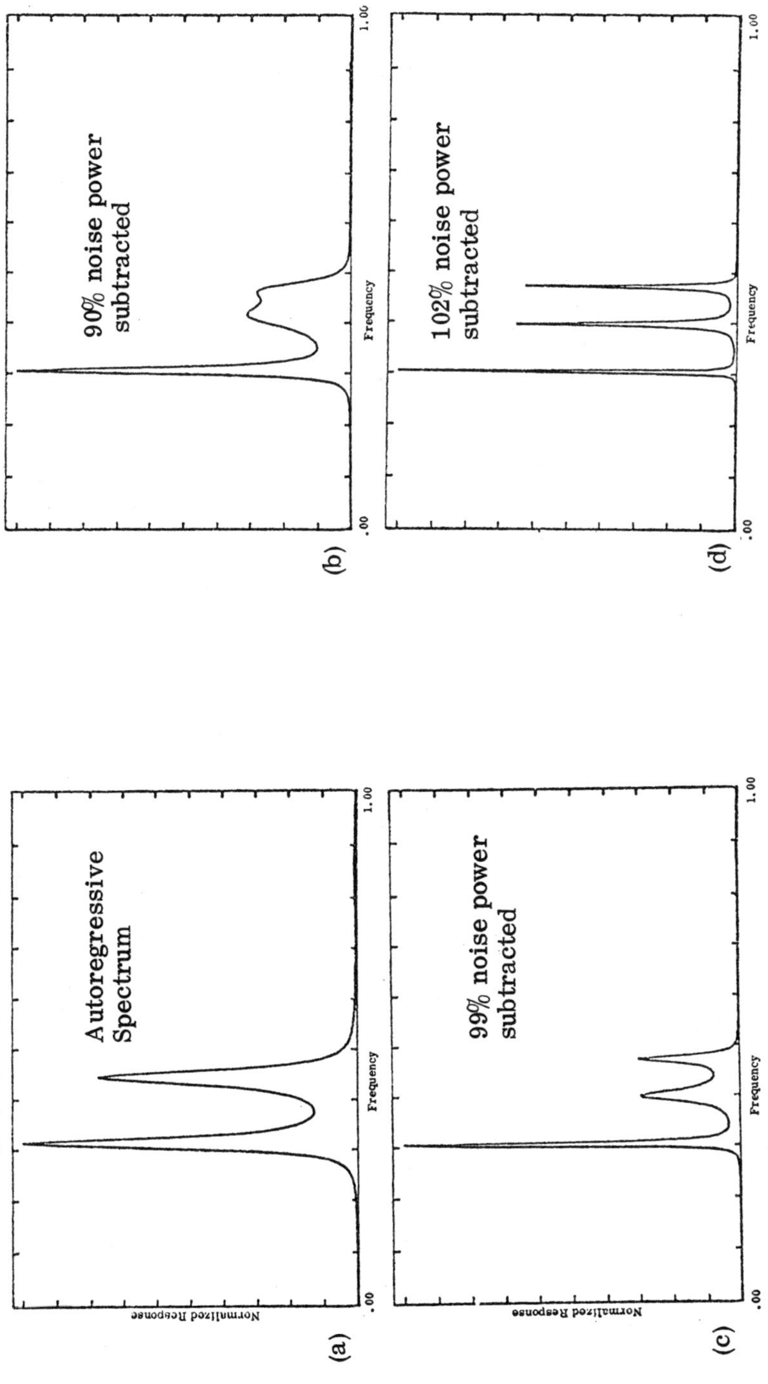

Fig. 16 IMPROVEMENT OF SPECTRAL RESOLUTION WITH PISARENKO DECOMPOSITON. Autocorrelation function at ten lags estimated by Blackman-Tukey technique applied to 300 data points. SNR = 10 dB.

Model of Sinusoids with Additive Noise

The form of the autocorrelation matrix produced by the time-series model in Fig. 14 can be now derived following the approach developed by Ulrych and Clayton (1975). The derivation yields Pisarenko's procedure for decomposing an autocorrelation matrix into the sum of sinusoids plus white noise.

By means of a trigonometric identity, any sinusoid can be modeled by a second-order difference equation

$$y(t) = d(1)y(t-1) + d(2)y(t-2) \tag{4.3}$$

that has the proper initial conditions. The roots of the polynomial, found

$$1 - d(1)Z - d(2)Z^2 = 0 \tag{4.4}$$

by taking the z-transform of both sides of Eq. (4.3), lie on the unit circle in the complex z-plane. The angular frequency $2\pi f_1$ of the sinusoid is the angle that the upper half-plane root subtends with the positive real axis. This model can be generalized to any number of sinusoids

$$y(t) = \sum_{k=1}^{p} d(k)y(t-k) \tag{4.5}$$

where p is twice the number of sinusoids present.

The model for n=p/2 sinusoids with additive noise is

$$q(t) = y(t) + n(t) \tag{4.6}$$

in which the noise is assumed to be white with the autocorrelation,

$$E[n(t)n(t-\ell)] = \sigma_n^2 \, \delta(\ell) \tag{4.7}$$

and is uncorrelated with the sinusoids so that

$$E[y(t)n(t-\ell)] = 0 \tag{4.8}$$

Spectral Decomposition

The sum of sinusoids can be expressed as the difference between the total output and the noise,

$$y(t) = q(t)-n(t) \tag{4.9}$$

Substituting into Eq. (4.5) obtains a new form of the model,

$$q(t) = \sum_{k=1}^{p} d(k)q(t-k) + n(t) - \sum_{k=1}^{p} d(k)n(t-k) \tag{4.10}$$

Written in matrix notation by defining the p-vector

$$D^T = [d(1), d(2), \ldots, d(p)]$$

and the p+1 vectors

$$\Gamma^T = [1, -d(1), -d(2), \ldots, -d(p)] = [1 \vdots -D]$$

$$\eta^T(t) = [n(t), n(t-1), \ldots, n(t-p)]$$

$$Q^T(t) = [q(t), q(t-1), \ldots, q(t-p)]$$

$$Y^T(t) = [y(t), y(t-1), \ldots, y(t-p)]$$

the new model [Eq. (4.10)] then becomes

$$Q^T(t)\,\Gamma = \eta^T(t)\,\Gamma \tag{4.11}$$

Premultiplying Eq. (4.11) by Q(t) and taking the expected value of both sides yields

$$E[Q(t)Q^T(t)]\,\Gamma = E[Q(t)\,\eta^T(t)]\,\Gamma \tag{4.12}$$

From Eq. (4.6), $Q(t) = Y(t) + \eta(t)$ and defining $C_{QQ} = E[Q(t)Q^T(t)]$, obtains

$$C_{QQ}\,\Gamma = E\,\{[Y(t) + \eta(t)]\,\eta^T(t)\}\,\Gamma \tag{4.13}$$

From Eq. (4.7) and (4.8), $E[\eta(t)\,\eta^T(t)] = \sigma_n^2\, I$ and $E[Y(t)\eta^T(t)] = 0$, respectively, and Eq. (4,13) then becomes an eigenvector equation,

$$C_{QQ}\Gamma = \sigma_n^2\,\Gamma \tag{4.14}$$

Given an autocorrelation matrix C_{QQ} formed by p/2 sinusoids plus white noise, the Pisarenko decomposition method solves Eq. (4.14) for the eigenvector of C_{QQ} corresponding to the smallest eigenvalue. If the order of the matrix is p+1 or greater, the smallest eigenvalue is the noise power σ_n^2 and, when normalized so that its first element is one, the corresponding eigenvector is equal to Γ. The frequencies of the sinusoids are found by solving

$$\Gamma^T Z = 0 \tag{4.15}$$

analogous to Eq. (4.4), where $Z^T = (z^0, z^1, z^2, \ldots, z^p)$. The p roots of Eq. (4.15) are denoted by $z_1, z_2, \ldots, z_i, \ldots, z_p$, where $z_i = \exp(j2\pi f_i)$; therefore, solving (4.15) is the same as determining where the Fourier transform of the elements of Γ^T equals zero.

Power Determination

When the frequencies of the n sinusoids are found, their amplitudes can be determined. Following Marple (1976), this is accomplished by solving the matrix form of Eq. (4.1),

$$FP = C_q \tag{4.16}$$

where the nxn matrix is

$$F = \begin{bmatrix} \cos(2\pi f_1) & \cos(2\pi f_2) \ldots & \cos(2\pi f_n) \\ \cos(2\pi f_1 2) & \cos(2\pi f_2 2) \ldots & \cos(2\pi f_n 2) \\ \vdots & \vdots & \vdots \\ \cos(2\pi f_1 n) & \cos(2\pi f_2 n) \ldots & \cos(2\pi f_n n) \end{bmatrix}$$

the n-vector of signal powers is

$$P^T = [P_{S1}, P_{S2}, \ldots, P_{Sn}]$$

where $P_{Si} = A_i^2/2$, and the n-vector of nonzero autocorrelation values is

$$C_q^T = [c(1), c(2), \ldots, c(n)]$$

The solution of Eq. (4.16) for the signal power is obtained from

$$P = F^{-1} C_q \tag{4.17}$$

The noise power is then

$$\sigma_n^2 = c(0) - \sum_{i=1}^{n} P_{Si} \tag{4.18}$$

which is redundant because it was also obtained from Eq. (4.14). This redundancy, however is used in the next section where the autocorrelation function must be estimated from the data.

Power and Frequency Estimation

In the above discussion, it was assumed that the autocorrelation matrix C_{QQ} was perfectly known and that the time series was generated by sinusoids with additive white noise. In most practical cases, however, the autocorrelation function must be estimated from the data. As a result of statistical errors in the standard autocorrelation-estimation procedures (such as Blackman-Tukey or Bartlett methods), there is no guarantee, even if the generating process is sinusoids with additive noise, that the Fourier transform of the autocorrelation-matrix eigenvector corresponding to the smallest eigenvalue will have zeros [that Eq. (4.15) will have roots on the unit circle)].

There are many possible procedures for using roots located near the unit circle to determine the sinusoid frequencies. Two procedures that work are based on the fact that, for an input signal consisting of pure sinusoids in the absence of noise, the PD and AR spectra are the same (Pisarenko, 1973). This equivalence suggests removing the noise contribution to the autocorrelation matrix and performing an AR spectrum analysis to obtain the Pisarenko spectrum. From Eq. (3.16), the noise from a perfectly known autocorrelation matrix can be removed by forming a new noisefree autocorrelation matrix.

$$\tilde{C} = C - \sigma_n^2 I \tag{4.19}$$

where σ_n^2 is the smallest eigenvalue of C. Inserting $\tilde{C}$ into the AR coefficient-determing Eq. (3.6) yields

$$(C - \sigma_n^2 I) B = 0 \tag{4.20}$$

because σ_v^2 vanishes in the noisefree case. The solution vector B thus obtained is the same as the solution Γ to the Pisarenko eigenvector problem Eq. (4.14)

$$(C - \sigma_n^2 I) \Gamma = 0 \tag{4.21}$$

because the equations are the same.

To avoid singular $\tilde{C}$ matrices, one approach is to subtract a large portion but not all of the effect of the noise by forming an autocorrelation matrix

$$\tilde{C} = C - 0.9\, \sigma_n^2 I \tag{4.22}$$

and finding the AR spectrum corresponding to $\tilde{C}$. This procedure yields the Pisarenko-decomposition spectrum as the fraction of noise subtracted approaches unity; in Figs. 15 and 16, it is shown to improve significantly the resolution and sinusoid-frequency measurement capability of AR spectrum analysis in the presence of additive noise.

A second procedure, proposed by Marple, iteratively determines an acceptable fraction of the noise power to subtract. The AR spectrum of an estimated autocorrelation function is first examined, the frequencies of any strong peaks are then determined, and the powers P_{Si} of the sinusoids are estimated according to Eq. (4.17). These P_{Si} are used in Eq. (4.18) to obtain an estimate of the noise power

$$\hat{\sigma}_n^2 = c(0) - \sum_{i=1}^{n} P_{Si}$$

A small fraction of this estimate (perhaps 10 percent) is subtracted off the diagonal of the autocorrelation matrix, the AR spectrum of C - 0.1 σ_n^2 I is determined, and the iteration begins again. After each AR spectrum analysis, new spectral peaks may appear and may be included in the power-estimation matrix F. At each iteration, a residual noise power is formed, equal to the difference between C(0) and the sum of the noise power actually subtracted (0.1 σ_n^2) plus the estimated signal powers. The iteration terminates (and takes the previous step as its final result) when the residual noise power obtained during an iteration is larger than the one obtained in the previous iteration.

Table 1 lists the improvements obtained by applying the Marple procedure to 24 simulations of two sinusoids in white noise at the AR resolution limit. Even before noise-power subtraction is applied, the signal-power and frequency estimation accuracies obtained by applying peak-positiion estimation and Eq. (4.17) to the AR spectrum are less than 15 percent in error from their actual values. The power-measurement accuracy is vastly better than approximations based on observing the AR spectral-peak magnitudes that may vary over a 100:1 range as a function of frequency; note also that, at the limits of resolution, the area under the peaks, reported by Lacoss (1971) to be proportional to the signal power, is difficult to measure. Compared to the straightforward use of Eq. (4.17) in the AR spectrum, the Marple technique increases the average frequency-measurement accuracy and the geometric mean of the two signal-power measurement accuracies by approximately one order of magnitude.

5. FUTURE TRENDS

Fourier spectral analysis is not about to be replaced. Its range of applications is pervasive, and its optimality for the detection of sinusoids in white noise by matched filtering is well known; it can also provide the maximum-likelihood frequency estimator for a single sinusoid in white gaussian noise when the observation window is long compared to the

Table 1. ACCURACY OF AUTOREGRESSIVE PISARENKO DECOMPOSITION FREQUENCY AND POWER ESTIMATES AT THE POINT OF AR RESOLUTION LIMITS FOR TWO SINUSOIDS IN NOISE AT 0 dB. Four nonzero autocorrelation lags were estimated from N=64 data points by the Burg algorithm. Frequencies of the two sinusoids were close to and centered about 0.5, or one-half the Nyquist folding frequency.

Estimation	Lower Frequency	Standard Deviation	Upper Frequency	Standard Deviation	White Noise	Standard Deviation
	Frequency					
AR Error[1]	0.0377	0.0164	0.0358	0.0176	--	--
PD Error[2]	0.0028	0.0080	0.0049	0.0140	--	--
Improvement Factor	13.5	2.0	7.3	1.3	--	--
	Power					
Actual Power	1.0	--	1.0	--	1.0	--
AR Power Estimate[1,3]	0.8892	0.3617	0.9517	0.3866	--	--
Error	0.1108	--	0.1483	--	(1.0)	--
PD Power Estimate[2,3]	0.9704	0.3100	0.9974	0.3748	0.2238	0.0657
Estimate Error[2,3]	0.0296	--	0.0026	--	0.7762	--
Improvement Factor	3.7	--	57	--	1.3	--

1. From AR spectrum peaks.
2. From AR spectrum peaks, Marple iteration.
3. Equation (4.17).

sinusoid's period (Viterbi, 1966). Discrete Fourier analysis to characterize cyclic phenomena continues to be appropriate when the cycle is known.

Fourier spectral analysis, however, is in the process of being augmented by new techniques that assume alternative and often more realistic time-series generation models. The advantages of applying these new models to estimation problems in one or more dimensions are beginning to be recognized. Beneficial applications have been reported in such fields as radio astronomy (Wernecke and D'Addario, 1976), geology (Claerbout, 1976), electroenchphalogography (Gersch, 1970), and speech synthesis (Atal and Hanauer, 1971).

Other still newer models are emerging. For Example, Ulrych and Clayton (1975) have observed that the time-series model of sinusoids with additive white noise has an important alternative interpretation. Equation (4.10) may be interpreted as a special form of a feedback-feedforward filter excited by white noise, also called an autoregressive moving-average (ARMA) model (Fig. 17). Generalizations of this model are likely to be developed for processing signals in the practical sonar acoustic environment of colored noise (noise not spectrally flat). Other methods for dealing with colored noise have been proposed by Widrow et al (1976).

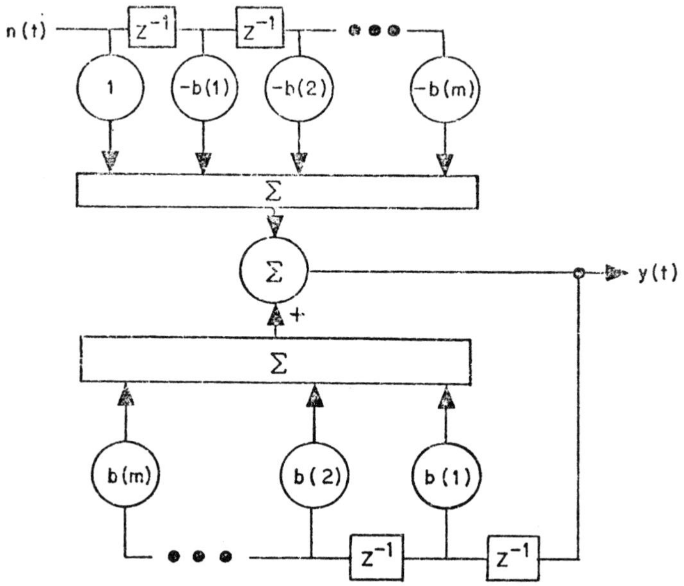

Fig. 17. ALTERNATIVE AUTOREGRESSIVE MOVING-AVERAGE MODEL FOR GENERATION OF SINUSOIDS WITH ADDITIVE WHITE NOISE.

Adaptive computational methods are also being devised for processing high data rates (Griffiths, 1975); (Widrow et al, 1975). Computational methods for extracting more information from a limited data set have already been developed for the autoregressive model (Burg, 1975). Similar techniques for other models may help to better determine the number and properties of signals in the time series analyzed in Fig. 16.

BIBLIOGRAPHY

Anderson, N. (1974). "On the Calculation of Filter Coefficients for Maximum Entropy Spectral Analysis, " Geophysics, 39, pp. 69-72.

Atal, B. S. and S. L. Hanauer (1971). "Speech Analysis and Synthesis by Linear Prediction of the Speech Wave, " J. Acoustical Soc. America, (Part 2), pp. 637-655.

Bartlett, M. S. (1950). "Periodogram Analysis and Continuous Spectra," Biometrika, 37, pp. 1-16.

Blackman, R. B. and J. W. Tukey (1959). The Measurement of Power Spectra from the Point of View of Communications Engineering, Dover Press, New York.

Bryson, A. E. and Y. Ho (1969). Applied Optimal Control, Blaisdell, Waltham, Mass.

Burg, J. P. (1967). "Maximum Entropy Spectral Analysis" presented at the 37th Meeting of the Society of Exploration Geophysicists in Oklahoma City.

Burg, J. P. (1968). "A New Analysis Technique for Time Series Data," presented at the NATO Advanced Study Institute on Signal Processing with Emphasis on Underwater Acoustics.

Burg, J. P. (1975). "Maximum Entropy Spectral Analysis," Ph.D. Thesis, Stanford University, Stanford, Calif.

Chen, W. Y. and G. R. Stegen (1974). "Experiments with Maximum Entropy Power Spectra of Sinusoids," J. Geophys. Res., 79, pp. 3014-3022.

Claerbout, J. F. (1976). Fundamentals of Geophysical Data Processing, McGraw-Hill Book Co., New York.

Gersch, W. (1970). "Spectral Analysis of EEG's by Autoregressive Decomposition of Time Series," Mathematical Biosciences, 7, pp. 205-222.

Griffiths, L. J. (1975). "Rapid Measurement of Digital Instantaneous Frequency," IEEE Trans on Acoustics, Speech, and Signal Processing, ASSP-23, pp. 207-222.

Lacoss, R. T. (1971). "Data Adaptive Spectral Analysis Methods," Geophysics, 36, 4, pp. 661-675.

Marple, S. L. (1976). Engineer Thesis, Dept. of Electrical Engineering, Stanford University, Stanford, Calif., in preparation.

Mohammed, A., and R. G. Smith. (1975). "Data Windowing in Spectral Analysis," Defense Research Establishment Atlantic (DREA) Report 75/2, Dartmouth, Nova Scotia.

Oppenheim, A. V. and R. W. Schafer (1975). Digital Signal Processing, Prentice-Hall, Englewood Cliffs, N. J.

Pisarenko, V. F. (1973). "The Retrieval of Harmonics from a Covariance Function," Geophysics, J. R. Astr. Soc., pp. 347-366.

Schaerf, M. C. (1963). "Estimation of the Covariance and Autoregressive Structure of a Stationary Time Series," Ph.D. Thesis, Stanford University, Stanford, Calif.

Schuster, A. (1898). "On the Investigation of Hidden Periodicities with Application to a Supposed 26 Day Period of Meteorological Phenomena," Terr. Magn., 3, pp. 13-41.

Ulrych, T. J. (1972). "Maximum Entropy Power Spectrum of Truncated Sinusoids," J. Geophys. Res., 77, B, pp. 1396-1400.

Ulrych, T. J. and T. N. Bishop (1975). "Maximum Entropy Spectral Analysis and Autoregressive Decomposition," Reviews of Geophysics and Space Physics, 13, 1.

Ulrych, T. J. and W. R. Clayton (1975). "Time Series Modeling and Maximum Entropy," presented at the IUGG Congress, Grenoble; to be published in Earth and Planetary Interiors, 1976.

Van den Bos, A (1971). "Alternative Interpretation of Maximum Entropy Spectral Analysis," IEEE, Trans. on Information Theory, pp. 493-494.

Viterbi, A. J. (1966). Principles of Coherent Communication, McGraw-Hill Book Co., New York.

Wernecke, S. J. and L. R. D'Addario (1976). "Maximum Entropy Image Reconstruction," to appear in IEEE Trans. Computers.

Widrow, B., J. Glover, J. McCool, J. Kaunitz, C. Williams, R. Hearn, J. Zeidler, E. Dong, and R. Goodlin (1975). "Adaptive Noise Cancelling: Principles and Applications," Proc. IEEE, 63, 12, pp. 1692-1716.

Widrow, B., J. Glover, J. McCool, and J. Treichler (1976). "Response to Letter from D. W. Tufts" (Tuft's letter entitled "Adaptive Line Enhancer and Spectral Analysis"), to appear in Proc. IEEE.

DISCUSSION

Comment : C. VAN SCHOONEVELD

(1) In your paper you concentrated on the case of separate sinewaves. Would you agree that the Pisarenko technique could also be applied to the case of a random signal with an all-pole power-density spectrum, superposed on a white-noise spectrum?

(2) In the case of sinewaves, what happens if the estimated coefficients of the autoregressive model correspond to poles outside the unit circle? Would it perhaps be advantageous to add some exponential weighting to the time-series data before analysis, in order to foresee the estimated poles to lie inside the unit circle?

Reply : O.L. FROST

(1) Yes, the Pisarenko technique may be extended to the case of a general all-pole plus white-noise power-density spectrum by solving equation [4.15] for zeros that may be off the unit circle, yielding damped sinusoids plus white noise as the spectral components.

(2) If poles are found outside the unit circle then the observed process evidently has been found to contain components that are growing in amplitude. Whether such components should be suppressed from the analysis probably depends on the use to which the analysis will be put.

Comment : B. PICINBONO

There are two very different problems in spectral analysis:

(i) Power spectrum estimation, which is the estimation from observed data of the power spectrum of a signal in general random. The power spectrum $\gamma(\nu)$ is, by definition, the Fourier transform of the correlation function, and the only assumption is that the signal observed is a sample of finite duration of a stationary process defined on an infinite duration.

(ii) Estimation of parameters of the spectrum, which is particularly the case if we know a priori that the considered signal has a parametric representation as, for example, a sum of finite number of sinusoids or an autoregressive representation.

The comparison presented in the paper is essentially based on a criterion of a resolution of sinusoidal signals and in this case it is clear that the classical methods of spectral analysis are not very convenient. It would be interesting to compare the three methods presented in the case of a stochastic signal with continuous spectrum; the possible criterion of comparison could be the bias and variance of the estimator.

Reply : O.L. FROST

Yes. In order to compare classical methods of power spectrum estimation with the two newer methods, the point of view presented in the paper was that estimation of the power spectrum by means of a discrete Fourier transform on a finite amount of observed data is equivalent to estimation of parameters of the spectrum, which are powers of sinusoids evenly spaced in frequency. Another comparison presented in the paper was based on the two-sinusoid resolution of the spectral analysis techniques. There are many other possible criteria for evaluation of the techniques, such as spectral stability, frequency measurement accuracy, and signal detectability, and it is clear that a technique which is best under one criterion will not necessarily be best under another.

INTRODUCTION TO DETECTION AND ESTIMATION

B. PICINBONO

Laboratoire des Signaux et Systèmes
E.S.E. Plateau du Moulon. 91190 GIF/YVETTE (France)

ABSTRACT. We present a general introduction to detection problems and estimation problems. In the first section we present shortly the two points of view, which are different. We show that in the case of Gaussian noise there is a strong connexion between detection and estimation and we give some examples. In particular we present some ideas on adaptive detection which needs estimation of a reference noise alone which is defined and used to give an example of detection receiver.

1. INTRODUCTION

It is very difficult to present in one lecture a general introduction to detection and estimation problems and also some recent advanced results. Indeed there is a lot of excellent books on detection with several hundred of pages, and it is absolutely impossible to give shortly an idea of all the problems concerned by the title. Consequently we will arbitrarily select some problems of interest in such a way that the lecture is coherent. But evidently we will not give all the details concerning points which can be found in classical books [1][2].

G. Tacconi (ed.), Aspects of Signal Processing, Part 1, 163-179. *All Rights Reserved.*

After a short description of classical theories of detection and estimation the lecture attemptsto present some recent results concerning the connexion between detection and estimation and particularly in order to discuss some results concerning adaptive detection which is actually a very large and important problem with only a few explicit results.

2. DETECTION AS BINARY HYPOTHESIS TESTING

2.1. Notations

We start from an observation x, vector in a N-dimensional space. Th number N can become very large or even infinite if the observation is a continuous stochastic process (s.p.). But in all the first calculations we will suppose that N is finite.
From this observation we must decide if it belongs to one of two hypothesis H_0 or H_1. Evidently we restrict the problem to binary tests only to simplify the presentation.
We suppose that H_0 and H_1 are simple, i.e. that the probability densities $p_{0_1}(x)$ of the observation under H_{0_1} are known.
The detection problem is solved if we have found a strategy $\Phi(x)$ $x \in R^N$. This function has only values o or 1 and if $\Phi(x) = 1$ or o we decide H_1 or H_0.The test function (or strategy) splits the observation space in two regions, R_0 and R_1.
In order to obtain $\Phi(x)$ we must introduce a detection criterion.

2.2. Bayes strategy

In this strategy we suppose that the a priori probabilities of H_0 or H_1 are known (π_0, $\pi_1 = 1-\pi_0$). In many problems, and particularly radar or sonar problems it is difficult to introduce such assumption. Moreover we suppose that is possible to introduce with a physical meaning the cost function C_{ij} which is the cost of the decision i if H_j is true. It is also sometime difficult to justify such costs, and this point will be discussed in the

following.
The Bayes strategy $\Phi^B(x)$ is the strategy for which the mean cost C is minimum. This cost is evidently defined by

$$C = \int \Sigma_{ij}\ C_{ij}\ \Phi_i(x)\ \pi_j\ p_j(x)\ dx \qquad (2.1)$$

where $\Phi_1(x) = \Phi(x)$ and $\Phi_o(x) = 1 - \Phi(x)$. We can immediately simplify this expression by introducing the false-alarm probability α

$$\alpha = \int \Phi(x)\ p_o(x)\ dx \qquad (2.2)$$

and the detection probability

$$\beta = \int \Phi(x)\ p_1(x)\ dx. \qquad (2.3)$$

After very simple calculation we can write C as

$$C = C_o - \pi_1\ \gamma_1(\beta - k\ \alpha) \qquad (2.4)$$

where $k = \pi_o\ \gamma_o / \pi_1\ \gamma_1$ and

$$\gamma_o = C_{1o} - C_{oo} \quad \text{and} \quad \gamma_1 = C_{o1} - C_{11}\ .$$

In equation (2.4) only α and β are depending on Φ.
Thus the Bayes strategy Φ_k^B is the strategy which gives the maximum of $\beta - k\alpha$ for a given k. But this expression can be written

$$B = \beta - k\alpha = \int \Phi(x)\ [p_1(x) - k\ p_o(x)]\ dx \qquad (2.5)$$

and the maximum is evidently obtained if $\Phi(x) = 1$ when $p_1(x) - k\ p_o(x) > o$. Thus we have the classical result

$$\Phi_k^B(x) = 1 \leftrightarrow L(x) = \frac{p_1(x)}{p_o(x)} \geq k. \qquad (2.6)$$

The function L(x) is the likelihood ratio and the Bayes strategy leads to a likelihood ratio test.

2.3. Neyman Pearson strategy

The Bayes strategy is depending on k defined by Eq. (2.4). It is clear that in many problems k has no physical meaning because we do not know the π_i and the C_{ij}. Thus we need another strategy. In this strategy we try to find the test $\Phi^{NP}(x)$ which satisfies.

$$\alpha \leq \alpha_o$$
$$\beta \text{ maximum}$$

It is well known (see Appendix A) that

$$\Phi^{NP}_{\alpha_o} = \Phi^{B}_{k_o} \tag{2.8}$$

where k_o is such that

$$a(k_o) = \alpha_o , \tag{2.9}$$

and

$$a(k) = \alpha\left[\Phi^{B}_{k}\right]$$

i.e. the false alarm probability of the test Φ^{B}_{k}.
Thus the N-P. strategy is directly connected to the B. strategy, and the test is evidently also a likelihood test.
This point is evidently due to the fact that L(x) is a sufficient statistic for our problem of hypothesis testing.

3. LINEAR ESTIMATION

We will very shortly summarize the classical results of linear estimation which are useful for the following. As previously we start from the observation x which is a zeromean second order vector. We want to estimate the vector y, which is also zeromean and second order, and we suppose that the covariance matrix

$$\Gamma_x \triangleq E[x\ x'] \ , \quad \Gamma_{xy} \triangleq E[x\ y'] \tag{3.1}$$

are known.
The estimate of y is linear and can be written

$$\hat{y} \triangleq E\ L\ [y|x] = R_{yx}\ x \tag{3.2}$$

and the problem is to calculate R_{yx} in order to obtain the minimum error in the quadratic mean sense. Let us introduce the error of estimation z defined by

$$z = y - \hat{y} \ . \tag{3.3}$$

The error matrix is defined by

$$\varepsilon^2 \triangleq E\ [z\ z'] \ . \tag{3.4}$$

It is well known that the best estimate in the quadratic mean sense is defined by the projection theorem and can be written

$$\hat{y} = P_{roj}\ [y|x] \leftrightarrow z \perp x. \tag{3.5}$$

This equation is equivalent to the Wiener-Hopf equation which can be written

$$E\,[\hat{y}\; x'] = E\,[y\; x'] \qquad (3.6)$$

or by using Eq. (3.1)

$$R_{yx}\,\Gamma_{xx} = \Gamma_{yx}\,. \qquad (3.7)$$

If the matrix Γ_{xx} is regular we obtain

$$R_{yx} = \Gamma_{yx}\,\Gamma_{xx}^{-1}\,. \qquad (3.8)$$

In this case the mean square error can be written

$$\varepsilon^2 = E\,[(y - \hat{y})\,(y - \hat{y})'] = \Gamma_{yy} - \Gamma_{\hat{y}y}$$

$$= \Gamma_{yy} - \Gamma_{yx}\,\Gamma_{xx}^{-1}\,\Gamma_{yx}'\,. \qquad (3.9)$$

We can simplify this equation by using Eq. (3.7) and write

$$\varepsilon^2 = \Gamma_{yy} - R_{yx}\,\Gamma_{xx}\,R_{yx}'\,. \qquad (3.10)$$

The vector z which is orthogonal to x is sometimes called innovation, and has interesting properties.

4. APPLICATION TO THE GAUSSIAN CASE

In this section we suppose that the noise is Gaussian, with zero-mean value and covariance matrix Γ. The probability distribution of this noise is evidently

$$p(x) = \alpha \exp - \frac{1}{2}\,(x'\,\Gamma^{-1}\,x)\,. \qquad (4.1)$$

Let us consider detection problems in two different cases.

4.1. Deterministic signal

The signal is non random and represented by the vector s. Thus the probability distribution in presence of signal is $p_1(x) = p(x - s)$ and the likelihood ratio defined by (2-6) is

$$L(x) = p(x - s)/p(x)$$

$$= \exp - \frac{1}{2}\,(s'\,\Gamma^{-1}\,s - 2\,s'\,\Gamma^{-1}\,x)\,. \qquad (4.2)$$

To detect a signal we have to compare this function to a threshold Evidently it is equivalent to use the logarithm of this function, and the receiver can be written as

$$s(x) = s' \Gamma^{-1} x \gtrless \lambda \quad . \qquad (4.3)$$

Thus the optimal receiver is linear on the observation x and is called the matched filter. In the case of white noise we have evidently $s(x) = x'\ x$.

For this calculation we have supposed that in Eq. (4.2) $s' \Gamma^{-1} s$ is finite. If that is not the case we are in a case of singular detection.

It is easy to explain this situation. If Γ has no inverse, that means that at least one of the r.v. $\{x_i\}$ can be expressed in terms of the other, $x_i = \Sigma_{j \neq i} \lambda_j x_j$. If we have for the signal $s_i \neq \Sigma_{j \neq i} \lambda_j s_j$, it is evident that $(x_i - \Sigma_{j \neq i} \lambda_j x_j)$ is a function of the observation which is o in the case of noise alone and different from o in presence of signal which allows a detection without error. This example explains why the condition of singular detection concerns the signal and the noise together and is not a condition on the noise only.

We can give a spectral interpretation of the previous results, valid in the case of stationary noise and signal long compared to the correlation time of the noise. The Eq. (4.3) can be written $\Lambda(x) = u'\ x$, $u' = s' \Gamma^{-1}$. That corresponds to a linear filter with the gain

$$G(\nu) = s^{*}(\nu)/\gamma(\nu) \qquad (4.4)$$

where $s^{*}(\nu)$ is the Fourier transform of the signal and $\gamma(\nu)$ the spectrum of the noise. That is the standard expression of the matched filter. Moreover the output signal to noise ratio can be written

$$d^2 = s' \Gamma^{-1} s = \int \frac{|s(\nu)|^2}{\gamma(\nu)} d\nu \qquad (4.5)$$

which allows us to interpret in the frequency domain the condition of singular detection.

The matched filter has very interesting properties, which are not studied in this short paper [3].

4.2. Gaussian signals

Now let us suppose that the signal is zeromean and Gaussian. As the sum of two zeromean Gaussian s.p. has the same property, we deduce that $p_o(x)$ and $p_1(x)$ have the structure of Eq. (4.1) with (α_o, Γ_o) and (α_1, Γ_1). By taking the logarithm of the likelihood ratio whe obtain easily that the test is defined by

$$\Lambda(x) = x'\ A\ x \gtrless \lambda \qquad (4.6)$$

where

$$A = \Gamma_o^{-1} - \Gamma_1^{-1}. \qquad (4.7)$$

Eq. (4.6) shows easily that the optimum receiver is a quadratic form of the observation x which is a classical result for the applications.

5. RELATIONS BETWEEN DETECTION AND ESTIMATION

Apparently there is no direct relation between the theories of detection and of estimation. Nevertheless it appears that in the Gaussian case the theory of detection uses the inverses of correlation matrices. This fact is evidently due to the structure of the gaussian law. But, without gaussian assumption, we have also inverse of correlation matrices in estimation theory. Thus it appears that common structures can be found. We will study more precisely this point.

5.1. Estimation and detection of Gaussian signal

Let us suppose that the signal and the noise are independent zeromean Gaussian s.p. with covariance Γ_S and Γ_B. The matrix A of Eq. (4.7) can be written

$$A = \Gamma_B^{-1} - \Gamma_{S+B}^{-1} . \tag{5.1}$$

Now let us introduce the matrix R_S which is used on the problem of linear estimation of the signal s from the observation S + B. From Eq. (3.8) this matrix can be written

$$R_S = \Gamma_{S,\ S+B}\ \Gamma_{S+B}^{-1} . \tag{5.2}$$

As S and B are independent, we deduce

$$\Gamma_{S,\ S+B} = \Gamma_S \quad \text{and} \quad \Gamma_{S+B} = \Gamma_S + \Gamma_B . \tag{5.3}$$

Thus we can write $R_S = (\Gamma_{S+B} - \Gamma_B)\ \Gamma_{S+B}^{-1}$ which gives

$$\Gamma_{S+B}^{-1} = \Gamma_B^{-1} - \Gamma_B^{-1}\ R_S \tag{5.4}$$

and the matrix A from Eq. (5.1) becomes

$$A = \Gamma_B^{-1}\ R_S . \tag{5.5}$$

With this expression the quadratic form appearing in the test $\Lambda(x)$ of Eq. (4.6) becomes

$$\Lambda\ (x) = x'\ \Gamma_B^{-1}\ R_S\ x \tag{5.6}$$

$$= x'\ \Gamma_B^{-1}\ \hat{s} \tag{5.7}$$

where $\hat{s}$ is the best linear estimate of s from the observation s+B.

This expression is interesting because it is exactly the same as for a deterministic signal (matched filter) in which we replace the known signal s by its estimation $\hat{s}$ obtained from observation x. Evidently the receiver is still quadratic, but can be decomposed in two systems : 1. linear estimation, $\hat{s} = R_s\ x$ 2. Matched filtering, $x'\ \Gamma_B^{-1}\ \hat{s}$. Thus the estimation procedure allows to present a strong connection between detection of deterministic or random signals in gaussian noise [4].

5.2. Case of white noise [5]

We suppose now that the noise is a discrete white noise, i.e. a sequence of independent and stationary random variables, (r.v.). In this case we will see that it is possible to use a causal estimation instead of the general estimation considered previously. Let us consider the observation betweem the times instants t_1 and t_n. The likelihood ratio can be written as

$$L(x^1_{\ n}) = L(x^1_{n-1})\ \frac{p_1(x_n|x^1_{\ n-1})}{p_0(x_n|x^1_{\ n-1})} \qquad (5.8)$$

where x'_n is the sequence $x_1, \ldots, x_n$, and $p_{\substack{1\\o}}$ the conditional probabilities.

We have assumed that the noise is white, which gives

$$p_0(x_n|x^1_{n-1}) = p_0(x_n) = (2\pi\sigma_B^2)^{-1/2}\ \exp -\frac{1}{2}\ \frac{x_n^2}{\sigma_B^2}\ . \qquad (5.9)$$

The signal + noise is evidently non white, and we know that the conditional probability $p(x_n|x^1_{n-1})$ is still gaussian whith mean and variance given by

$$E[x_n|x^1_{n-1}] \triangleq \hat{x}_n \qquad (5.10)$$

$$\tilde{\sigma}_n^2 = E[(x_n - \hat{x}_n)^2]\ . \qquad (5.11)$$

In these expressions $\hat{x}_n$ is the causal linear estimate of x_n in terms of $x_1, \ldots, x_{n-1}$ and $\tilde{\sigma}_n^2$ is the variance of this estimate. If n is large enough, $\hat{x}_n$ is the causal estimate of the signal + noise at time instant n in terms of all the past. and $\tilde{\sigma}^2$ becomes independent of n because of the stationarity of the signal and noise.

Thus we can write

$$p_1(x_n|x^1_{n-1}) = (2\pi\sigma^2)^{-1/2}\ \exp -\frac{1}{2}\ \frac{(x_n - \hat{x}_n)^2}{\tilde{\sigma}^2}\ . \qquad (5.12)$$

By taking the logarithm of (5-8) we obtain easily

$$\Lambda(x'_n) = \Lambda(x'_{n-1}) + \lambda_n \tag{5.13}$$

with

$$\lambda_n = \text{Log}\,\frac{\sigma_B}{\tilde{\sigma}} + \frac{1}{2}\left[\frac{x_n^2}{\sigma_B^2} - \frac{(x_n - \hat{x}_n)^2}{\tilde{\sigma}^2}\right].$$

$$= c + \frac{1}{\tilde{\sigma}^2}\left[x_n\,\hat{x}_n - \frac{1}{2}\,\hat{x}_n^2 - \left(1 - \frac{\tilde{\sigma}^2}{\sigma_B^2}\right) x_n^2\right] . \tag{5.14}$$

In this expression x_n is the observation and $\hat{x}_n$ the causal estimation of the x_n in terms of all the past. But $x_n = s_n + b_n$, which gives evidently $\hat{x}_n = \hat{s}_n + \hat{b}_n$. Moreover the noise is white and independent of the signal. Thus we can conclude

$$\hat{x}_n = \hat{s}_n, \tag{5.15}$$

where $\hat{s}_n$ is the causal estimation of the signal in terms of the observation $x_n = s_n + b_n$. In conclusion the only term in λ_n which depends on the observation can be written

$$t_n(x_n) = \frac{1}{\tilde{\sigma}^2}\left[x_n\,\hat{s}_n - \frac{1}{2}\,\hat{s}_n^2 - \left(1 - \frac{\tilde{\sigma}^2}{\sigma_B^2}\right) x_n^2\right] . \tag{5.16}$$

Now it is interesting to compare this expression to that obtained in the case of deterministic signal in white gaussian noise. From Eq. (4.2) we see easily that $\Lambda(x'_n)$ can be written as Eq. (5.13) and

$$t'_n = \frac{1}{\sigma_B^2}\left[x_n\,s_n - \frac{1}{2}\,s_n^2\right] \tag{5.17}$$

We see that there is a difference between (5.15) and (5.16) which is the term $(1 - \sigma^2/\sigma_B^2)\,x_n^2$. This terms disappears if we use a continuous white noise, but this kind of noise, with an infinite power, is certainly not physically realisable experimentally.

Finally we can notice that Eqs. (5.13) and (5.14) are not recursive equationsto construct the likelihood ratio. Indeed $\hat{s}_n$ needs in general the knowledge of $x_1, \ldots, x_{n-1}$. But we know that if s_n is a Gaussian process with a markovian representation, the Kalman-Bucy algorithm allows in to obtain a recursive method to calculate $\hat{s}_n$, and therefore $\Lambda(x_n)$.

6. ADAPTIVE DETECTION

6.1. Introduction . Reference noise alone

All the previous structures are determined by the assumptions concerning the signal and the noise, and particularly by the Gaussian assumption and the knowledge of the correlation functions.

It is clear that this situation is not at all common in practice and it is necessary to conceive and construct systems which are optimal and can be adaptive, i.e. in general have a structure which do not need too much assumptions about the signal and the noise.
It is very difficult to present a general theory concerning simultaneously optimality and adaptivity. A tentative in this direction was presented in a previous NATO Advanced Study [6][7]. In these papers was particularly studied the interest for the detection to use an automatic gain control (A.G.C.) before the matched filter in the case of a noise with stable spectral properties but with strong variations of instantaneous power. Many other studies were presented in the same directions, particularly for systems with multiple sensors [8][9].
One important problem which arises in the use of A.G.C. system is the eventual pollution by a strong signal considered as a case of noise. Thus it is necessary to introduce a system which has a A.G.C. insensible to the signal, and can be called a reference noise alone, (R.N.A.). We will give some preliminary ideas about such a problem.

6.2 Interpretation of the matched filter

At first we will show that a R.N.A. can appear in the classical theory of matched filter presented in section 4. We have seen that the optimal receiver computes

$$\Lambda(x) = s' \Gamma^{-1} x = \sigma' x \qquad (6.1)$$

and compares the result to a threshold.
The signal s and observation x are vectors of R^N, and in this space we can apply the projection theorem as

$$x = x_s + x_\perp \qquad (6.2)$$

where $x_\perp$ is the projection of x in the subspace $E_\perp$ of R^N orthogonal to s. We have evidently

$$x_s = \frac{s' x}{s^2} . s \qquad (6.3)$$

and

$$x_\perp = x - \frac{s' x}{s^2} s. \qquad (6.4)$$

The decomposition (6.2) is applicable for any vector x of R^N. In particular we have evidently

$$s = s + 0 \qquad (6.5)$$

which means that $E_\perp$ is a R.N.A. space in which there is never signal. Thus all the procedure of estimation which are performed from $x_\perp$ are also R.N.A. The decomposition as (6.2) can also be applied to σ and b which are written as

$$\sigma = \sigma_s + \sigma_\perp \tag{6.6}$$

$$b = b_S + b_\perp \quad . \tag{6.7}$$

The correlation matrix $\Gamma_B = E[b\ b']$ can also be partitioned in the form

$$\Gamma_B = \begin{pmatrix} \Gamma_\perp & \gamma \\ \gamma' & \sigma^2 \end{pmatrix} \tag{6.8}$$

in which

$$\Gamma_\perp = E[b_\perp\ b_\perp'] \quad ; \gamma = E[b_\perp\ b_S] \ ; \sigma^2 = E\ [b_S^2]. \tag{6.9}$$

Thus the test function defined by Eq. (6.1) can also be written as

$$\Lambda(x) = \sigma_\perp'\ x_\perp + \sigma'_s\ x_s \tag{6.10}$$

where the vector σ is defined by

$$\Gamma_B\ \sigma = s. \tag{6.11}$$

Using (6.5), (6.6.) and (6.8), this equation can be written as

$$\Gamma_\perp\ \sigma_\perp + \Gamma_{\perp s}\ \sigma_s = o \tag{6.12}$$

which gives

$$\sigma_\perp = -\ \Gamma^{-1}\ \Gamma_{\perp S}\ \sigma_S \tag{6.13}$$

Thus the test $\Lambda(x)$ becomes

$$\begin{aligned} \Lambda(x) &= \sigma_S'\ x_S - \sigma_S'\ \Gamma_{S\perp}\ \Gamma_\perp^{-1}\ x_\perp \\ &= \sigma_S'(x_S - \hat{x}_S) \end{aligned} \tag{6.14}$$

where

$$\hat{x}_S = \Gamma_{S\perp}\ \Gamma_\perp^{-1}\ x \quad . \tag{6.15}$$

By comparing with Eq. (3.1) and (3.7), we see that $\hat{x}_s$ is the best linear estimation of the noise component b_s by using the observation x which is a R.N.A.

Now let us calculate the component σ_s. We have evidently

$$\sigma_S = \frac{s'\ \sigma}{s^2}\ .\ s = \frac{s'\ \Gamma^{-1}\ s}{s^2}\quad s = \frac{d^2}{E}\ s \tag{6.16}$$

where E is the energy of the signal s and d^2 the signal to noise ratio at the output of the matched filter. In conclusion, the test can be written

$$\Lambda(x) = \frac{d^2}{\sqrt{E}} (x_s - \hat{x}_s) \qquad (6.17)$$

which is equivalent to (6.1), but with another interpretation. A block diagram of this system is shown in Fig. 1.

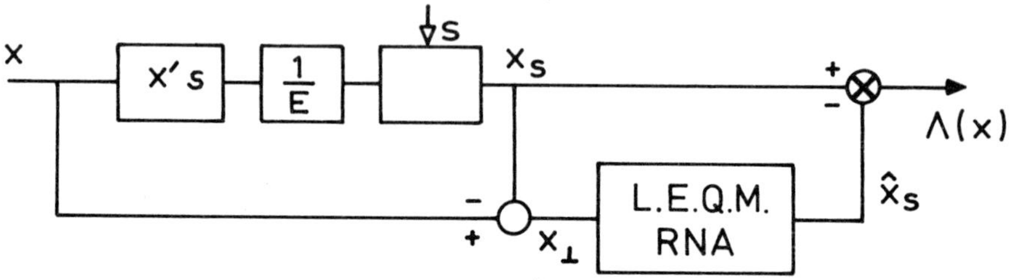

Fig. 1 : Representation of the matched filter.

In this block diagram we see that $\vec{x}' . \vec{s}$ is the matched filter used for the detection of $\vec{s}$ in the case of white noise (see Eq. (6.1)).
In this case x_s and x are independent, and we have evidently $\hat{x}_s$ = o. Thus the system is reduced to the classical matched filter in white noise.
But if the noise is non white we see that this fact is only used to calculate the component of the noise in the direction s, and this component is evidently a R.N.A. Moreover we can very simply interpret the cases of singular detection. If from $x_\perp$, i.e. $b_\perp$ it is possible to estimate without error $\hat{x}_s$, i.e. $\hat{b}_s$, we see that $\Lambda(x)$ = o if there is no signal and $\Lambda(x) \neq$ o if there is a signal, which gives a perfect detection of this signal.

6.3. Estimation of the power with a R.N.A.

To construct an A.G.S. which is not perturbed by the presence of a strong signal, it is sufficient to estimate the power of the noise by using not all the observation x, but only x with is a R.N.A., so that its value is independent of the presence or absence of the signal.
It is well known that there is two good estimates of the power of a random signal, depending on the a priori knowledge of this signal. The first estimate is evidently

$$\hat{P} = \frac{1}{n} x' x. \qquad (6.18)$$

But if we have an a priori knowledge about the normalized correlation $\tilde{\Gamma}$ of x, there is an optimal estimator of the power which can be written

$$\hat{P}_o = \frac{1}{n} x' \tilde{\Gamma}^{-1} x. \qquad (6.18')$$

The physical meaning of this estimator is that with $\tilde{\Gamma}^{-1}$ it is possible to obtain from correlated samples x uncorrelated samples which allows a best estimation of the power. It is clear that $\hat{P}_o$ is equivalent to $\hat{P}$ in the case of a white noise.
If we hope an estimate which is a R.N.A., it is clear that we will use in Eqs. (6.18) and (6.18') $x_\perp$ instead of x. Thus the first estimate, valuable in the case of a white noise can be written

$$\hat{P} = \frac{1}{m} [x_\perp' x_\perp] \qquad (6.19)$$

with m = n-1, because x is on a subspace of R^n.
By using Eq. (6.4) we obtain

$$\hat{P} = \frac{1}{m} [x' x - \frac{(s' x)^2}{s' s}] \qquad (6.20)$$

and it is clear that

$$\hat{P}(b) = \hat{P}(s + b) \qquad (6.21)$$

which is a characteristic property of a R.N.A.
Now let us consider the case of the optimal estimator $\hat{P}_o$. In this case Eq. (6.18') can be written

$$P_o = \frac{1}{m} x_\perp' \Gamma_\perp^{-1} x_\perp \qquad (6.22)$$

which is by construction a R.N.A. As for Eq. (6.20) it is interesting to give an expression of P_o in terms of the total observation. The calculation is presented in Appendix B and the result is

$$x_\perp' \tilde{\Gamma}_\perp^{-1} x_\perp = x' \tilde{\Gamma}^{-1} x - \frac{(s' \tilde{\Gamma}^{-1} x)^2}{s' \tilde{\Gamma}^{-1} s} \quad . \qquad (6.23)$$

On this last expression we verify very simply that an equation like to (6.21) is still true.

6.4. Adaptive matched filter

Let us suppose that the noise has stable spectral properties, defined by $\tilde{\Gamma}$, but strong variations of power. It is possible to show that an optimal receiver is on the form [10]

$$T(x) = \frac{s' \tilde{\Gamma}^{-1} x}{[x' \tilde{\Gamma}^{-1} x]^{1/2}} \gtrless t \qquad (6.24)$$

It is clear that the numerator is the classical matched filter,

(6.1), and the denominator an estimation of the power. Thus the system appears as a regulation of the output of the matched filter by an estimation of the fluctuating power. This method is evidently analog to a A.G.C. before the matched filter.
But let us now consider the case of strong signal. If x = k s, we see that

$$T(x) = (s' \tilde{\Gamma}^{-1} s)^{1/2} = d \qquad (6.25)$$

which is independent of k. Thus there is no difference, is absence of noise, between a small or a large signal, because of the A.G.C. This situation is not satisfactory and leeds to the idea of a A.G.C. which is a R.N.A.
Now let us consider the other receiver T'(x) defined by

$$T'(x) = \frac{s' \tilde{\Gamma}^{-1} x}{[\tilde{x}'_{\perp} \tilde{\Gamma}_{\perp}^{-1} \tilde{x}_{\perp}]^{1/2}} \gtrless t \qquad (6.26)$$

The numerators of T(x) and T'(x) are the same (N), and the denominator are different (D, D'). But from Eq. (6.23) we see that

$$D'^2 = D^2 - \frac{N^2}{d^2} \qquad (6.27)$$

where d^2 is as previously $s' \tilde{\Gamma}^{-1} s$. Thus we can write

$$T'(x) = \frac{T(x)}{\left[1 - \frac{1}{d^2} T^2(x)\right]^{1/2}} \qquad (6.28)$$

One can see that T'(x) is a monotone function of T(x) if $T^2 < d^2$, which is true as noticed previously. We conclude that T(x) and T'(x) have exactly the same properties concerning the detection. But T'(x) is more interesting than T(x) because if x = k s we obtain $T' \to \infty$ instead of T = d.
This example shows very simply the interest of a R.N.A. in this problem of adaptive detection.
In practice the situation is not so simple because the signal is a function on time, and the situation described previously is only true at the time instant where we observe the output of the matched filter. Nevertheless some analog structures can be found which give similar results.

APPENDIX A

Bayes and Neyman-Pearson strategies

Let us consider the class $C_{\alpha o}$ of tests Φ defined by

$$\Phi \in C_{\alpha o} \leftrightarrow \alpha[\Phi] \leqslant \alpha_o \qquad (a.1)$$

where α is defined by (2.2) and introduce the function b(k) defined by

$$b(k) = \alpha[\Phi_k^B] \qquad (a.2)$$

where Φ_k^B is defined by (2.6).
If there exists a k_o such that

$$b(k_o) = \alpha_o \qquad (a.3)$$

thus

$$\Phi_{\alpha o}^{N.P} = \Phi_{ko}^B \qquad (a.4)$$

Indeed, $\forall\ \Phi$ we have

$$\beta(\Phi) - k_o\ \alpha(\Phi) \leq \beta(\Phi_{ko}^B) - k_o\ \alpha(\Phi_{ko}^B) \qquad (a.5)$$

because, according to (2.5), the Bayes strategy Φ_{ko}^B gives the maximum of $\beta - k_o\ \alpha$. Using (a.2) and (a.3) we deduce that $\forall\ \Phi$

$$\beta(\Phi) \leq \beta(\Phi_{ko}^B) - k_o\ [\alpha_o - \alpha(\Phi)] \qquad (a.6)$$

Now $\forall\ \Phi \in C_{\alpha o}$ we have from (a.1)

$$\beta(\Phi) \leq \beta(\Phi_{ko}^B) \qquad (a.7)$$

which gives Eq. (2.8).
Now the only thing to show is that Eq. (a.3) can be true for any α_o. This point is due to the fact that

$$b(k) = \int \Phi_k^B(x)\ p_o(x)\ dx \qquad (a.8)$$

where Φ_k^B is defined by (2.6). Thus b(k) has the structure of a repartition function which secures Eq. (a.3).

APPENDIX B

Let us consider a regular and symetric matrix M. The matrix

$$M^* = M - \frac{M\ s\ s'\ M}{s'\ M\ s} \qquad (b.1)$$

is a R.N.A. because

$$M^*(s + b) = M^*b. \qquad (b.2)$$

Indeed we see directly from (b.1) that $M^*\ s = o$.
We deduce that there exists a matrix M operating in the subspace $E_\perp$ such that

$$x'\ M^*\ x = x'_\perp\ M_\perp\ x_\perp\ . \qquad (b.3)$$

To find $M_\perp$ we use the partition

$$M = \begin{pmatrix} M_1 & m \\ m' & c \end{pmatrix} \qquad x = \begin{pmatrix} x_\perp \\ x_s \end{pmatrix} \tag{b.4}$$

By calculating the two terms of Eq. (b.3) we find easily

$$M_\perp = M_1 - \frac{1}{c}\, m\, m' . \tag{b.5}$$

We see that this procedure can be applied to Eq. (6.23) where $M = \tilde{\Gamma}^{-1}$. Thus the problem is to obtain for $\tilde{\Gamma}^{-1}$ the terms corresponding to M_1, m and c. For this purpose we start from the fact that $\tilde{\Gamma}^{-1}$ is the inverse of $\tilde{\Gamma}$ which can be written as (6.8), with the only difference that we have a normalized covariance matrix which is

$$\tilde{\Gamma} = \begin{pmatrix} \tilde{\Gamma}_\perp & \gamma \\ \gamma' & 1 \end{pmatrix} \tag{b.6}$$

We write $\tilde{\Gamma}^{-1}$ on the same form with B_1, b, c and after a simple calculation starting $\tilde{\Gamma}\,\tilde{\Gamma}^{-1} = I$ we obtain

$$\tilde{\Gamma}^{-1} = \alpha \begin{pmatrix} \frac{1}{\alpha}\,\Gamma_\perp^{-1} + \lambda\,\lambda' & -\lambda \\ -\lambda' & 1 \end{pmatrix} \tag{b.7}$$

with

$$\lambda = \Gamma_\perp^{-1}\,\gamma \tag{b.8}$$

$$\alpha = (1 - \gamma'\,\Gamma^{-1}\,\gamma)^{-1} \tag{b.9}$$

By using Eqs. (b.5) and (b.7) we obtain that if $M = \tilde{\Gamma}^{-1}$,

$$M_\perp = (\tilde{\Gamma}^{-1})_\perp = \tilde{\Gamma}_\perp^{-1} \tag{b.10}$$

which shows completely Eq. (6.23).

REFERENCES

[1] C.W. HELSTROM : Statistical theory of signal detection, Pergamon, 1968.

[2] H.L. VAN TREES : Detection, estimation and modulation theory. Wiley, 1968.

[3] I. BAR-DAVID : A sample path property of matched-filter outputs with applications to detection and estimation. I.E.E.E. Trans. Inf. Thery 22, 225, 1976.

[4] T. KAILATH : correlation detection of signals perturbed by a random channel. I.E.E.E. Trans. Inf. Theory IT 6, 361, 1960.

[5] See Ref. [2], t 3, p. 19.

[6] G. VEZZOSI : What is optimality for an adaptive detection system. NATO Advanced Study on Signal Processing, 1973, Academic Press.

[7] G. VEZZOSI : Detection d'un signal dans un bruit de loi incertaine. Thèse de Doctorat d'Etat, Université de Paris-Sud, 1976.

[8] H. MERMOZ : Ecueils et diversités des traitements adaptatifs d'antenne. Ann. Telecomm. 28, 244, 1973.

[9] H. MERMOZ : Modularité du traitement adaptatif d'antenne. Ann. Telecomm. 29, 43, 1974.

[10] See Ref. (6), on table III and also
G. VEZZOSI and B. PICINBONO : Détection d'un signal certain dans un bruit sphériquement invariant. Structure et comparaison des différents récepteurs. Ann. Télécomm. 27, 95, 1972.

DISCUSSION

Comment : J.W.R. GRIFFITHS

Could the lecturer explain why his model shows the "estimate and plug" model to be optimum? What is the difference for the case considered by Dr W.S. Hodgkiss who showed such a system to be sub-optimum?

Reply : B. PICINBONO

In general it is not possible to use a Bayer criterion if there is some uncertain parameter without probability distribution. In many cases one uses the classical structure and replaces the uncertain parameter by its estimation, but it is not possible to show that such a receiver is optimal. The receiver considered in the lecture is optimal in the class of receivers with constant false alarm probability. But that is a particular property of the problem considered with uncertainty on the power. The general problem is extensively studied in Ref [7] of my paper.

MODELLING AND DETECTION

Ph. de Heering

SACLANT ASW Research Centre
La Spezia, Italy

ABSTRACT. This review paper first presents some general concepts about the meaning of modelling a physical process; attention is then restricted to the modelling of the radar/sonar target as seen by the receiver, for which some fluctuation models are presented. Detection performance curves are presented and discussed. The paper concludes with an example of application of a detection model to a specific communication problem.

1. INTRODUCTION

This paper concentrates on a few detection models, more from an applicative point of view than from a theoretical one to give an idea about the operation, advantages, and limitations of such models. The selection is arbitrary, although some of the models presented have been extensively used in Sonar.

2. MATHEMATICAL MODELLING

The complexity of nature calls for partial descriptions that are sometimes mathematical in character and then generally called mathematical models. The relation between nature and a mathematical model can be represented schematically as in Fig. 1. Here are some qualities necessary or desirable for such a model [1,2,3]:

a. Self-consistency.
b. Consistency with existing data.
c. Consistency with possible ideal physical mechanisms.
d. Have a minimum number of degrees of freedom (or independent

G. Tacconi (ed.), Aspects of Signal Processing, Part 1, 181-201.

input measurements).
e. Be adapted to the user, in that it can predict useful parameters from input measurements accessible to the user.

a. and c. give confidence that the model is also valid outside of the regions where it has been verified; d. has to do with the predictive power of the model.

3. NEYMAN-PEARSON STRATEGY FOR ECHO DETECTION [4]

The concepts of Neyman-Pearson strategy, likelihood ratio detector, sufficient statistic, detection and false alarm probabilities have been introduced at this Institute by Picinbono [5].

It will be sufficient to stress a few important ideas useful for the remainder of this paper.

A likelihood ratio detector is seen (Fig. 2) to be composed of two components in series: a data processor and a threshold device. The data processor performs a particular mapping of the multi-dimensional observation space into a one-dimensional space (the likelihood ratio), thereby extracting from the observation all the information necessary to perform the detection decision. The output of the data processor, or any monotonically increasing function of it, can be characterized by a probability of detection P_d and a probability of false alarm P_{fa}, both functions of a threshold

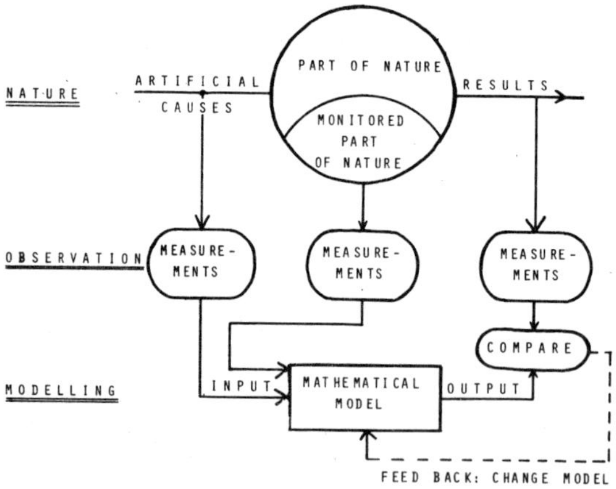

Fig. 1. Mathematical model and nature.

that can be considered as a parameter (Fig. 3). The threshold device operation does not depend on the data or the transition probabilities $P(\underline{v}|H_1)$, $P(\underline{v}|H_o)$, but only on the a <u>priori</u> probabilities $P(H_1)$, $P(H_o)$ and the Bayes costs. When these are not available (Neyman-Pearson strategy), the detector operation can be entirely characterized by the functional relationship between P_d and P_{fa}.

4. SOME SIMPLE MODELS FOR TARGET ECHO AND BACKGROUND, AND THE RELATED DETECTOR STRUCTURE [4,6]

4.1 Background

The hypothesis-testing approach to the detection problem is presented together with the example of detection of an exactly known signal. This approach is also used in the other examples. We consider here the detection of a signal against a background of white gaussian stationary noise of known spectral density N_o. This assumption applies to sufficiently narrow band echoes observed in sufficiently short time intervals. Further, the coloured stationary background noise can often be reduced to the case of the white background noise by means of appropriate filtering (prewhitening) if the observation interval is long enough. This background model is the one adopted in the remainder of the paper.

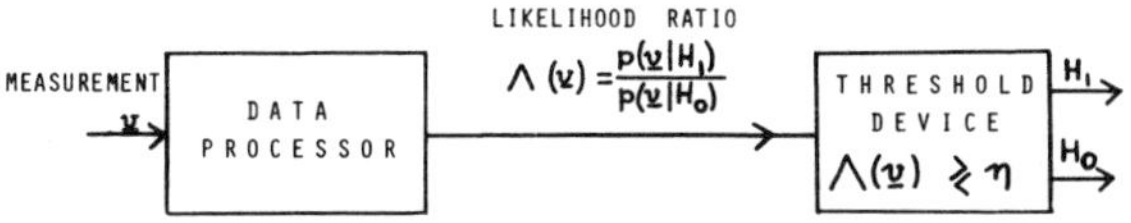

Fig. 2. Likelihood ratio detector.

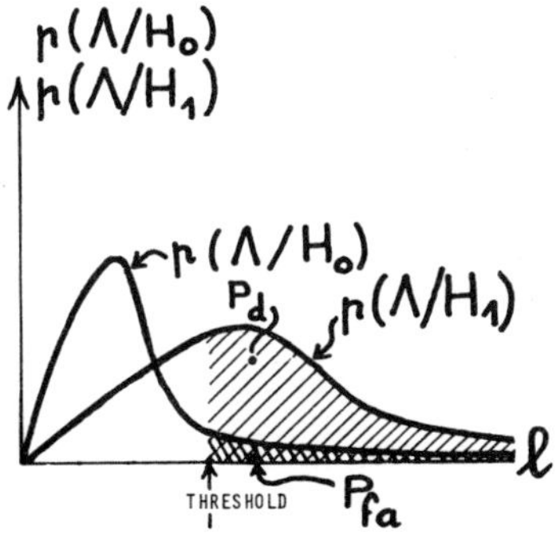

Fig. 3. Definition of P_d, P_{fa}.

4.2 Exactly known signal

Let H_1 = the hypothesis that "signal plus noise is present"
H_o = the hypothesis that "noise only is present"

$v(t)$ = the received waveform
$s(t)$ = the (exactly known) signal of duration T, starting at $t = 0$ and of bandwidth W
$n(t)$ = the (stationary white gaussian) noise waveform bandlimited in bandwidth W

Then

$$\text{under } H_1 \qquad v(t) = s(t) + n(t) \tag{4.1}$$

$$\text{under } H_o \qquad v(t) = n(t) \tag{4.2}$$

The <u>likelihood ratio detector</u> structure is most easily derived if one considers the received signal sampled at 1/2W intervals.

The vectors representing the sampled waveforms are written: $\underline{v} = \{v_i\}$, $\underline{s} = \{s_i\}$, $\underline{n} = \{n_i\}$ $i = 1,2,\ldots, 2WT$. Thus

$$H_1 : \quad \underline{v} = \underline{s} + \underline{n} \tag{4.3}$$

$$H_o : \quad \underline{v} = \underline{n} \tag{4.4}$$

We write the conditional joint probability densities of the observation vector $\underline{v}$ under the hypotheses H_o and H_1 respectively:

$$p(\underline{v}|H_o) = \frac{1}{(2\pi)^{WT} (N_oW)^{2WT}} \exp\left(\frac{-\sum_{i=1}^{2WT} v_i^2}{2WN_o}\right) \tag{4.5}$$

$$p(\underline{v}|H_1) = \frac{1}{(2\pi)^{WT} (N_oW)^{2WT}} \exp\left(\frac{-\sum_{i=1}^{2WT} (v_i-s_i)^2}{2WN_o}\right) \tag{4.6}$$

Both densities are written as products because of the statistical independence between samples of the noise waveform.

The likelihood ratio $\Lambda(\underline{v})$ is the ratio $p(\underline{v}|H_1)/p(\underline{v}|H_o)$

$$\Lambda(\underline{v}) = \frac{\exp\left(-\sum_{i=1}^{2WT} (v_i-s_i)^2/2WN_o\right)}{\exp\left(-\sum_{i=1}^{2WT} v_i^2/2WN_o\right)} \tag{4.7}$$

$$\Lambda(\underline{v}) = \exp\left(2 \sum_{i=1}^{2WT} \frac{v_i s_i}{2WN_o} - \sum_{i=1}^{2WT} \frac{s_i^2}{2WN_o}\right) \tag{4.8}$$

Since $\sum_{i=1}^{2WT} \frac{s_i^2}{2WN_o} \simeq \frac{E}{N_o}$ where E is the signal energy

$$\Lambda(\underline{v}) = \exp\left(+2 \sum_{i=1}^{2WT} \frac{v_i s_i}{2WN_o} - \frac{E}{N_o}\right) \tag{4.9}$$

In (4.9), we recognize that $\Lambda(\underline{v})$ is a monotonically increasing function of the summation term in the exponent, which contains all the useful information about the received waveform. This term is the output of the data processing part of the receiver:

$$\Lambda'(\underline{v}) = \frac{1}{2W} \sum_{i=1}^{2WT} (v_i s_i) \tag{4.10}$$

We recognize in (4.10) that the data processing part of the receiver correlates the received waveform with the known signal. This cross-correlation is then compared with a threshold. The receiver is schematically represented in Fig. 4, where the correlator can also be realized as a matched filter.

The receiver performance in terms of detection probability P_d and false alarm probability P_{fa} is easily found by recognizing that the test statistic (4.10) is a gaussian random variable under both H_o and H_1. Since both the noise samples and the signal-plus-noise samples (if taken at the rate 2W) are jointly gaussian

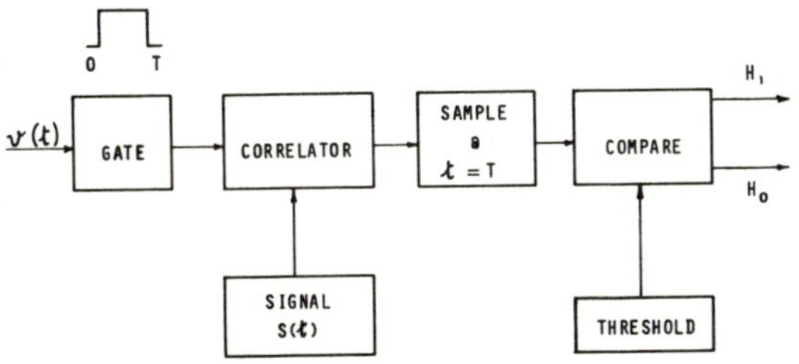

Fig. 4. Likelihood ratio receiver.

random variables, then by definition all their linear combinations are gaussian random variables.

It is easy to calculate the mean and standard deviation of $\Lambda'(\underline{v})$ under the two hypotheses. While $\Lambda'(\underline{v}) = 0$ under H_o, the remaining quantities are all functions of the signal-to-noise ratio (SNR) E/N_o

$$H_o : \quad \overline{\Lambda'(\underline{v})} = 0 \tag{4.11}$$

$$\sigma\{\Lambda(\underline{v})\} = k(E/N_o) \tag{4.12}$$

$$H_1 : \quad \overline{\Lambda'(v)} = k(E/N_o) \tag{4.13}$$

$$\sigma\{\Lambda'(\underline{v})\} = k(E/N_o) \tag{4.14}$$

where σ is the standard deviation and k a constant multiplicative factor.

Performance curves can be derived by application of the tabulated error function and are presented in Fig. 5, where P_d is represented as a function of SNR with P_{fa} as a parameter. The value of P_{fa} is determined by the threshold in the receiver. This diagram entirely characterizes the signal processing part of the receiver.

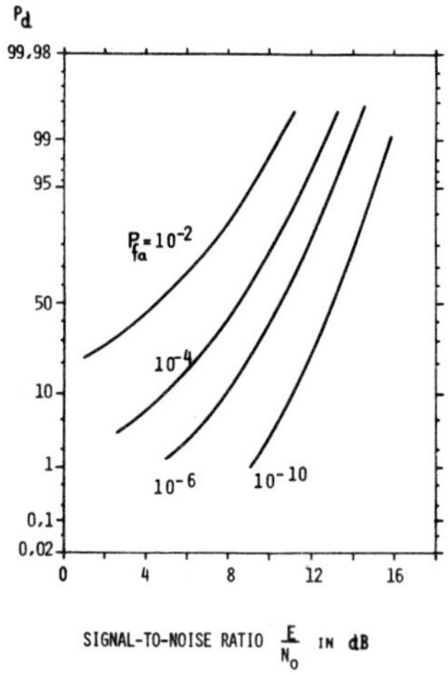

Fig. 5. Detection performance for an exactly known signal.

4.3 Narrow-band signal with known amplitude and unknown phase

The unknown phase is assumed to be uniformly distributed in [0, 2π]. This case is more realistic than the preceding one, since the phase of the received signal is practically always unknown. The derivation of the likelihood ratio detector is made by expressing the likelihood ratio as a function of the initial phase of the signal, then averaging over the interval [0, 2π] with a uniform probability density function for the phase. This calculation results [6] in the receiver structure of Fig. 6. A performance diagram is given in Fig. 7. Comparison with Fig. 5 shows that the loss due to not knowing the initial phase of the echo is of the order of 2 dB at low SNR and decreases to about 0.5 dB at high SNR.

The correlator envelope detector configuration is the narrow-band approximation to a more general detector (for wide and narrow band) that computes the correlation of the received signal with in-phase and in-quadrature components of the transmitted signal, and then computes the quadratic sum of the correlations.

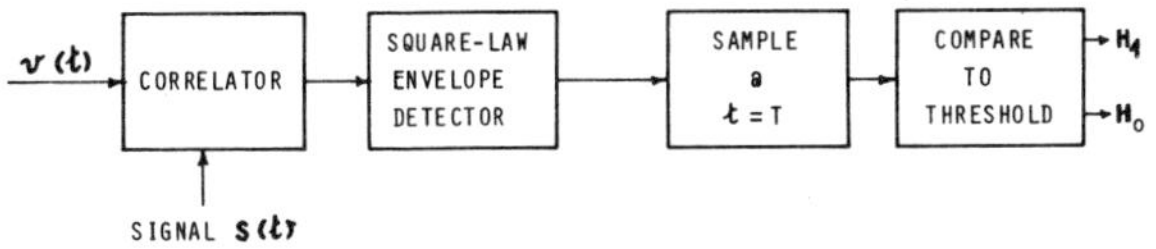

Fig. 6. Likelihood ratio receiver for signal with unknown phase.

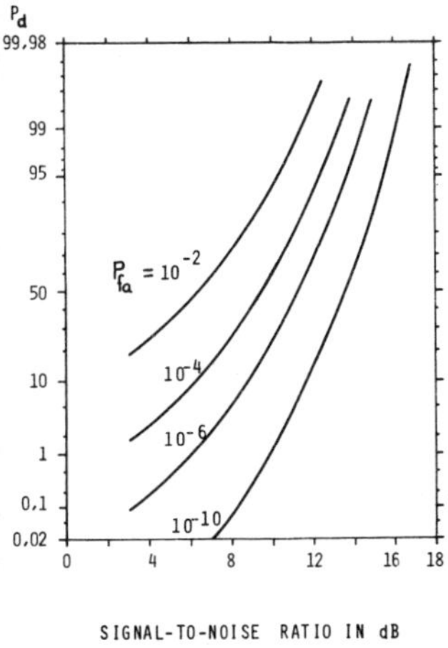

Fig. 7. Detection performance for signal with unknown phase.

4.4 Rayleigh fluctuating signal (with random phase)

When the unresolved target echo results from many scatterers of comparable size adding vectorially with random phases, then the echo amplitude is Rayleigh distributed (and the echo energy is thus exponentially distributed). This can be seen as follows. Vectorial addition of the scatterers in an in-phase, in-quadrature coordinate system yields a resultant vector whose projection on any axis is a gaussian random variable (by the Central Limit Theorem). For reasons of symmetry, the projections on all axes have the same probability density, which shows that the joint density of the in-phase, in-quadrature components is cylindrically symmetric. For two jointly gaussian random variables, this is equivalent to their being uncorrelated, therefore statistically independent. Their root-mean sum is thus [7] Rayleigh distributed and the sum of their squares is exponentially distributed.

The received signal energy E has thus an exponential p.d.f. (also called χ^2 with two degrees of freedom)

$$p_{\bar{E}}(E) = \frac{1}{\bar{E}} \exp -\left(\frac{E}{\bar{E}}\right) \tag{4.15}$$

and the phase of the received signal is again assumed to have a uniform density in the interval $[0, 2\pi]$.

It is shown in [6] that the structure of the optimum receiver for a signal of uniform phase is independent of the amplitude p.d.f. It follows, in particular, that the detector of Fig. 6 is also optimal for a Rayleigh fluctuating signal. The detector performance is obtained by averaging the performance for a non-fluctuating target with unknown phase (Fig. 4) with an exponential weighting function for the SNR as examplified on Fig. 8. The performance curves (Fig. 9) then have the simple expression

$$P_{fa} = (P_d)^{\frac{\bar{E}}{N_o}+1} \tag{4.16}$$

because of the exponential density of the test statistic.

Comparison of the curves in Figs. 7 and 9 shows the effect of echo fluctuation on detection performance. It can be seen that for $P_d > 30\%$ there is a fluctuation loss that increases with P_d; for $P_d < 30\%$ there is a small fluctuation gain. The reason for this can be seen with reference to Fig. 8. When the mean SNR is high, the weighting of the non-fluctuating echo performance curve is done mostly on P_d values smaller than those corresponding to $\bar{E}/N_o$; this causes the loss. When the mean SNR is low, the

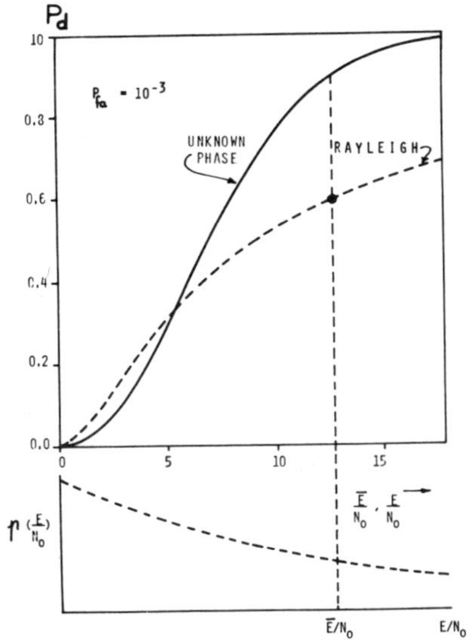

Fig. 8. Calculation of receiver performance for Rayleigh signal.

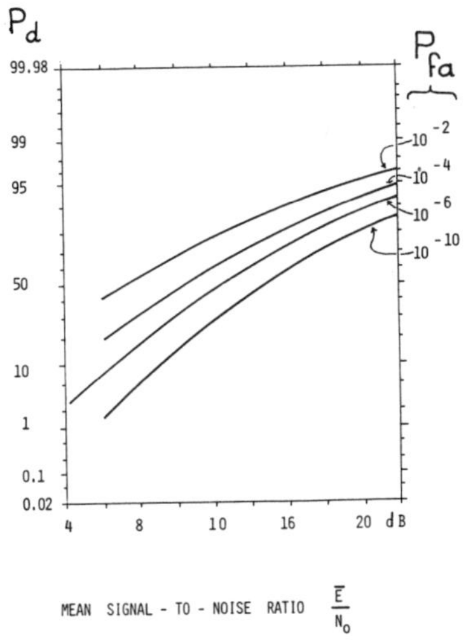

Fig. 9. Detection performance for Rayleigh signal.

weighting is performed mainly on P_d values higher than those corresponding to $\overline{E}/N_0$; this causes the gain. The Rayleigh fluctuation is a particular case of the Swerling I and II fluctuation models.

5. SWERLING I AND II FLUCTUATION MODELS [2,6,8]

5.1 Introduction

When a Rayleigh fluctuating target (Sect. 4.4) is illuminated by a pulse train (and not a single pulse as was assumed in the previous chapter), the echo energy from the successive train pulses can fluctuate. Swerling I and II fluctuation models consider two extreme types of fluctuation. Swerling I model assumes that the echo energy variation is slow compared with the duration of the pulse train, so that all pulse train echoes have equal energy. Swerling II model assumes that the echo energy variation is fast compared with the pulse-to-pulse interval, so that the energies of the pulse train echoes are statistically independent. In both cases, the initial phase of each of the echoes in the train is taken as independent from the initial phases of all other echoes of the train with a uniform distribution between 0 and 2π.

A physical mechanism giving an exponential density of a single echo energy has been described above (Sect. 4.4). The assumption that the initial phases of the pulse train echoes are random and independent comes from the fact that small changes in target orientation and range generally cause large variations in the phase of the received echo.

Swerling II fluctuation models are also applicable [9] when using pulses whose range resolution is high compared with the target radial extension. This case will be studied in more detail in Sect. 5.4.

5.2 Optimal receiver

The optimal receiver structure for both Swerling I and Swerling II fluctuation models is shown in many references [4,6,8] to be that of Fig. 10.

For ease of implementation, this receiver is often simplified into the one shown in Fig. 11, which is equivalent in performance within a fraction of a decibel [6]. This last receiver will be studied in the rest of the chapter. One should note that the receiver in Fig. 11 is similar to the one in Fig. 6 (Rayleigh fluctuation target). The present receiver processes each pulse train echo like the Rayleigh target receiver. This is followed by incoherent recombination of the energy contributions of all the train pulses.

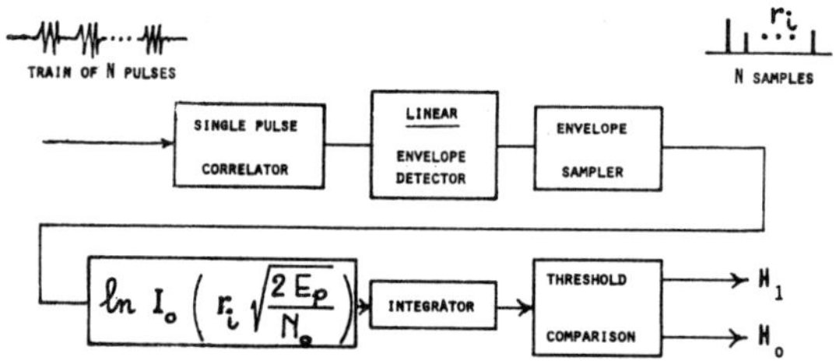

Fig. 10. Optimum detector for Swerling I and II fluctuating signal.

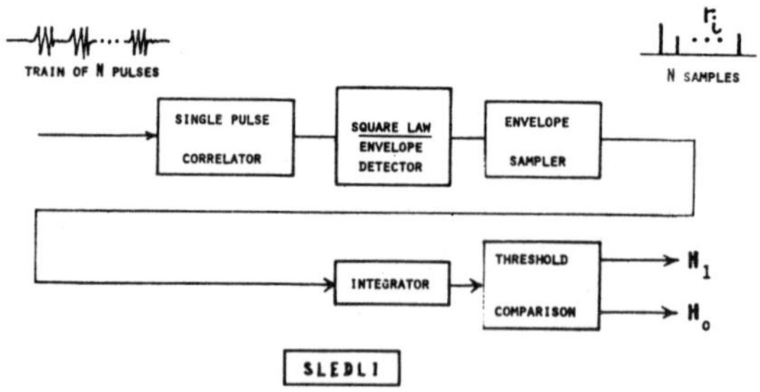

Fig. 11. Near-optimum detector for Swerling I and II fluctuating signal.

5.3 Performance

The false alarm probability of the receiver in Fig. 11 is easy to calculate as a function of the threshold, if one observes that under hypothesis H_0 the test statistic is χ^2 with 2N degrees of freedom.

For both Swerling I and II models P_{fa} is given by

$$P_{fa} = \exp(-d) \sum_{i=0}^{N-1} \frac{d^i}{i!} \tag{5.1}$$

where N is the number of pulses in the pulse train,
d is a parameter proportional to the value of the threshold.

This is simply a series expansion of the χ^2 distribution function with 2N degrees of freedom.

The derivation of P_d for the Swerling I fluctuation model can be found in [6]. Some representative performance curves are given in Figs. 12, 13 and 14 [N = 1, 10, 100].

The P_d performance of the detector of Fig. 11 in the Swerling II fluctuation model is immediately derivable from the χ^2 distribution expressions since the test statistic is then also χ^2 with 2N degrees of freedom. The expression is functionally the same as (5.1).

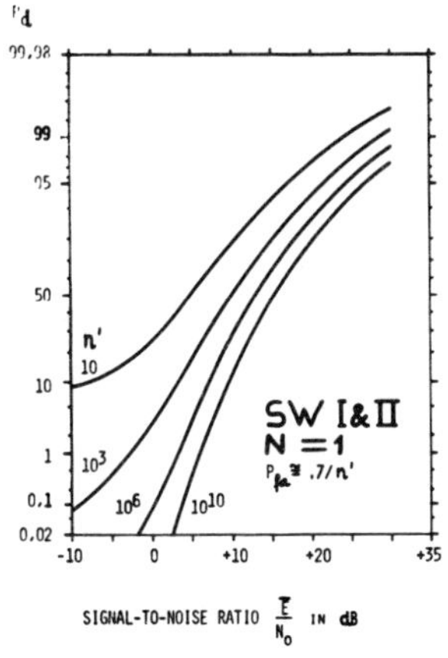

Fig. 12.

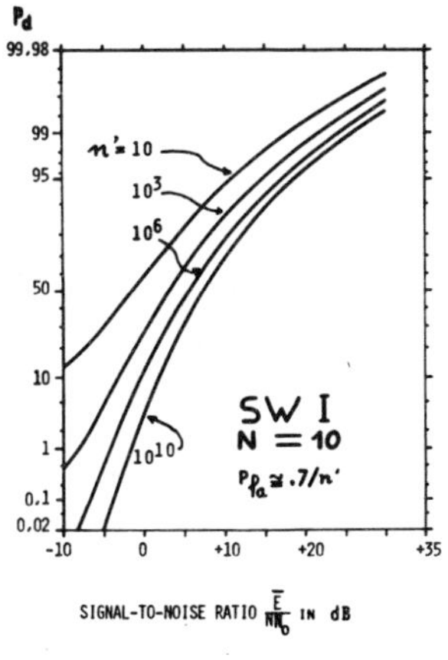

Fig. 13.

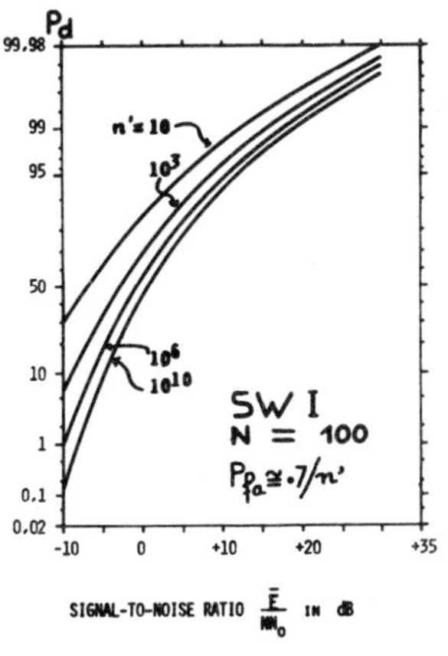

Fig. 14.

Detection performance for Swerling signal.

$$P_d = \exp(-Cd) \sum_{i=1}^{N-1} \frac{(Cd)^i}{i!} \qquad (5.2)$$

where $C = \dfrac{1}{1 + \dfrac{\bar{E}/N}{N_o}}$ (reciprocal of the mean signal plus noise-to-noise ratio per pulse)

and E/N is the mean signal energy per pulse.

Selected performance curves are given in Figs. 15 and 16 [N = 10, 100]. For N = 1 , both models correspond to the same echo fluctuation (Rayleigh target).

5.4 Comparison of Swerling I and II performance curves: 'Integration loss'

The performance curves of Figs. 13 and 14 (Swerling I) as compared with those of Figs. 15 and 16 (Swerling II) show a striking performance difference, at the higher signal-to-noise ratios, in favour of the Swerling II model due to the fluctuation suppression effect present in the Swerling II model detector; for higher values of N the test statistic resembles more and more that of a non-fluctuating pulse train echo.

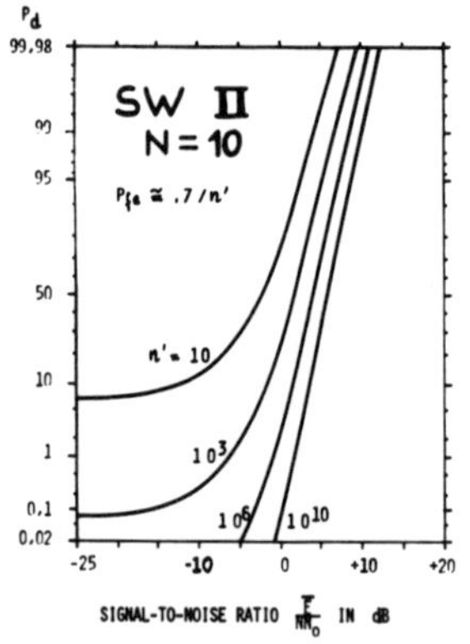

Fig. 15.

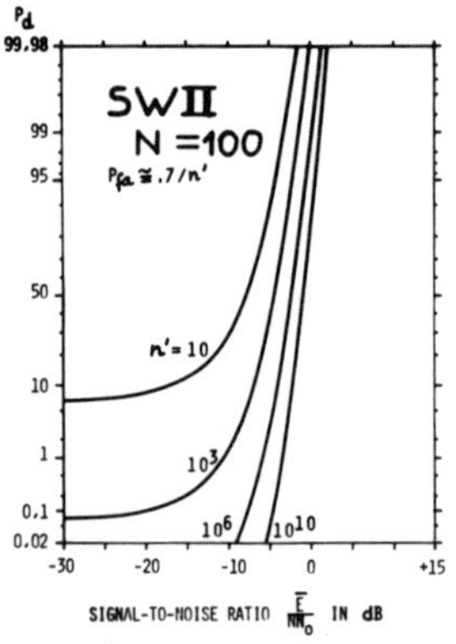

Fig. 16.

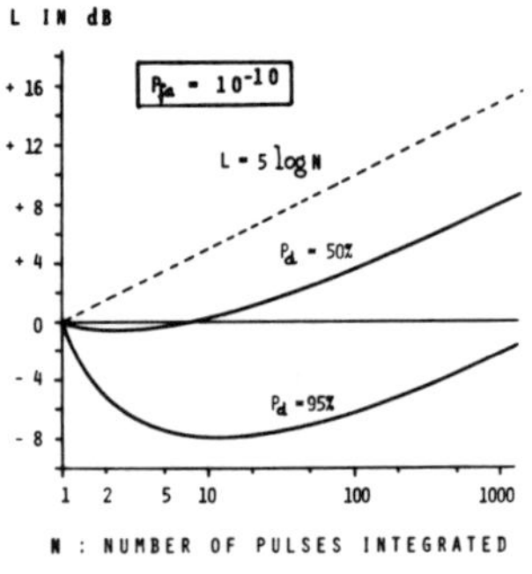

Fig. 17.

Figs. 15 and 16. Detection performance for Swerling signal.

Fig. 17. Integration loss for Swerling II fluctuating signal.

An important concept pertinent to the detection of echo trains is the so-called 'Integration loss'. It is a measure of the performance difference in the reception of the two following signals:

Signal a : one pulse of mean energy $\bar{E}$.

Signal b : N incoherent pulses, each of mean energy $\bar{E}/N$.

Then the 'Integration loss' L is defined as the ratio of the noise power densities N_{01}, N_{ON} that produce the same P_d and P_{fa} for signals a. and b.

$$L(P_d, P_{fa}, N) = 10 \log \frac{N_{01}}{N_{ON}} \tag{5.3}$$

It is expressed in decibels and is in general a function of P_d, P_{fa} and N.

In order to calculate L, we rewrite it in the form

$$L = 10 \log \left[\frac{(E/N)/N_{ON}}{(E/N_{01})/N} \right] \tag{5.4}$$

or

$$L = [SNR]_N - [SNR]_1 + 10\log N \tag{5.5}$$

where $[SNR]_N$ and $[SNR]_1$ are the signal-to-noise ratios necessary to get a specified P_d and P_{fa} with signals b. and a. respectively.

A plot of L for selected values of P_d and P_{fa} is given in Fig. 17, (Swerling II fluctuation model).

It is useful to notice that in the Swerling II model for smaller values of N and higher values of the SNR, the integration loss is negative; that is, it is actually a gain. In such cases the fluctuation suppression effect is greater than the SNR loss caused by splitting the echo energy among N pulses.

5.5 Detection of an extended target in noise

Van der Speck [9] has applied the Swerling II fluctuating model formalism to a somewhat different situation : the detection of an extended target in noise by a broad-band pulse. He considers a transmission of energy E_T and bandwidth W. The target has a range extension L. The target echo contains on average a fraction r of the transmitted energy so that

$$\bar{E} = r\, E_T \tag{5.6}$$

Since the transmission of bandwidth W has a time resolution of 1/W, the target can be thought of as being constituted by N resolution cells in which

$$N = 2LW/c \tag{5.7}$$

where c is the wave propagation velocity in the medium considered.

The signal energies in all resolution cells are modelled as independent, exponentially-distributed random variables, with the mean energy per cell equal to $\bar{E}/N$.

The noise is assumed white, gaussian with spectral power density N_0. The mean SNR is thus a function of W.

$$\overline{SNR} = \frac{\bar{E}}{NN_0} = \frac{\bar{E}\ c}{2LW\ N_0} \tag{5.8}$$

This model corresponds exactly to the Swerling II model hypotheses where the role of the number N of pulses in the pulse train is played by the number N of resolution cells in the target.

It follows that the best bandwidth to use under the model hypotheses can be deduced from "integration loss" plots such as the one in Fig. 17.

In the context of the present problem, this plot should be interpreted as follows. The horizontal axis represents the number N of resolution cells in which the modelled target is resolved by the wideband pulse. N is related to the bandwidth of the transmitted pulse and the (supposed known) target range extension by (5.7). The vertical axis represents the extra energy required to retain a specific performance in terms of P_d and P_{fa}.

6. OTHER FLUCTUATION MODELS

The detection models presented in the preceding sections are but a small fraction of all the models that have been investigated for the radar and sonar targets.

Other models include, in the χ^2 family, apart from the classical Swerling III and IV fluctuation models [8], models and solutions for partially correlated pulse trains [9,10] and generalized χ^2 distribution models [2]. Other families of fluctuations statistics have also been investigated (for the square-law envelope detector and integrator) such as the log-normal family [12] and the Rice family [2].

7. APPLICATION OF A DETECTION MODEL TO A COMMUNICATION PROBLEM [13]

7.1 Motivation and problem definition

Radar/Sonar detection models are also applicable in one-way communication situations, as will be examplified by the idealized problem treated in this chapter. One assumes that the fluctuations are produced by a communication channel, rather than by a target.

We are concerned here with the following situation. A binary (YES/NO) message has to be transmitted from a transmitter to a receiver by means of a particular message fromat that insures very low message false alarm probability and high message secrecy. Hereafter we study the transmission performance under simplifying assumptions that make possible the application of a simple fluctuation model.

7.2 Message definitions

A message is constituted by n digits transmitted consecutively. A digit is incoded into k frequency bands, the digit elements, of which, for the digit to be valid, f have to be occupied by a tone pulse, and e have to empty (Fig. 18).

A digit can take $k = \frac{k!}{f!\ e!}$ values ; (7.1)

A message has $c = \left[\frac{k!}{f!\ e!}\right]^n$ possible states. (7.2)

We restrict our attention to only one of all possible message states being taken as valid. This state we will call a valid message state.

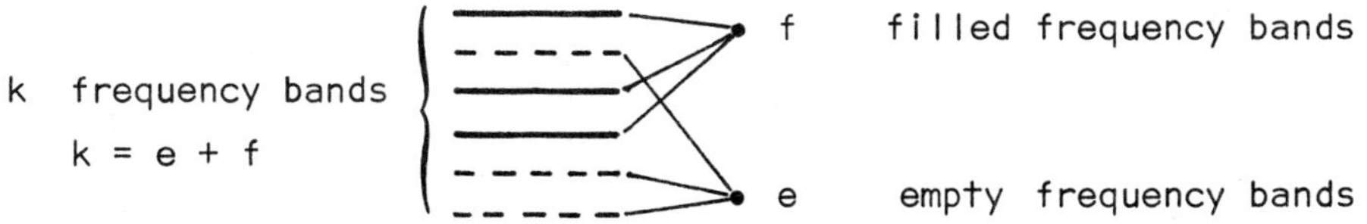

Fig. 18. Digit encoding.

7.3 Fluctuations model

Each digit element in the message is Rayleigh fluctuating, independent from all other digit elements, and with a common mean. The background is white gaussian stationary noise. Thus, with P_d and P_{fa} referring to the digit elements, one has (Sect. 4.4)

$$P_{fa} = (P_d)^R \tag{7.3}$$

where $R = \bar{E}/N_0 + 1$. The threshold is related to P_{fa} by

$$P_{fa} = \exp(-d) \tag{7.4}$$

7.4 Message probabilities of detection and false alarm

The <u>message probability of detection</u> P_d^M is the probability that all the consecutive digits of a transmitted message will be decoded correctly. The <u>message probability of false alarm</u> P_{fa}^M is the probability that (with no message transmitted) the background will be erroneously decoded as the valid message.

The two following equations hold, because of the postulated independence between the digit elements:

$$P_d^M = \left[(P_d)^f \quad (1-P_{fa})^e \right]^n \tag{7.5}$$

$$P_{fa}^M = \left[(P_{fa})^f \; (1-P_{fa})^e \right]^n \tag{7.6}$$

Substituting (7.3) into (7.5) yields the performance equations with P_{fa} as the parameter:

$$P_d^M = \left[(P_{fa})^{f/R} \; (1-P_{fa})^e \right]^n \tag{7.7}$$

$$P_{fa}^M = \left[(P_{fa})^f \quad (1-P_{fa})^e \right]^n \tag{7.8}$$

Elimination of P_{fa} between (7.7) and (7.8) in order to get an explicit relation between P_d^M and P_{fa}^M seems impossible. Equation (7.8) does not always have real solutions for P_{fa} in the [0, 1] interval corresponding to a given P_{fa}^M. If it has such solutions, it can be shown, by taking the ratio of (7.7) over (7.8), that the smallest solution in P_{fa} always yields the greatest P_d^M.

Some performance curves are given in Fig. 19. It can be seen that

P_d^M tends to the asymptotic value $(1-P_{fa})^{e\,n}$ as the SNR increases to infinity. The performance curves have the somewhat unusual feature that for certain choices of the signal parameters and of the SNR, P_d^M decreases as P_{fa}^M is increased. This is, of course, due to our definitions.

7.5 Choice of the best operating point

The choice of the receiver operating point (that is the choice of P_{fa} or the threshold choice) is made by placing an upper bound F on P_{fa}^M

$$P_{fa}^M \leq F \tag{7.9}$$

and maximizing P_d^M as a function of P_{fa} in (7.7) subject to this bound.

Simple algebra yields the following results:

Let $P_{fa}(F)$ stand for the smallest real solution between 0 and 1 of $P_{fa}^M = F$ [in (7.6)]. Then if $P_{fa}(F) \geq \frac{f/R}{f/R+e}$ or does not exist, then

$$P_{fa\ OPT} = \frac{f/R}{f/R + e} \tag{7.10}$$

where $P_{fa\ OPT}$ is the digit element false alarm probability that yields the largest P_d^M and satisfies (7.9). If $P_{fa}(F) \leq \frac{f/R}{f/R+e}$ then

$$P_{fa\ OPT} = P_{fa}(F) \tag{7.11}$$

$P^M_{d\ OPT}$ and $P^M_{fa\ OPT}$ are found be replacing P_{fa} by $P_{fa\ OPT}$ in (7.5) and (7.6) respectively.

The procedure for choosing the receiver P_{fa} can now be summarized as follows:

- Choose $\overline{SNR}_{MIN}$ (i.e. the minimum $\overline{SNR}$ at which the receiver must operate) and F (the maximum tolerable P_{fa}^M.
- Calculate $P^M_{d\ OPT}$ for $\overline{SNR}_{MIN}$
- The detection threshold is found from $P_{fa\ OPT}$ by (6.4).

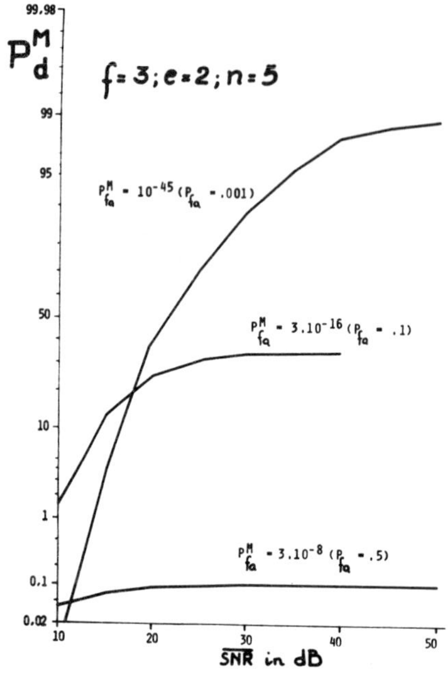

Fig. 19. Some performance curves.

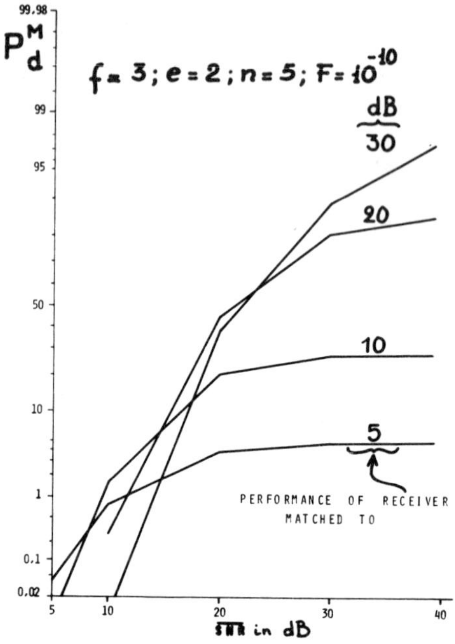

Fig. 20. Performance of receiver with matched P_{fa}.

This threshold choice will be the best possible at $\overline{SNR}_{MIN}$ and will yield a P_d^M that increases with $\overline{SNR}$.

An example is given in Fig. 20 for $f = 3$, $e = 2$, $n = 5$, $F = 10^{-10}$.

7.6 Uncertainty in the received frequencies

Unknown doppler and other reasons may cause the received frequency to be uncertain within certain frequency limits.

We thus assume that, due to this effect, the bandwidth of each of the digit element's receivers has to be N times larger than the bandwidth W of the original signal. To keep the same receiver performance will require a higher signal energy (for constant N_o); this increase can equally well be considered as a SNR loss. Consider the receiver in Fig. 21, hereafter called the split-band receiver (SBR). Each of its parallel branches contains a filter matched to a replica of the transmitted signal frequency shifted by an integer multiple of the original signal bandwidth W.

Together, the N receiver branches thus cover the bandwidth NW. (Remark: *The case of a received signal straddling two adjacent branches is not explicitely considered, since for tone pulses this corresponds to a relatively small (≃3 dB) SNR loss. One could consider a receiver with, say 2N or 3N branches, partially overlapping in frequency.)*

For the SBR to be equivalent to the original receiver, both must have the same P_d and P_{fa}. The P_d in each of the branches of the SBR has to be the same as that of the original receiver, whereas the P_{fa} of each of the branches of the SBR has to be approximately N times smaller.

Applying (4.15) gives the desired relation

$$\begin{bmatrix}\text{SNR loss due to SBR with}\\ \text{N channels in parallel}\end{bmatrix} = \log\left(1 - \frac{\log N}{\log P_{fa} - \log P_d}\right) \qquad (7.12)$$

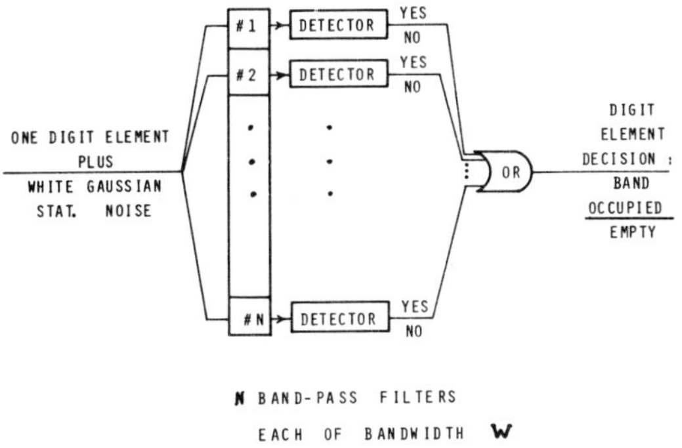

Fig. 21. Split-band receiver.

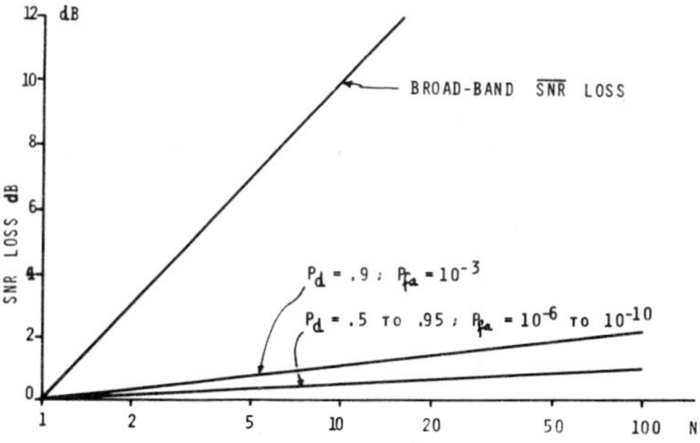

Fig. 22. SNR loss for split-band receiver.

This relation is plotted on Fig. 22 for selected values of P_d and P_{fa}. It can be seen that the loss is much less than the "broad-band" SNR loss (i.e. 10 log N). The SBR is not the optimal receiver, because it does not use all the available information, but it is believed to be reasonably close to the optimum.

8. CONCLUSION

Several detection models have been presented in this paper. They represent a biased sampling of the numerous target models available for detection, the bias being notably towards simplicity and sonar applications. The fundamental concepts used are of a general nature and may therefore be applied to a wide variety of models.

ACKNOWLEDGEMENTS. Many thanks are due to E.H. Hug and L. Meier, both of the SACLANT ASW Research Centre, for helpful suggestions concerning the manuscript.

REFERENCES

1. Slepian, D. On bandwidth. Proceedings IEEE 64, 1976: 292-300
2. Swerling, P. Survey of echo fluctuation impact on detection, NATO UNCLASSIFIED. In Vettori, G. and Hug, E. eds. Submarine Echo Properties, SACLANTCEN Conference Proceedings CP-18, Pt 2, NATO CONFIDENTIAL. La Spezia, Italy, SACLANT ASW Research Centre, 1976: pp. 13.1 - 13.45.
3. Hug, E.H. SACLANT ASW Research Centre, La Spezia, Italy, Private communication.
4. Van Trees, H.L. Detection, Estimation and Modulation Theory. New York, Wiley, 1968.
5. Picinbono, B. Introduction to detection and estimation. In *Present publication:* pp. 163-179. 1976.
6. Di Franco, J.V. and Rubin, W.L. Radar Detection. Englewood Cliffs, N.J., Prentice-Hall, 1968.
7. Papoulis, A. Probability, Random Variables and Stochastic Processes. New York, McGraw-Hill, 1965.
8. Swerling, P. Probability of detection for fluctuating targets. IRE Trans. Information Theory IT-6, 1960: 269-308.
9. Spek, G.A. van der. Detection of a distributed target. IEEE Trans. Aerospace & Electronic Systems AES-7, 1971: 922-931.
10. Swerling, P. Detection of fluctuating pulsed signals in the presence of noise. IRE Trans. Information Theory IT-3, 1957: 175-178.
11. Shwartz, M. Effects of signal fluctuations in the detection of pulsed signals in noise. IRE Trans. Information Theory IT-2, 1956: 66-71.

12. Heidbreder, G.R.H. and Mitchell, R.L. Detection probabilities for log-normally distributed signals. IEEE Trans. Aerospace & Electronic Systems AES-3, 1967: 5-13.
13. Heering, Ph. de. Theoretical analysis of a secure communication system. SACLANTCEN unpublished document. La Spezia, Italy, SACLANT ASW Research Centre, 1975.

DISCUSSION

Comment : J.W.R. GRIFFITHS

In the model, if the resolution cell is decreased in size there is a possibility of the original requirement of the target echo being composed of many scatterers being violated. What happens under these conditions?

Reply : P. DE HEERING

Yes, indeed, when resolution increases there comes a moment when the model is no longer valid; the condition for validity being notably that each resolution cell contains enough scatterers so that the target strength of each cell be Rayleigh.

Comment : L. PESCATORI

With reference to your example of a communication system, is the choice of the particular type of coding dictated solely by security requirements, or does in depend on propagation reasons as well? Finally, do you think that a general procedure could be established to design signal for information transmission via underwater links?

Reply : P. DE HEERING

In answer to your first question, the type of coding is dictated mainly by security requirements. With regard to the second, I do not think that a general procedure exists in the general case; it all depends on how well one knows the channel scattering function.

BAYES OPTIMAL VERSUS ESTIMATE AND PLUG ARRAY PROCESSOR PERFORMANCE WHEN THERE IS DIRECTIONAL UNCERTAINTY*

W. S. Hodgkiss
Naval Undersea Center, San Diego, California, U.S.A.

L. W. Nolte
Duke University, Durham, North Carolina, U.S.A.

ABSTRACT. Detection performance of four candidate receiver structures for the signal known except for direction (SKED) array problem is investigated. Included are the Bayes optimal detector, two estimate and plug structures, and a fixed estimate structure. Estimators considered are the maximum likelihood (ML) and maximum a posteriori (MAP).

1. INTRODUCTION

An appealing approach to an array detection problem where uncertain parameters exist is first to estimate these parameters, then plug them into the conditional likelihood ratio as if they were known exactly. When a "good" estimation scheme is used, the structure illustrated in Figure 1 appears to be operating in an optimal fashion. The underlying assumption is that there is a canonical "optimum" processor which consists of the parameters known solution (i.e. the conditional likelihood ratio, $\Lambda(\underline{z}|\underline{\theta}_1,\underline{\theta}_0)$). Should uncertain parameters be present under either or both hypotheses as denoted by the vectors $\underline{\theta}_1$ and $\underline{\theta}_0$, the processor must adapt to or learn these parameters.

Two problems which have received considerable attention in the literature are: (1) detecting the presence or absence of a Gaussian signal imbedded in a Gaussian noise field and (2) detecting the presence or absence of a known form signal imbedded in a

* This work was supported by the Office of Naval Research (Code 222).

G. Tacconi (ed.), Aspects of Signal Processing, Part 1, 203-212. *All Rights Reserved.*

Gaussian noise field. Solutions assuming all parameters known originally were derived by Bryn [1] for the first and Mermoz [2] for the second. In the consideration of these two problems when uncertain parameters exist, the processors of Bryn or Mermoz typically are assumed as implementations of the conditional likelihood ratio in the figure below.

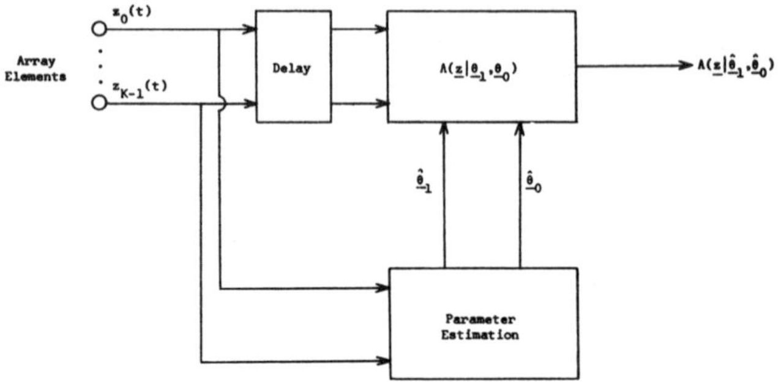

Figure 1. Estimate and Plug Array Processor.

Although the array processor structure in Figure 1 is intuitively appealing, it is not clear that piecing together techniques which may be optimal locally (i.e. a "good estimation scheme joined with a solution optimal when all parameters are known) will yield global optimality when the overall goal is good detection performance.

In this paper, performance of the processor which is to decide presence or absence of a signal known except for direction (SKED) in Gaussian noise which is independent from sensor to sensor will be discussed. ROC (receiver operating characteristic) curves for fou candidate receiver structures are compared--the Bayes optimal detector, two estimate and plug structures, and a fixed estimate structure. Estimators considered are the maximum likelihood (ML) and maximum a posteriori (MAP).

2. PROBLEM FORMULATION AND THE CONDITIONAL LIKELIHOOD RATIO

An array of K uniformly spaced elements on a straight line is assumed with the zeroth element being the right-most sensor. The processors will be asked to decide between the two mutually exclusive and exhaustive hypotheses H_1 that the time waveforms observed at the elements consist of signal plus noise and H_0 that they consist of noise alone.

A vector of real time waveforms is observed on the interval $(-T_0/2, T_0/2)$

$$\underline{z}(t) = [z_0(t),\ldots,z_{K-1}(t)]^T \quad (1)$$

where the subscript denotes the array element. Assuming the random processes are bandlimited to L $\omega_0/2\pi$ Hz. ($\omega_0 = 2\pi/T_0$), let

$$z_k(n) = \left(\frac{1}{T_0}\right)^{1/2} \int_{-T_0/2}^{T_0/2} z_k(t) \exp(-jn\omega_0 t)dt \quad (2)$$

$$\underline{z}(n) = [z_0(n),\ldots,z_{K-1}(n)]^T \quad (3)$$

and $$\underline{z} = [\underline{z}(0)^T,\ldots,\underline{z}(L)^T]^T. \quad (4)$$

In this way, the time waveforms observed at the K elements are mapped into a K (L+1) dimensional vector. Note that the Fourier coefficients for a single frequency index n and all K elements are grouped together.

When noise alone is observed, the array element outputs are sample functions of zero mean, stationary Gaussian random processes. The noise processes are assumed to be spatially uncorrelated (i.e., independent from sensor to sensor) with bandlimited power spectral density functions $N(\omega)$ which are assumed to be the same at each element. Choosing an observation period sufficiently long so that the function $N(x\omega_0)$ is relatively smooth with respect to increments in ω_0

$$p(\underline{z}|H_0) = \prod_{n=0}^{L} \left\{\frac{1}{\pi^K|\underline{Q}(n)|} \exp[-\underline{z}(n)^*\underline{Q}(n)^{-1}\underline{z}(n)]\right\} \quad (5)$$

where $$\underline{Q}(n)^{-1} = N^{-1}\underline{I} \quad (6)$$

and $$N = N(n\omega_0). \quad (7)$$

When signal plus noise is observed, the array element outputs consist of time delayed versions of a deterministic signal added to the noise processes described above. The bandlimited signal is represented by the Fourier coefficients $b_0(n)$ at the zeroth array element. Again, choosing an observation period sufficiently long so that the function $N(x\omega_0)$ is relatively smooth with respect to increments in ω_0

$$p(\underline{z}|\omega_0\tau,H_1) = \prod_{n=0}^{L} \left\{\frac{1}{\pi^K|\underline{Q}(n)|}\right.$$

$$\left.\cdot \exp[-(\underline{z}(n)-b_0(n)\underline{u}(n,\tau))^*\underline{Q}(n)^{-1}(\underline{z}(n)-b_0(n)\underline{u}(n,\tau))]\right\} \quad (8)$$

where $\underline{u}(n,\tau)^* = [1, \exp(jn\omega_0\tau), \ldots, \exp(j(K-1)n\omega_0\tau)]$. (9)
Uncertainty in the signal location is reflected in the parameter τ which is the time delay of the signal between adjacent elements.

The conditional likelihood ratio is by definition the ratio of (8) and (5)

$$\Lambda(\underline{z}|\omega_0\tau) \triangleq \frac{p(\underline{z}|\omega_0\tau,H_1)}{p(\underline{z}|H_0)} . \tag{10}$$

The expression in (10) can be broken into its single frequency constituents as follows:

$$\Lambda(\underline{z}|\omega_0\tau) = \prod_{n=0}^{L} \Lambda(\underline{z}(n)|\omega_0\tau) \tag{11}$$

where $\Lambda(\underline{z}(n)|\omega_0\tau) \triangleq \dfrac{p(\underline{z}(n)|\omega_0\tau,H_1)}{p(\underline{z}(n)|H_0)}$

$$= \exp[2\mathrm{Re}\{\underline{z}(n)^*\underline{Q}(n)^{-1}b_0(n)\underline{u}(n,\tau)\} - |b_0(n)|^2\underline{u}(n,\tau)^*\underline{Q}(n)^{-1}\underline{u}(n,\tau)]. \tag{12}$$

Making the appropriate substitutions from (5) to (9) into (12)

$$\Lambda(\underline{z}(n)|\omega_0\tau) = \exp[-\frac{K}{N}|b_0(n)|^2+2\,\mathrm{Re}\{\frac{1}{N}z_0(n)^*b_0(n)\}]$$

$$\cdot \exp[2\mathrm{Re}\sum_{\ell=1}^{K-1} A_\ell \cos(\ell n\omega_0\tau + B_\ell)] \tag{13}$$

where

$$A_\ell \cos B_\ell = \mathrm{Re}[G_\ell]$$
$$A_\ell \sin B_\ell = -\,\mathrm{Im}[G_\ell] \tag{14}$$

and

$$G_\ell \triangleq \frac{1}{N} z_\ell(n)^* b_0(n) . \tag{15}$$

3. DIRECTIONAL UNCERTAINTY

The array processor sees location uncertainty reflected in terms of an uncertain time delay of the directional signal source between adjacent elements. And, in turn, this corresponds to an uncertain phase delay in the frequency domain where the processing is actually carried out. Thus, location uncertainty will be summarized by the probability density function $p(\omega_0\tau)$. Specifically [3, 4]

$$p(\omega_0\tau) = \frac{L}{2\pi I_0(A_0)} \exp[A_0 \cos(L\omega_0\tau+B_0)] , \quad \frac{-\pi}{L} \le \omega_0\tau \le \frac{\pi}{L} \tag{16}$$

where $A_0 \ge 0, -\pi \le B_0 \le \pi$, and $I_0(\cdot)$ is the modified Bessel function of order zero. For all cases where performance is reported here, L = 8 and B_0 = 0. As Figure 2 indicates, varying the parameter A_0 from zero to infinity models a wide range of uncertainty from diffuse to highly localized. The array elements are assumed one half wavelength apart at frequency $L\omega_0/2\pi$ Hz. Thus, (16) corresponds to a location uncertainty over $\pm 90^0$ in physical angle from broadside to the array or $\pm\pi$ in phase at frequency $L\omega_0/2\pi$ Hz. The parameters A_0 and B_0 should not be confused with similarly denoted parameters in (13)-(14).

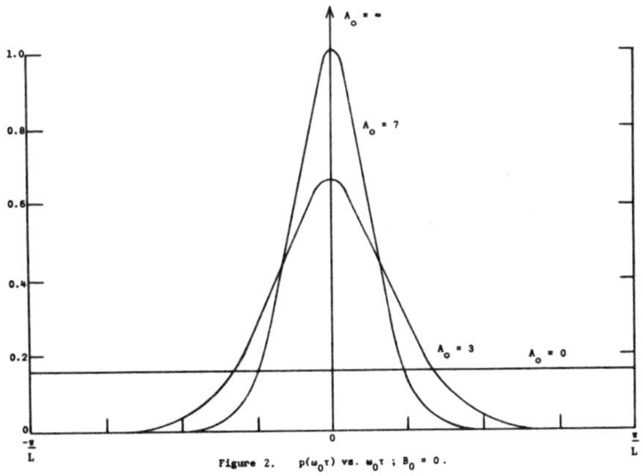

Figure 2. $p(\omega_0\tau)$ vs. $\omega_0\tau$; $B_0 = 0$.

4. PROCESSOR STRUCTURES

Four candidate receiver structures for the SKED problem will be considered. The environmental uncertainty in which each operates is described by the probability density function $p(\omega_0\tau)$.

With the Bayes optimal detector, the processor has a priori knowledge of the environmental uncertainty. The optimal test statistic is formed by averaging the conditional likelihood ratio over this knowledge of the uncertain parameter

$$\Lambda(\underline{z}) = \int_{\omega_0\tau} \{ \prod_{n=0}^{L} \Lambda(\underline{z}(n)|\omega_0\tau)\} \, p(\omega_0\tau)d\omega_0\tau. \tag{17}$$

The two estimate and plug processors have the structure illustrated in Figure 1. The maximum likelihood estimate is obtained by determining the global maximum of the function $p(\underline{z}|\omega_0\tau,H_1)$. Since $p(\underline{z}|H_0)$ is independent of the uncertain parameter, an equivalent computation is to determine the global maximum of the conditional likelihood ratio

$$\max_{\omega_0\tau} \Lambda(\underline{z}|\omega_0\tau) \rightarrow \hat{\omega_0\tau}_{ML}. \tag{18}$$

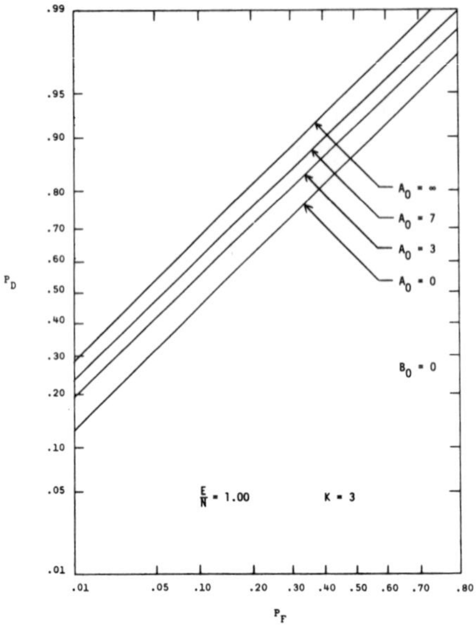

Figure 3. Performance of the Optimal SKED Array Processor.

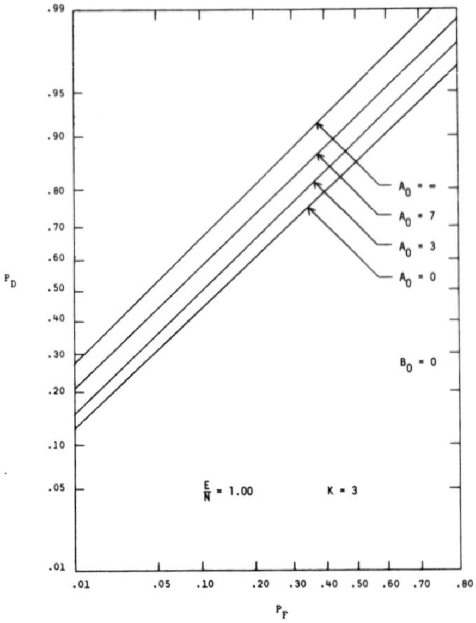

Figure 4. Performance of the Suboptimal SKED Array Processor. MAP Estimate

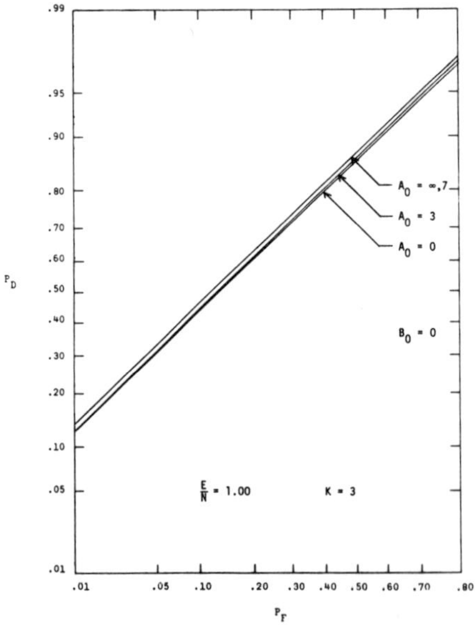

Figure 5. Performance of the Suboptimal SKED Array Processor. ML Estimate

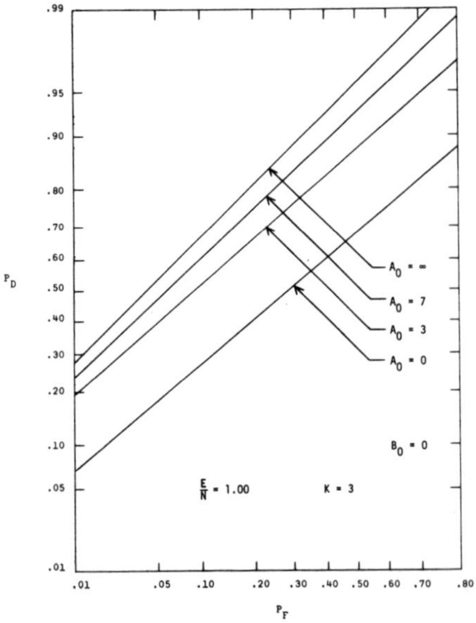

Figure 6. Performance of the Suboptimal SKED Array Processor. Fixed Estimate ($\omega_0\tau = 0$)

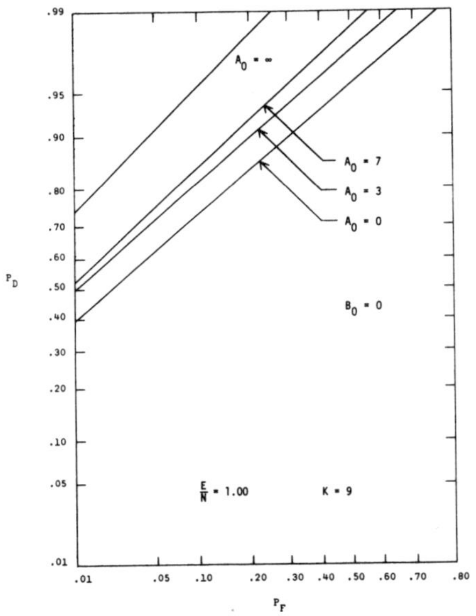

Figure 7. Performance of the Optimal SKED Array Processor.

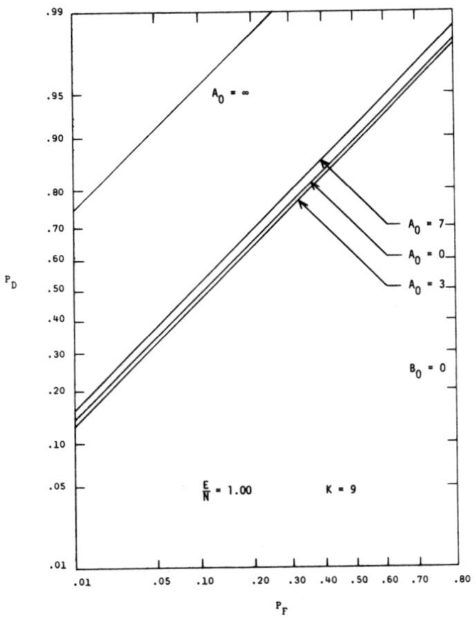

Figure 8. Performance of the Suboptimal SKED Array Processor. MAP Estimate

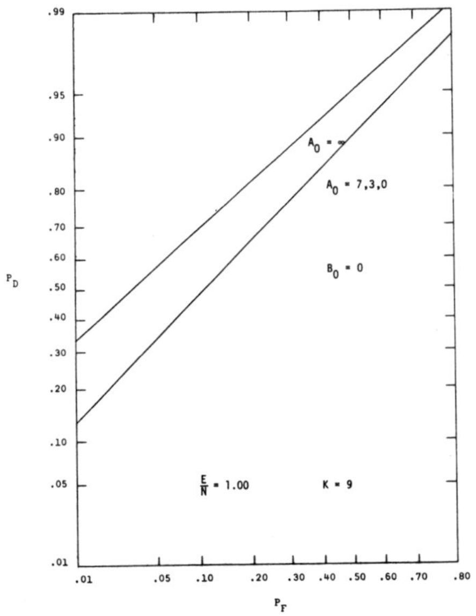

Figure 9. Performance of the Suboptimal SKED Array Processor. ML Estimate

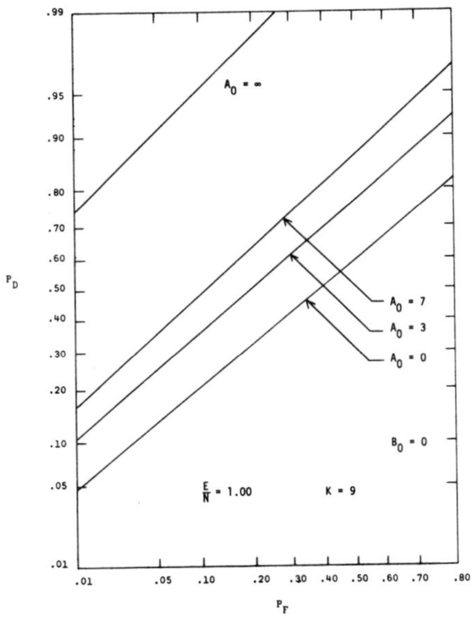

Figure 10. Performance of the Suboptimal SKED Array Processor. Fixed Estimate ($\omega_0\tau = 0$)

The maximum a posteriori estimate is determined by utilizing knowledge of the environmental uncertainty with Bayes' rule to form the a posteriori probability density function. The MAP estimate corresponds to the global maximum of the density $p(\omega_0\tau|\underline{z},H_1)$. Since $p(\underline{z}|H_0)$ is independent of the uncertain parameter, an equivalent computation is to determine the global maximum at the conditional likelihood ratio multiplied by the a priori knowledge

$$\max_{\omega_0\tau} \Lambda(\underline{z}|\omega_0\tau)\, p(\omega_0\tau) \rightarrow \hat{\omega_0\tau}_{MAP} . \tag{19}$$

The last processor to be considered also is cast in the estimate and plug structure. The parameter estimate, however, is not data dependent and is always set to zero ($\hat{\omega_0\tau} = 0$). Note that $\hat{\omega_0\tau} = 0$ corresponds to the peak of the environmental uncertainty probability density function.

5. PERFORMANCE

Detection performance in terms of the ROC (receiver operating characteristic) curve is presented in Figures 3-10 for the four array processor structures. Each ROC is labeled by the A_0 value corresponding to a particular level of environmental uncertainty. The array sizes investigated were for three and nine elements as denoted by the parameter K. A single frequency signal is assumed at $L\omega_0|2\pi$Hz. Its energy over the observation interval is given by $E = 2\underline{b}_0(L)^*\underline{b}_0(L)$. The Gaussian noise which is independent from sensor to sensor has a power spectrum which is white and bandlimited to $L\omega_0|2\pi$Hz. Its spectral height is denoted by N.

Figures 3 and 7 correspond to the Bayes optimal detector for the SKED problem. Note that increasing location uncertainty ($A_0 \rightarrow 0$) leads to a greater degradation in performance the larger the array size.

Performance of the suboptimal SKED array processor which utilizes the MAP estimate is reported in Figures 4 and 8. The three element ROC is similar to that of the optimal detector, but a dramatic performance loss from optimal has occurred for the nine element array processor.

Figures 5 and 9 correspond to the maximum likelihood, estimate and plug structure. No a priori knowledge is available to the processor. When k = 3, the ROC's for all levels of environmental uncertainty cluster about the $A_0 = 0$ ROC of the optimal detector. Performance for the nine element array processor again shows a dramatic drop from that of the optimal detector.

Performance of the suboptimal array processor which always plugs a fixed estimate ($\hat{\omega_0\tau} = 0$) into the conditional likelihood ratio is shown in Figures 6 and 10. The resulting processor is a realization of that derived by Mermoz [2] with a front end consisting of a beamformer looking broadside to the array. For both the three and nine element processors, significant performance degradation is suffered from optimal for all levels of uncertainty other than $A_0 = \infty$. These ROC's give an indication of how sensitive performance is to a mismatch between an estimate of signal source location and the actual parameter value. Additional work on array processor sensitivity to mismatch has been pursued by Cox [5].

6. SUMMARY

This paper has investigated the performance of four candidate receiver structures for the signal known except for direction (SKED) array problem. Included were the Bayes optimal detector, two estimate and plug structures, and a fixed estimate structure. Estimators considered were the maximum likelihood (ML) and maximum a posteriori (MAP).

Although an estimate and plug approach to an array detection problem where uncertain parameters exist is appealing, it has been shown that performance on the ROC curve can suffer greatly when such an ad hoc structure is implemented. The effects of a suboptimal configuration were shown to be significantly more severe the larger the array size.

REFERENCES

1. F. Bryn, "Optimum Signal Processing of Three-Dimensional Arrays Operating on Gaussian Signals and Noise," J. Acoust. Soc. Am., Vol. 34, No. 3, pp. 289-297, 1962.
2. H. Mermoz, "Filtrage Adapte et Utilization Optimale d'une Antenne, "Proceedings of the NATO Advanced Study Institute on Signal Processing with Emphasis on Underwater Acoustics, Grenoble, France, 14-26 September, 1964; results are summarized in: C. W. Horton, Signal Processing of Underwater Acoustic Waves, Washington: U.S. Government Printing Office, 1969.
3. M. A. Gallop, and L. W. Nolte, "Bayesian Detection of Targets of Unknown Location," IEEE Trans. on Aerospace and Electronic Systems, Vol. AES-10, No. 4, pp. 429-435, 1974.
4. W. S. Hodgkiss and L. W. Nolte, "Optimal Array Processor Performance Tradeoffs Under Directional Uncertainty," IEEE Trans. on Aerospace and Electronic Systems, Vol. AES-12, No. 5, 1976.

5. H. Cox, "Resolving Power and Sensitivity to Mismatch of Optimum Array Processors," J. Acoust. Soc. Am., Vol. 34, No. 3, pp. 771-785, 1973.

DISCUSSION

Comment : P.G. CABLE

Since the ML and MAP estimators converge in probability to the quantities being estimated as the number of samples used in the estimates grows without limit, should not the two estimate-and-plug processors approach the performance of the Bayes processor?

Reply : W.S. HODGKISS

The answer to this question depends upon which samples are being increased. More time samples (while leaving K fixed) will result in suboptimal processor performance approaching that of the Bayesian processor. An increase in the number of spatial samples (i.e. array size, K) results in a widening gap between optimal and suboptimal processor performance. Although the estimate of signal source location is getting better in this second case, the conditional likelihood ratio (or space-time matched filter) exhibits a sensitivity to mismatch between the true and estimated parameter value that gets worse even faster.

Comment : C.N. PRYOR

Can you describe the structure of the Bayes optimal processor in the $A_0 = 0$ case, and whether it simply did a non-coherent combination of the element outputs?

Reply : W.S. HODGKISS

No, the structure cannot be determined without solving an integral that is intractable above about two elements.

A SEQUENTIAL LIKELIHOOD RATIO DETECTION SYSTEM*

Hugh A. Reeder

Tracor, Inc., Austin, Texas, U.S.A.

ABSTRACT. A general framework for a computer system that accepts and analyzes the large quantity of data generated by a modern sonar system has been developed. The system combines a simple tracking algorithm with a statistical decision test based on Wald's Sequential Likelihood Ratio (SLR) procedure. The computer algorithms have been applied to both active and passive sonar systems and to combinations of systems. Performance results and a hardware implementation are discussed to illustrate the SLR processor's capabilities.

1. STATEMENT OF THE PROBLEM

Two of the major problems faced by modern sonar systems are the consolidation and presentation of a large quantity of data and the ability of the sonar operator to effectively assimilate and process these data. Potentially, a modern sonar system is capable of delivering thousands of channels of information to the operator who in turn, even in an alert state, cannot process all of this information. Moreover, operators do not typically perform in an alert manner when required to search for extended periods of time, especially when the incidence of contacts is low. The result of this is that targets may go undetected for longer than necessary, and when detected, the resultant time available for classification and tracking is reduced--possibly

* This work has been sponsored by the Exploratory Development Office of the U. S. Naval Sea Systems Command (NAVSEA 06H1).

G. Tacconi (ed.), Aspects of Signal Processing, Part 1, 213-220.

to an extent that seriously degrades the ASW system's performance. Our approach to the solution of this problem is to develop a computer processing system which can handle the vast quantity of data and in so doing operate in a near optimum manner by virtue of a ping-to-ping integration and tracking algorithm.

2. DESCRIPTION OF THE SLR PROCESSOR

The fundamental problem of statistical decision theory is that of choosing one of several possible hypotheses by utilizing information gained from the measurement of some quantity. A great deal of generality can be included in defining the algorithm to optimally carry out this procedure.

We specialize the problem to a choice between two simple hypotheses: H_o that a given observation is noise, and H_1 that it is signal. This type of hypothesis testing has been treated extensively in the literature [1,2]. A functional known as the likelihood ratio has been shown to play a central role. The likelihood ratio L(x) is defined by

$$L(x) \equiv p_1(x|H_1)/p_o(x|H_o) \; ,$$

where $p_1(x|H_1)$ and $p_o(x|H_o)$ are the probability density functions associated with the hypotheses H_1 and H_o, respectively. Broadly, the common statistical decision tests use this likelihood ratio with variations based on the amount of other information available, such as prior probabilities and costs of wrong decisions. For this paper we will consider only one test--the sequential likelihood ratio test [3] that makes no assumptions about prior probabilities or costs of incorrect decisions. It does require the fixing of probabilities of wrong decisions.

Typically we have a succession of echo returns or passive updates resulting in a sequence--$\underline{x} = (x_1, x_2, \ldots, x_n)$--of samples. A likelihood ratio is formed by using the joint probability density functions. We assume that the samples in the sequence are sufficiently separated in time that they are effectively independent; therefore, the joint probability density function is a product of individual probability densities. It is advantageous to use the logarithm of L(x). In many cases the log likelyhood ratio, $\ell(x)$, may be represented by a straight line in the region of interest. The joint log likelyhood ratio, $\ell(\underline{x})$, is determined by a simple summation. Since the logarithm is a monotonically increasing function, order relations for thresholding are preserved.

It is desired for the hypothesis H_1 to be any signal; as a practical matter we replace H_1 by the hypothesis that the signal-plus-noise samples are from a population with signal-to-noise ratio ρ_o, called the design signal-to-noise ratio. In the sequential test this effectively provides 0.5 probability of detection or more for sample sequences from population with signal-to-noise ratios of $\rho' \leq \rho < \infty$ with $\rho' < \rho_o$ [3].

Two log likelihood ratio equations often used are

$$\ell(x) = \rho_o\, x/(1+\rho_o) - M \ln (1+\rho_o)$$

for gamma distributed outputs, e.g., Gaussian input with square-law detection and averaging for M independent samples, and

$$\ell(x) \approx \rho_o\, x - (\rho_o^2/2 + \tfrac{1}{2} \ln 2\pi\, \rho_o)$$

for Rayleigh-Rice outputs, e.g., sine wave plus Gaussian noise followed by an envelope detector. These log likelihood ratio equations have proved to be adequate in many cases.

The sequential test uses two thresholds, T_L and T_D, based on the errors of incorrect decision. These may be determined by the very good approximations given in [3]. If the joint likelihood ratio is below the lower threshold, T_L, the null hypothesis, H_0, that the sequence is noise, is accepted and the processing stops. If the ratio is above T_D, the alternate hypothesis, H_1, that the sequence is signal, is accepted. In the original formulation by Wald the testing would stop also for this latter case; however, in our formulation the testing continues but a maximum value is imposed on the log likelihood ratio. If the ratio is between T_L and T_D, the testing continues without either hypothesis being accepted.

In sonar applications one does not know uniquely how to make a single "next" observation. To solve this problem a simple tracking algorithm was developed and multiple linkages allowed. The tracking algorithm may be based on simple linear extrapolation of the last two observed sample coordinates, e.g., range, bearing, and Doppler, or on smoothed estimates using weighted averages of the observed sample coordinates. In any case a tracking window that varies with the number of observed samples is established about the predicted sample coordinates. The window sizes are based on expected measurement errors and target maneuverability. The object of the tracking algorithm is to eliminate obviously incorrect linkages rather than obtain precise track parameters.

To account for multiple linkages, the joint likelihood ratio must be altered to

$$L'(\underline{x}) = p_1(\underline{x}|H_1)P_1/p_0(\underline{x}|H_0)P_0 \quad ,$$

where P_1 is the probability that given linkage is a target under H_1 and P_0 is a probability that it is noise under H_0. We assume only one target is in any tracking window and its position is uniformly distributed in the window. If the window contains N resolution cells under H_1, the probability P_1 is then equal to 1/N and P_0 is equal to 1.0. Therefore,

$$L'(\underline{x}) = L(\underline{x})/N .$$

When the likelihood ratio is large, it may combine with a noise sample to form a significant, but spurious, track. Heuristically, to prevent this situation, large likelihood ratios are allowed to link only with the largest likelihood ratio in the tracking window.

The SLR processor has been implemented on large scale digital computers and a minicomputer system. The basic information is contained after each ping in a track file. This file contains the predicted location of each track, the last observed position, a track window index, and the log likelihood associated with the track. When additional data becomes available from the signal processor, it is normalized in time and space and then thresholded. This thresholding eliminates samples that are most likely of no interest. These thresholded data are first used to create tracks based on a single update, allowing tracks to begin on any update. Secondly, the locations of these data are compared with predicted locations contained in the track file. If an observed location and a predicted location agree within the tolerance defined by the window index, a linkage is considered. A new joint log likelihood ratio is calculated using the previous and new log likelihood ratios and the track correction. If the results pass the lower decision test, a new track unit is created and the predicted location updated. Independent of this process, the previous tracks are extrapolated and their log likelihood ratios degraded. If the resulting ratios pass the lower decision test, the track is stored in the new track file, allowing strong tracks to continue when a few track samples are missed. Since one track may be reproduced by different program sections, the new track file is searched and duplicates eliminated. Finally, an output display is generated from the new track file by subjecting the log likelihood ratios to the upper decision threshold test. For the next update the previous present track file is used as a track history. Using this process it is only necessary to retain information from the last update and the accumulated results on the present update--reducing the computer memory demands.

This general structure has been used to process data from an active low Doppler processor with data indexed by bearing and range; an active high Doppler processor indexed by bearing, range, and Doppler; a passive broadband processor indexed by bearing; and a passive spectrum analyzer processor indexed by bearing and frequency.

It is possible that two or more sonar processors may be in operation at the same time. Typically, a broadband and spectrum analyzer processor may be operated simultaneously to cover different types of target signals. This information is displayed independently and combined by an observer. Since the SLR processor generates likelihood ratios, the different processor outputs have been transformed into a common space and may be combined, giving a joint measure of the likelihood that a particular search direction contains a target. This joint measure may then be used to alert the operator, who in turn examines the detailed output for further classification and localization. If independence is assumed between channels, the joint measure is easily calculated by

$$\ell_J(\underline{x},\underline{y}) = \max[\ell_{SA}(\underline{x}) + \ell_{BB}(\underline{y}), \ell_{SA}(\underline{x}), \ell_{BB}(\underline{y})] \ ,$$

where $\ell_{SA}(\underline{x})$ is the output of the spectrum analyzer SLR, and
$\ell_{BB}(\underline{y})$ is the output of the broadband SLR.

The maximum over the combined and individual channels is taken to maintain detectability when only one channel is effective in detecting the target.

3. PERFORMANCE RESULTS

Such a combined system has been implemented in a hardware laboratory test bed. The equipment consists of an analog tape drive for 12 channels of data input, a spectrum analyzer, broadband detectors, analog-to-digital interface, two Data General 2/10 minicomputers each with 16K words of core, a 131K word disc, CRT displays, and a nine track digital magnetic tape to record data output. This system accomplishes real-time normalization and SLR processing for 12 beams of spectrum analyzed and broadband data. In addition, it combines their outputs and generates various displays. Typically, about one thousand tracks are processed each update although up to 4000 may be processed.

The principal performance results have been obtained from simulated data that consists of noise generator outputs with injected sine waves to simulate tonals and scaled background noise to simulate broadband energy. A variety of input signal-to-noise ratios for both tonals and broadband were generated and processed by the system. As an example of the performance of a single channel processor consider Fig. 1. This shows the probability of exceeding a decision threshold (probability of accepting noise as signal-plus-noise is 10^{-3}) as a function of input signal-to-noise ratio and SLR update number (2, 5, and 10 updates). Also shown are the results of non-SLR or conventionally processed data. The results are typical for SLR processed

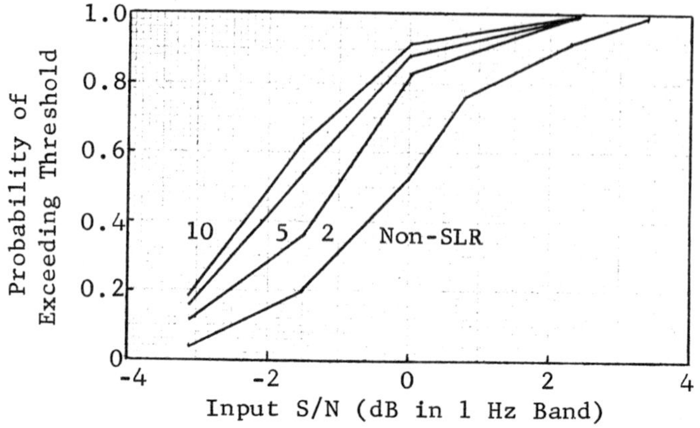

Fig. 1. Performance of SLR and non-SLR processed data for spectrum analyzer data ($P_{FA} = 10^{-3}$).

data--most of the processing gain is obtained in a few updates. The amount of gain is limited by the internal thresholding causing missing track samples and the multiple linkage track corrections. This is the cost of automatic tracking of dynamic targets.

As an example of combined processing, consider Fig. 2 which shows the probability of the combined processor output (CP) crossing a decision threshold after five updates for a constant tonal input signal-to-noise ratio and varying broadband input signal-to-noise ratio. Also shown are the results for the

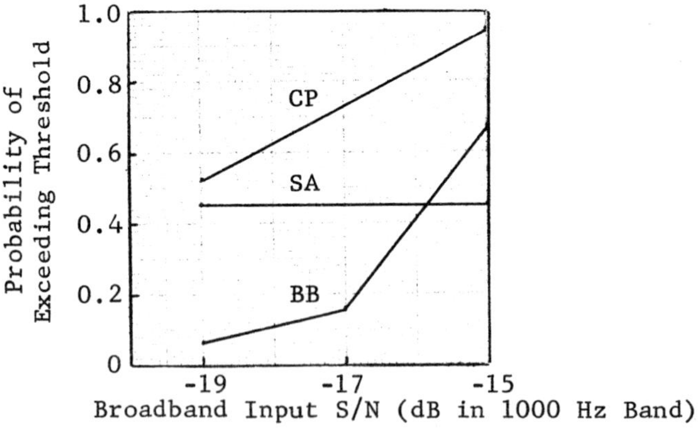

Fig. 2. Performance of combined processor after 5 updates for a constant tonal input signal-to-noise ratio of 0 dB (1 Hz band) ($P_{FA} = 10^{-3}$).

single channel SLR processor results for the spectrum analyzer (SA) SLR processor and the broadband (BB) SLR processor. The combined processor clearly dominates the individual channels which in turn dominate the conventionally processed data. Overall, significant gains may be achieved by combined processing when marginal signals are present at the input to each processor.

4. SUMMARY

The SLR algorithms have provided consistent and uniform automatic search performance on quantities of sonar data per unit time that would exceed the ability of an operator to perform undegraded search. They provide a means of substantially reducing (nominally a factor of 1000:1) the data rate that is to be presented to a search operator with no significant loss in detectability. Also, the algorithms exhibit a desirable robustness to deviations in input statistics.

Finally, the Sequential Likelihood Ratio processor, with its decision and tracking procedures, operates much like an alert operator but without the variability of an operator who is susceptible to fatigue, subjectivity, boredom, poor training, and a host of other deterrents to ideal, time-invariant detection performance. More important than this perhaps, is the fact that the information handling capacity of this computer process is far in excess of that of the operator and is also subject to expansion as computer technology improves whereas the capacity of the operator is unlikely to be expanded or improved by any significant amount in spite of advances in display technology.

REFERENCES

1. Carl Helstrom, Statistical Theory of Signal Detection, Pergamon Press, New York, 1960.
2. John C. Hancock and Paul A. Wintz, Signal Detection Theory, McGraw Hill Book Company, New York.
3. A. Wald, Sequential Analysis, John Wiley & Sons, Inc., New York, 1948.

DISCUSSION

Comment : J.N. MAKSYM

Can you comment on the relative effectiveness of various pulse types when used with your tracking system; in particular, did you find FM slide more effective than pulses yielding high resolution in doppler? Were you able to use high doppler resolution in doppler for tracking.

Reply : H.A. REEDER

For the systems studied, the FM slide gave better performance due to better processing gain and range resolution. The doppler system studied did not have high resolution; therefore it was not possible to use doppler effectively in tracking. With a high-resolution doppler sonar it would be advantageous to use doppler in tracking.

Comment : C. VAN SCHOONEVELD

Could you give us an idea of the processing gain of your tracking system.

Reply : H.A. REEDER

The gain realizable is a complex function of several variables, such as the design signal-to-noise ratio, track window size, initial threshold, and up-date number. In example seven [Fig 1], 0.5 probability of detection is achieved at a little over 2 dB, less input signal-to-noise ratio, with the SLR processor after ten up-dates than with the conventional non-SLR processor.

ESTIMATION OF THE TIME POSITION OF A PULSE IN NOISE

A. R. Pratt, B.Sc., Ph.D.,

Department of Electronic and Electrical Engineering,
University of Technology, Loughborough, U.K.

1. INTRODUCTION

The work reported in this paper was motivated by interest in delay or range estimation made on sonar displays of limited dynamic range. Such displays have characteristics which approximate to those of infinite clippers. This paper considers the problem of delay estimation after the received signal has been processed by a clipper. It is shown that an estimation procedure may be based on the mid-point between an upward and the next downward transition of a decision threshold by the signal and noise waveform. A suitable method for the mathematical analysis of this problem has been found which permits direct computation of the variance of the local estimate of delay and which shows the estimate to be unbiased given certain conditions. Some of the mathematical details are presented in the appendix. Finally, a comparison is made between the variance of this estimator and the Cramér-Ráo lower bound.

2. SYSTEM MODEL

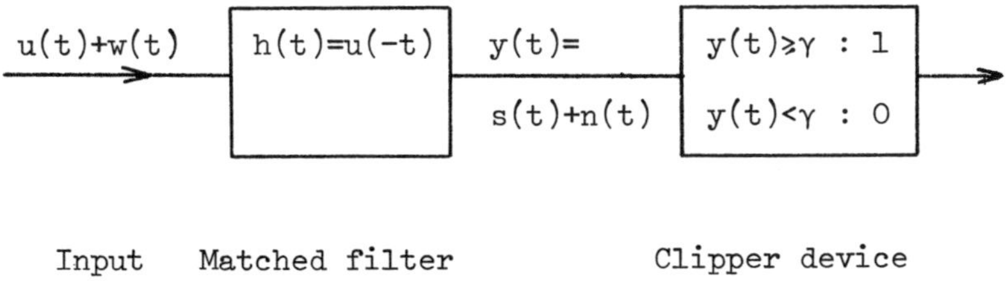

Figure 1 : System Model and Clipper Action.

G. Tacconi (ed.), Aspects of Signal Processing, Part 1, 221-228. *All Rights Reserved.*

The received signal and white noise waveform, $u(t)+w(t)$, is passed through a matched filter as shown in figure 1. The resulting waveform, $y(t)$, is processed by a clipper. The transitions of the output of the clipper at times t_1 and t_2 coincide in time with the transitions by $y(t)$ of the decision level γ. Positional estimates are made of the peaks of $y(t)$ by computing the mid-points between the transitions of the clipper output.

In order to compute the variance of the mid-point, t_m, between the transitions at t_1 and t_2 it is necessary to find the probability density function for t_m with respect to some datum which we choose as the peak of $s(t)$. The required distribution may be obtained from a rotational transformation of the joint probability density function, $P_{+-}(t_1,t_2,|\gamma,s)$, for an upward transition of γ at time t_1 and a downward transition of γ at t_2 by the signal and noise process, $y(t)$. Mathematical details concerning the derivation of $P_{+-}(t_1,t_2|\gamma,s)$ are to be found in the appendix.

We now perform the rotational transformation in the t_1, t_2 plane to obtain a joint density function for the mid-point time, t_m, defined from the t_1, t_2, datum and the time delay, t_d, between t_1 and t_2. The Jacobian of this transformation is unity so that:-

$$P(t_m,\ t_d|\gamma,s)=P_{+-}(t_1,t_2|\gamma,s) \qquad \text{...(2.1)}$$

where $t_m=\frac{1}{2}(t_1+t_2)$, $t_d=t_2-t_1$

The density function of the mid-point between an upward and downward transition is easily obtained by integrating over all possible values of time delay t_d,

$$P(t_m|\gamma,s) = \int_0^\infty P(t_m,t_d|\gamma,s)\ dt_d \qquad \text{...(2.2)}$$

It is thus possible to find the variance of t_m in the usual way

$$\sigma^2_{t_m} = \int_{-\infty}^{\infty} P(t_m|\gamma,s)\ t_m^{\ 2}\ dt_m - E(t_m)^2 \qquad \text{...(2.3)}$$

It is also shown in the appendix that $P(t_m|\gamma,s)$ is symmetrical about the axis $t_m=0$ and hence has zero mean. Thus, the estimate is unbiased.

3. RESULTS AND DISCUSSION

Computer programs have been written to calculate $P(t_m|\gamma,s)$ and hence the variance of the mid-point estimate. A Gaussian shape has been used for the signal at the output of the matched filter. This shape also corresponds to the autocorrelation function of the noise process $n(t)$. Specifically:-

$$s(t)= A \exp(-0.125t^2) \quad , \quad \text{and } E(n(t).n(t+\tau))= \exp(-0.125\tau^2)$$

A problem arises in attempting to compute the variance of the mid-point estimate, t_m, because at low signal to noise ratios, the signal and noise waveform does not always cross the threshold. This results in the area under the distribution $P(t_m|\gamma,s)$ being less than unity for signal to noise ratios less than 12dB for the threshold used ($2\sqrt{2}\sigma_n$). A plausible solution to this difficulty would be to scale $P(t_m^n|\gamma,s)$ so that the area under this probability distribution becomes unity. This corresponds to making estimates conditional on a signal present decision.

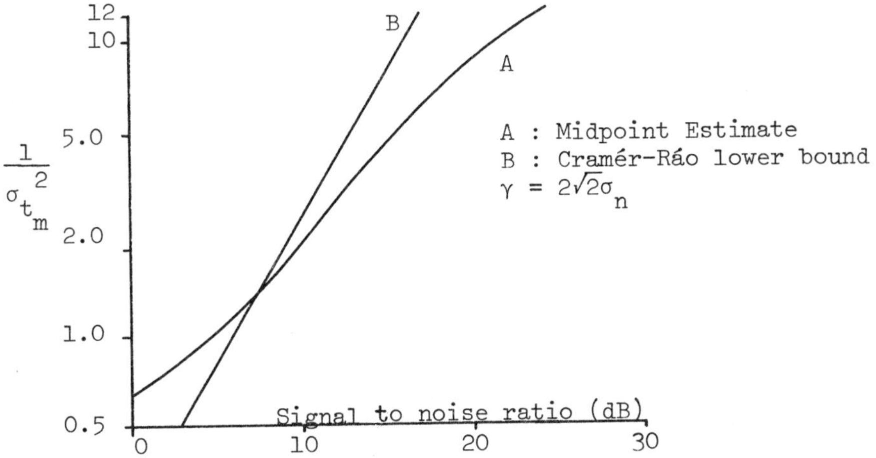

Figure 2 : Results - Midpoint Estimate and Cramér-Ráo Lower Bound.

As can be seen from the results plotted in figure 2(curve A), this predicts a variance lower than the Cramér-Ráo lower bound (curve B) at lower signal to noise ratios. Two mechanisms can be identified which might account for this behaviour. Firstly for the threshold level used, the signal and noise waveform does not always cross threshold for signal to noise ratios less than 12dB. When a threshold crossing does occur for low signal to noise ratios, it will occur near the peak of $y(t)$. The use of the peaks of $y(t)$ as estimators of the signal position corresponds to maximum likelihood estimation for which the Cramér-Ráo lower bound is normally a good approximation. Thus, as the system tends to estimate on the peaks of $y(t)$ at low signal to noise ratios, we may expect the clipped system behaviour to approach the lower bound. Supporting evidence for this hypothesis is that the probability with which y(t) crosses the decision level, γ, at the intersection of curves A and B is approximately 0.4.

The second factor which has an influence at low signal to nois ratios can be most simply explained by considering the signal to be absent. The remaining noise process, n(t), occasionally crosses the threshold producing false alarms and these will be uniformly distributed over the interval during which we seek the signal. The variance of these false alarm estimates is a function only of the length of this observation interval. This maximum variance appropriate to the observation interval used in computing the results in figure 2 corresponds to a value of $1/\sigma^2_{tm}$ of 0.3. Curve A must flatten to this level at low signal to noise ratios.

At high signal to noise ratios, insufficient use is made of the signal energy present since the threshold level, γ, is lower than the signal peak. We may, therefore, expect the variance of estimates made at high signal to noise ratio to have a larger variance than the Cramér-Ráo lower bound. In fact the results indicate that the variances diverge and that the estimator is not asymptotically efficient. Results, not shown here, for other thresholds are similar in shape to curve A but are displaced up or down parallel to the lower bound.

In conclusion this research indicates that acceptable estimation performance is available from clipper systems and presumably from P.P.I. or B-scan sonar displays at low signal to noise ratios. However, at high signal to noise ratios relatively poor estimation performance is likely to be available. Improved performance under these conditions could be obtained by using several different threshold levels, making position estimates from the highest threshold level crossed.

4. MATHEMATICAL APPENDIX

For simplicity we assume a baseband signal and normally distributed noise. From the system model of figure 1 we define:

$$y(t)=s(t)+n(t) \quad , \quad E(n(t).n(t+\tau))=s(\tau) \qquad \ldots(4.1)$$

Any time waveform and in particular the noise waveform, may be represented by the infinite Taylor series expansion:-

$$n(t+\Delta t)=n(t)+n'(t)\Delta t+n''(t)\frac{(\Delta t)^2}{2!}+n'''(t)\frac{(\Delta t)^3}{3!}+\ldots \qquad \ldots(4.2)$$

where n'(t) is the first derivative, with respect to time, of n(t) at t, n''(t) the second derivative, and so on. The statistics of the noise waveform, $n(t)$ may thus be represented by the joint probability distribution of all the derivatives of $n(t)$ at one point in time. The time varying probability distributions of the output of the matched filter, $y(t)$, may be found using equations (4.1) and (4.2) provided that both $s(t)$ and the probability distribution of $n(t)$ are known.

The probability density, $P_{+-}(t_1,t_2|\gamma,s)$, that $y(t)$ will pass through the level γ in the interval t_1 to t_1+dt_1 with positive slope and through the level γ in the interval t_2 to t_2+dt_2 with negative slope is a generalisation of a result obtained by S.O.Rice (2)(3) using the above signal and noise model(1).

$$P_{+-}(t_1,t_2|\gamma,s) = -\int_0^{\infty}\int_{-\infty}^{0} y_1' y_2' P(\gamma-s_1,\gamma-s_2,y_1'-s_1',y_2'-s_2')\,dy_1'\,dy_2' \quad \ldots(4.3)$$

where $P(n_1,n_2,n_1',n_2')$ is the probability density function for the four random variables:-

$$n_i=y_i-s_i=y(t_i)-s(t_i)\ ,\quad E(y(t_i))=s(t_i)\ ;\quad i=1,2$$

$$n_i'=y_i'-s_i'=y'(t)-s'(t_i)\ ,\quad E(y'(t_i))=s'(t_i)\ ;\quad i=1,2$$

In many situations of interest $P(n_1,n_2,n_1',n_2')$ will be jointly normal.

$$P(X)=\frac{1}{(2\pi)^2}\cdot\frac{1}{|K|^{\frac{1}{2}}}\cdot\exp\{-\tfrac{1}{2}X^TK^{-1}X\} \quad \ldots(4.4)$$

where $X^T=(n_1,n_1',n_2',n_2)$, the covariance matrix $K=E(XX^T)$, and $|K|$ is the determinant of the covariance matrix.

Without loss of generality, we may assume that the matched filter scales both the signal and noise waveform so that $y(t)$ has unit variance. In this case, the energy signal to noise ratio becomes the maximum value of $s^2(t)$.

Writing K_{rs} as the cofactor of element rs in K, substituting equation 4.4 in 4.3 and using the following symbol assignments, we may show after considerable manipulation:-

$$P_{+-}(t_1,t_2|\gamma,s)=\frac{A\cdot m^2}{(1-r^2)^2K_{22}^{\ 2}}\cdot J(h_1,-h_2,-r) \quad \ldots(4.5)$$

where: $m=E(n_1n_2)$, $m'=E(n_1n_2')$, $r=K_{23}/K_{22}$

$$A=\frac{1}{2\pi}\left[\frac{1-r^2}{|K|}\right]^{\frac{1}{2}}\exp\left[\frac{-1}{2(1-m^2)}\left[(\gamma-s_1)^2+(\gamma-s_2)^2-2m(\gamma-s_1)(\gamma-s_2)\right]\right]$$

$$a_1=\frac{m'}{1-m^2}\left[\gamma-s_2-m(\gamma-s_1)\right]\quad ;\ a_2=\frac{m'}{1-m^2}\left[m(\gamma-s_2)-\gamma+s_1\right]$$

$$z_i=\left[\frac{(1-r^2)K_{22}}{|K|}\right]^{\frac{1}{2}}\quad(y_i'-s_i'+a_i)\quad,\quad i=1,2$$

$$h_i=\left[\frac{(1-r^2)K_{22}}{|K|}\right]^{\frac{1}{2}}\quad(a_i-s_i)\quad,\quad i=1,2$$

$$J(h_1,h_2,r)=\int_{h_1}^{\infty}\int_{h_2}^{\infty}\frac{(z_1-h_1)(z_2-h_2)}{2\pi(1-r^2)^{\frac{1}{2}}}\exp\left[-\frac{(z_1^2+z_2^2-2rz_1z_2)}{2(1-r^2)^{\frac{1}{2}}}\right]dz_1dz_2$$

$J(h_1,h_2,r)$ is a standard integral(3) which may be found from tables of the volume of the bivariate normal distribution, $K(h_1,h_2,r)$.

To show that estimates obtained by the method set out above are unbiased, it will be sufficient to show that $P(t_m,t_d|\gamma,s)$ is invariant for a sign change in t_m. A relationship between t_1,t_2, t_m and t_d is (see equation 2.1):

$$t_1=\tfrac{1}{2}(t_m-t_d)\quad,\quad t_2=\tfrac{1}{2}(t_m+t_d)$$

We define two new time variables t_1' and t_2' which corresponds to t_1 and t_2 respectively but with negative t_m:-

$$t_1'=\tfrac{1}{2}(-t_m-t_d)=-t_2\quad,\quad t_2'=\tfrac{1}{2}(-t_m+t_d)=-t_1\qquad\ldots(4.6)$$

Thus the conditions for $P(t_m,t_d|\gamma,s)$ to be symmetrical about the $t_m=0$ axis may be restated in terms of $P_{+-}(\cdot)$ as follows:

$$P_{+-}(-t_2,-t_1|\gamma,s)=P_{+-}(t_1,t_2|\gamma,s)$$

The first step in the proof is to note that the covariance matrix, its determinant, cofactors and elements are dependent only upon t_d provided that the white noise process, $w(t)$, is stationary. These are unchanged by variations in t_m and will be subsequently ignored

Considering now the contributions of the signal, $s(t)$, to equation 4.5 under the transformation $t_1 \to t_1'$ and $t_2 \to t_2'$ recalling that $s(t)$ is the signal contribution to a matched filter output and therefore even, we find that

$$s(t_1) \to s(t_1') = s(t_2) \quad , \quad s(t_2) \to s(t_2') = s(t_1)$$

$$s'(t_1) \to s'(t_1') = -s'(t_2) \quad , \quad s'(t_2) \to s'(t_2') = -s'(t_1) \qquad \ldots (4.7)$$

Applying these relationships to the 'A' term in equation 4.5, we observe that s_1 and s_2 interchange positions only. The 'A' term is therefore unchanged. If we now consider $J(h_1, -h_2, -r)$ and apply the relationships in equation 4.7 to a_1, a_2 and h_1, h_2, we see that a_1 becomes $(-a_2)$; a_2 becomes $(-a_1)$; h_1 becomes $(-h_2)$ and finally h_2 becomes $(-h_1)$. Thus, the standard integral $J(h_1, -h_2, -r)$ changes to $J(-h_2, h_1, -r)$ as a result of a sign change in t_m. Examination of equation (4.5), reveals that the integral is invariant to interchanges in h_1 and h_2. As a result:-

$$P(t_m, t_d | \gamma, s) = P(-t_m, t_d | \gamma, s) \quad , \quad P(t_m | \gamma, s) = P(-t_m | \gamma, s)$$

and $\overline{t}_m = 0$

5. REFERENCES

1. A.R.PRATT, Some theoretical considerations concerning time statistics in signal detection, in Signal Processing ed by J.W.R.Griffiths et al, Academic Press 1973

2. S.O.RICE, Mathematic Analysis of Random Noise, BSTJ. Vol.24 sec 3 1945.

3. S.O.RICE, Distribution of the Duration of Fades in Radio Transmission, BSTJ Vol. 37 May 1958.

4. H.L.VAN TREES, Detection, Estimation and Modulation Theory Vol. 1 pp 273-287, Wiley 1968.

DISCUSSION

Comment : A. PLAISANT

Obviously the variance of the estimator tm depends on the statistical properties of input signal and noise. Have you run other examples as, for instance, s Swerling II type signal in a white Gaussian noise?

Reply : A.R. PRATT

The results shown in the paper are a constant amplitude signal, i.e. a completely known signal. However, since the Swerling II is a fast fading model, independent from pulse to pulse, it should be a straight forward exercise to weight the estimation variance by the probability density function for signal amplitude and plot a modified curve of estimator variance versus average signal-to-noise ratio. I would anticipate that estimation performance will improve at low signal-to-noise ratios, but would worsen at moderate and high signal-to-noise ratios.

MAXIMUM LIKELIHOOD DETECTION OF SIGNALS IN NOISE OF UNKNOWN LEVEL

P. G. Cable

Naval Underwater Systems Center
New London, Connecticut 06320

1. INTRODUCTION

There are many practical instances when the statistics of interference received by a signal detection system are subject to unknown and unpredictable variations. For example, this situation is encountered in reverberation (or clutter) limited active sonar (or radar) environments where both the level and the spectrum of the interference may be changing with time. In such a circumstance if the receiver employs conventional processing techniques (e.g., matched filtering and envelope detection) the probability of interference alone exceeding a fixed threshold at the receiver output can be extremely sensitive to small changes in the interference level. This paper is concerned with a type of adaptive threshold control device that, at the expense of introducing estimation noise into the detection procedure, maintains a specified false alarm probability P_F independent of changes in noise level.

Because the adaptive specification of a threshold depends upon a continual estimation of noise level, it is generally necessary that the noise background be varying slowly enough in time that independent samples of noise alone, from the same ensemble as the sample being tested for signal presence, are available in neighboring time increments (requirement of local stationarity). In the discussion that follows it is also supposed that the input to the receiving system has been filtered to remove noise outside the signal band and that the sampled outputs of the system filters are statistically independent. In order to restructure the originally nonparametric problem into a parametric form (although the parameters appearing in the probability density functions (PDFs) are unknown), it is further assumed that the noise is Gaussian.

G. Tacconi (ed.), Aspects of Signal Processing, Part 1, 229-250.

In the adaptive method under discussion, the values of the unknown parameters are estimated by separately maximizing the PDFs for the signal present hypothesis H_1 and signal absent hypothesis H_0 (maximum likelihood estimates). The likelihood ratio (LR) formed under these conditions (principle of maximum likelihood) can be used to obtain a generalized likelihood ratio (GLR) test. Following a brief, general statement of the hypothesis testing procedure, the maximum likelihood (ML) method is applied to practical examples of detection of phase incoherent signals and fading signals, where, in each instance, the noise level is unknown. In addition to the derivation of specific GLR tests, performance characteristics for the corresponding ML detectors are obtained analytically. These results partially summarize work previously reported in a series of unpublished documents [1-3]. Additional relevant work on this problem can be found in [4-6].

In the preceding discussion it was suggested that the assumption of local stationarity is critical to the analyses and results presented. To ascertain the consequences of nonstationarity, some new results on the effects of local trends in noise level on ML processor performance are also presented in this paper.

2. PROBLEM STATEMENT

Suppose that a detection system performs M observations of a received narrowband signal plus noise alone process centered on frequency Ω and having waveforms

$$\operatorname{Re}\{V_k(t)\exp(i\Omega t)\}, \qquad 1 \le k \le M, \tag{1}$$

where

$$V_k(t) = \begin{cases} S_k(t) + N_k(t), & H_1 \\ & \\ N_k(t), & H_0 \end{cases}, \quad 1 \le k \le M. \tag{2}$$

In (2), $S_k(t)$ and $N_k(t)$ are the complex envelopes of the received signal and Gaussian noise waveforms, respectively, for the k-th observation.

Further, it is assumed that N additional observations are available in which it is known that only noise is present. These latter waveforms are

$$\operatorname{Re}\{\tilde{V}_j(t)\exp(i\Omega t)\}, \qquad 1 \le j \le N, \tag{3}$$

where

$$\tilde{V}_j(t) = \tilde{N}_j(t) , \qquad 1 \leq j \leq N, \tag{4}$$

for H_1 and H_0.

Application of standard processing procedures results in specification of the assumed statistically independent complex quantities

$$\int dt\ F(t)V_k(t) \equiv Z_k = x_k + iy_k, \qquad 1 \leq k \leq M, \tag{5}$$

and

$$\int dt\ F(t)\tilde{V}_j(t) \equiv \tilde{Z}_j = \tilde{x}_j + i\tilde{y}_j, \qquad 1 \leq j \leq N \tag{6}$$

In the above, $F(t)$ is a filter that need not be optimum and the integrations are over the range of nonzero integrand. Because the noise is a Gaussian process, $\{Z_k\}$ and $\{\tilde{Z}_j\}$ are complex Gaussian random variables that are, by previous assumption, statistically independent of each other and of equal (but unknown) variance σ_0^2.

On the basis of the observation vector $\underline{R}$, where

$$\underline{R}^T = \left[x_1 y_1 \ldots x_M y_M \tilde{x}_1 \tilde{y}_1 \ldots \tilde{x}_N \tilde{y}_N\right] , \tag{7}$$

a decision is to be made between hypothesis H_1 (signal present in $\{x_k, y_k\}$) and H_0 (signal absent in $\{x_k, y_k\}$). The variance σ_0^2 associated with $\underline{R}$ is unknown, but it is desired that the processing of $\underline{R}$ ensure a specified P_F irrespective of the magnitude of σ_0^2.

3. ML DETECTION OF SIGNALS WITH UNKNOWN PHASE

Under hypothesis H_0, the PDF of $\underline{R}$, conditional on an assumed value of noise variance σ^2, is

$$P_0(\underline{R}|\sigma^2) = \prod_{k=1}^{M} \frac{1}{2\pi\sigma^2} \exp\left[-\frac{x_k^2 + y_k^2}{2\sigma^2}\right]$$

$$\cdot \prod_{j=1}^{N} \frac{1}{2\pi\sigma^2} \exp\left[-\frac{\tilde{x}_j^2 + \tilde{y}_j^2}{2\sigma^2}\right]. \quad (8)$$

The ML estimate, denoted by carats, of σ_0^2 under hypothesis H_0 is

$$2\hat{\sigma}^2 = \frac{\sum_{k=1}^{M} (x_K^2 + y_k^2) + \sum_{j=1}^{N} (\tilde{x}_j^2 + \tilde{y}_j^2)}{M + N}$$

$$= \frac{\sum_{k=1}^{M} |Z_k|^2 + \sum_{j=1}^{N} |\tilde{Z}_j|^2}{M + N}, \quad (9)$$

and is proportional to the total sample power.

Then substitution of (9) in (8) yields

$$\max_{\sigma^2} p_0(\underline{R}|\sigma^2) = \left(e\pi \frac{\sum_{k=1}^{M} |Z_k|^2 + \sum_{j=1}^{N} |\tilde{Z}_j|^2}{M + N}\right)^{-(M+N)} . \quad (10)$$

Under hypothesis H_1, the PDF of $\underline{R}$, conditioned on assumed values of signal means and σ^2, is

$$p_1(\underline{R}|\sigma^2, \underline{A}, \underline{\psi}) = \prod_{k=1}^{M} \frac{1}{2\pi\sigma^2} \exp\left[-\frac{(x_k - A_k \cos\psi_k)^2 + (y_k - A_k \sin\psi_k)^2}{\sigma^2}\right]$$

$$\cdot \prod_{j=1}^{N} \frac{1}{2\pi\sigma^2} \exp\left[-\frac{\tilde{x}_j^2 + \tilde{y}_j^2}{2\sigma^2}\right], \tag{11}$$

where, denoting ensemble averages on the noise by overbars, the means are

$$\bar{Z}_k = \bar{x}_k + i\bar{y}_k \equiv A_k \exp(i\psi_k) = \int dt\; F(t)\; S_k(t),$$

$$1 \leq k \leq M\;, \tag{12}$$

and

$$\underline{A}^T \equiv \left[A_1 \cdots A_M\right] \;, \; \underline{\psi}^T \equiv \left[\psi_1 \cdots \psi_M\right] \;. \tag{13}$$

The ML estimate of σ_0^2 under hypothesis H_1 is

$$2\hat{\sigma}^2 = \frac{\sum_{k=1}^{M} |Z_k - A_k \exp(i\psi_k)|^2 + \sum_{j=1}^{N} |\tilde{Z}_j|^2}{M+N}\,, \tag{14}$$

and from this there is obtained

$$\max_{\sigma^2} p_1(\underline{R}|\sigma^2, \underline{A}, \underline{\psi}) =$$

$$\left(e\pi \frac{\sum_{k=1}^{M} |Z_k - A_k \exp(i\psi_k)|^2 + \sum_{j=1}^{N} |\tilde{Z}_j|^2}{M+N}\right)^{-(M+N)} . \tag{15}$$

This function can be further maximized by choice of $\underline{A}$ and $\underline{\psi}$. If the signal amplitudes are not restricted to be equal*, the ML estimates of signal amplitudes and phases are

---

* For a discussion of the case where the amplitudes $\{A_k\}$ are assumed equal but unknown, see [2].

$$\hat{A}_k = |Z_k| \ , \qquad \hat{\psi}_k = \arg(Z_k) \ , \qquad 1 \le k \le M \ . \tag{16}$$

These give

$$\max_{\sigma^2, \underline{A}, \underline{\psi}} P_1(\underline{R}|\sigma^2, \underline{A}, \underline{\psi}) = \left(e\pi \frac{\sum_{j=1}^{N} |\tilde{Z}_j|^2}{M + N} \right)^{-(M+N)} . \tag{17}$$

A GLR test can now be constructed using (10) and (17), and is expressible as: Choose H_1 if

$$\frac{1}{M} \sum_{k=1}^{M} |Z_k|^2 \ge r \frac{1}{N} \sum_{j=1}^{N} |\tilde{Z}_j|^2 \ , \qquad M \ge 1 \ , \qquad N \ge 1 \ , \tag{18}$$

where $r(\geq 0)$ is a constant, and choose H_0 otherwise. The ML detector compares the sums of squared envelopes for potential signal samples with the summed squared envelope samples for noise.

The performance of the GLR test, (18), is evaluated in Appendix A. One form obtained for the probability of detection P_D is

$$P_D = 1 - \left(\frac{\alpha}{1+\alpha}\right)^M \exp\left(\frac{-d_T^2}{2} \frac{1}{1+\alpha}\right) \sum_{k=0}^{N-1}$$

$$\cdot \binom{M-1+k}{k} (1+\alpha)^{-k} \ {}_1F_1\left(-k; M; \frac{-d_T^2}{2} \frac{\alpha}{1+\alpha}\right) \ , \tag{19}$$

where ${}_1F_1$ is the confluent hypergeometric function and $\alpha \equiv rM/N$. The parameter of total power signal-to-noise ratio (SNR) $d_T^2/2$ is given by

$$\frac{1}{2} d_T^2 \equiv \sum_{k=1}^{M} \frac{1}{2} d_k^2 = \sum_{k=1}^{M} \frac{|\int dt F(t) S_k(t)|^2 / 2}{\frac{1}{2} \iint dt du F(t) F^*(u) R_{11}(t-u)} \ , \tag{20}$$

where

$$\overline{N_k(t) N_\ell^*(u)} = R_{k\ell}(t-u), \qquad 1 \le k, \ \ell \le M. \tag{21}$$

is the actual complex envelope noise crosscorrelation on the potential signal samples.

From (20) it can be observed that the exact way the total received energy is actually fractionalized in the M observations does not affect performance.

When d_T is set equal to zero in (19) there is obtained

$$P_F = 1 - \left(\frac{\alpha}{1+\alpha}\right)^M \sum_{k=0}^{N-1} \binom{M-1+k}{k} (1 + \alpha)^{-k} . \tag{22}$$

An alternative form for P_F discussed in Appendix A is given by

$$P_F = (1 + \alpha)^{-N} \sum_{k=0}^{M-1} \binom{N-1+k}{k} \left(\frac{\alpha}{1+\alpha}\right)^k . \tag{23}$$

Comparison of the two expressions for P_F indicates that (23) is computationally more efficient than (22) for $N > M$.

For the ML processor, as $N \to \infty$, the expressions (19), (22) and (23) for P_D and P_F tend to the forms obtained for envelope squared detection with known noise level [2], as indeed they must, for ML estimators are consistent.

Typical performance curves for the ML processor are shown in figures 1 and 2, where P_D is plotted versus d_T for $P_F = 10^{-4}$. It is apparent from an examination of these curves that ML processor performance steadily degrades as the signal energy fractionalization increases, that is, as M increases, for fixed d_T. This situation is somewhat alleviated as N becomes larger.

The improvement attainable in ML processor performance by increasing N is indicated in figure 3, which is a plot for $M = 1$ of required d_T in decibels to realize $P_D = .5$ versus N. The tic marks at the right edge of the plot show required d_T for $N = \infty$. Figure 3 also indicates that to achieve a given P_D for a specified d_T, N must be increased as P_F is lowered.

4. EXTENSION TO FLUCTUATING SIGNALS

It has been noted that the ML processor performance is unaffected by the manner in which the received signal energy is fractionalized. If the signal in the various observations fades in a random manner, the quantity $R_T \equiv d_T^2/2$ is a random variable; the average probability of detection is then given by averaging over (19) according to

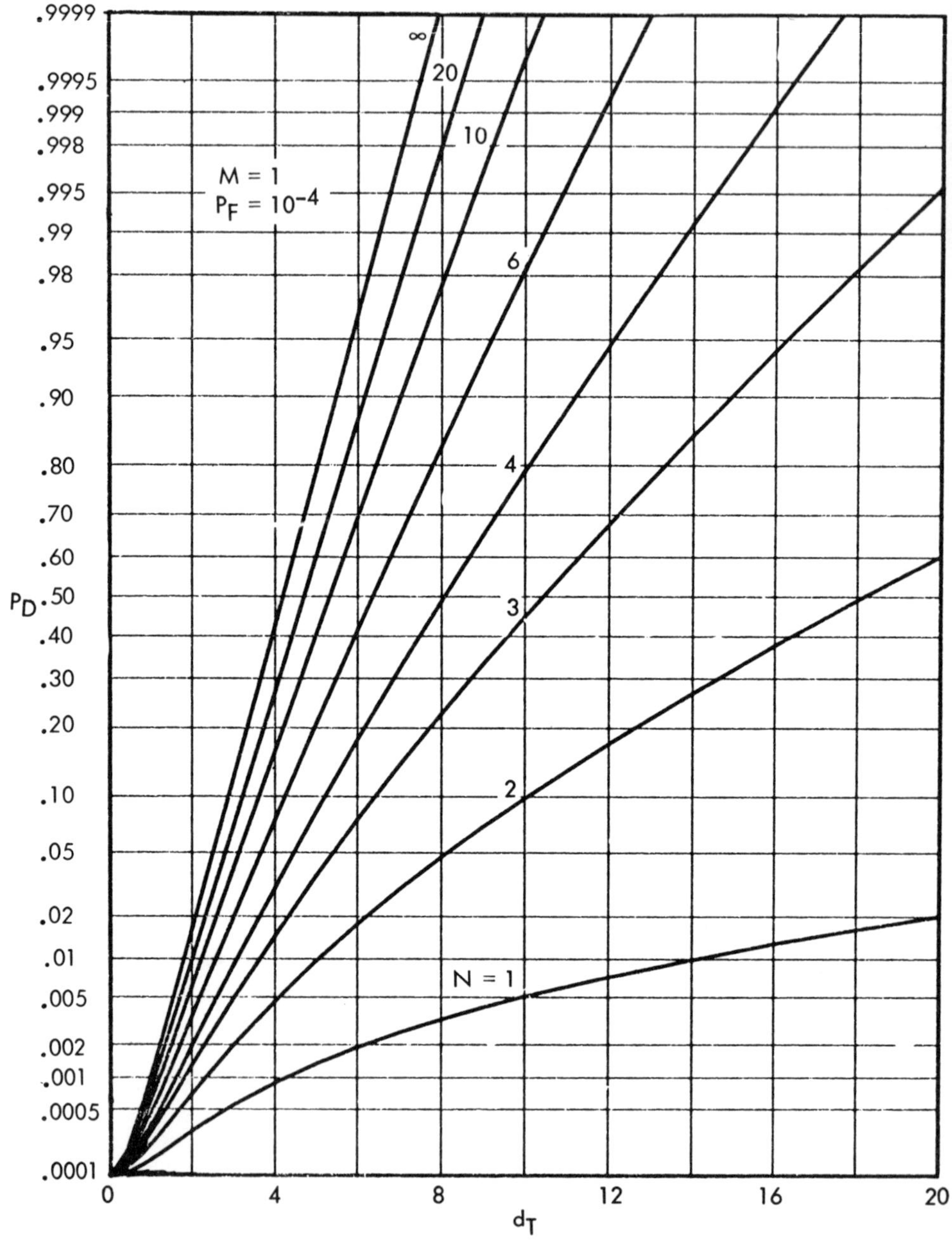

Figure 1. Detection Characteristics for ML Processor

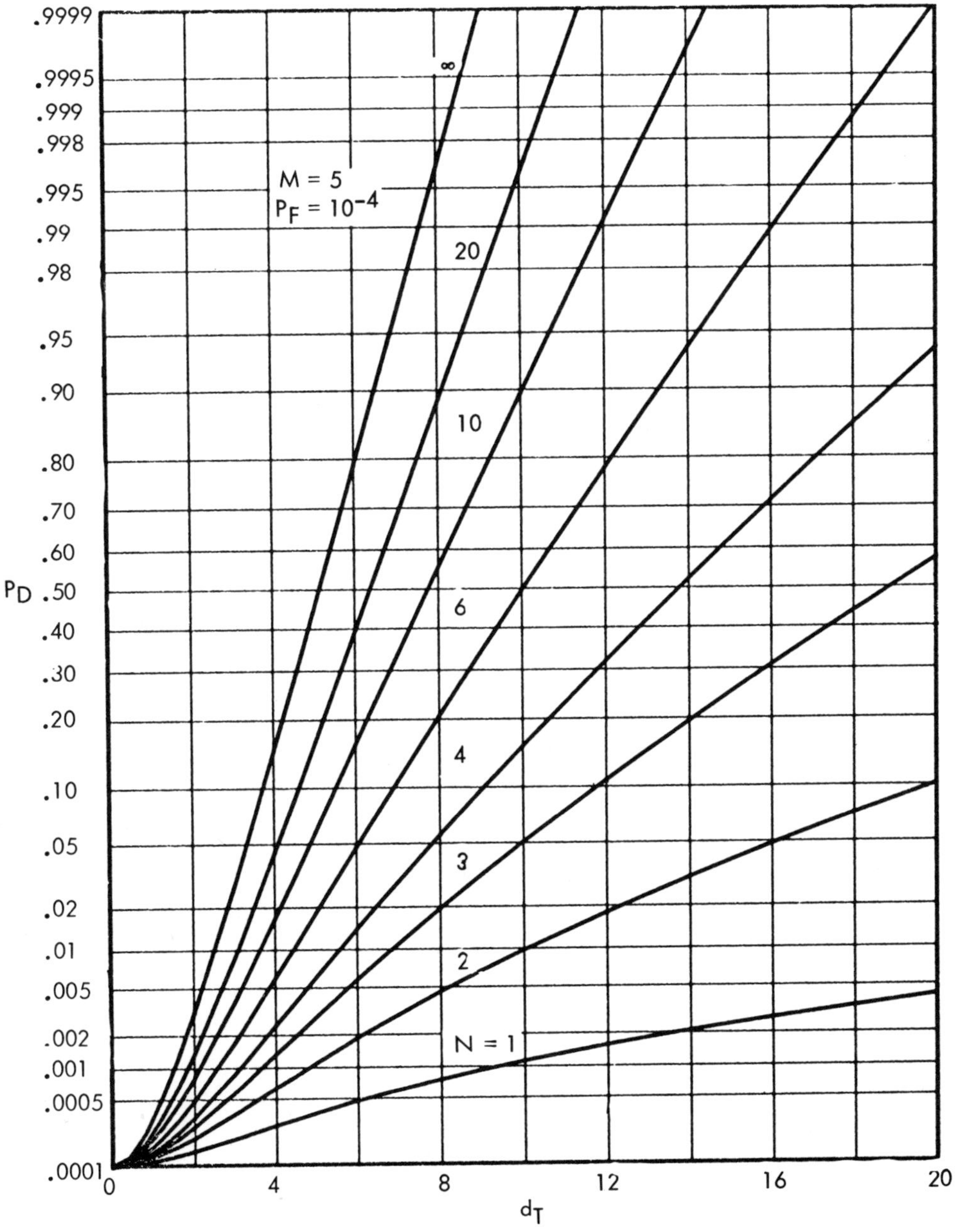

Figure 2. Detection Characteristics for ML Processor

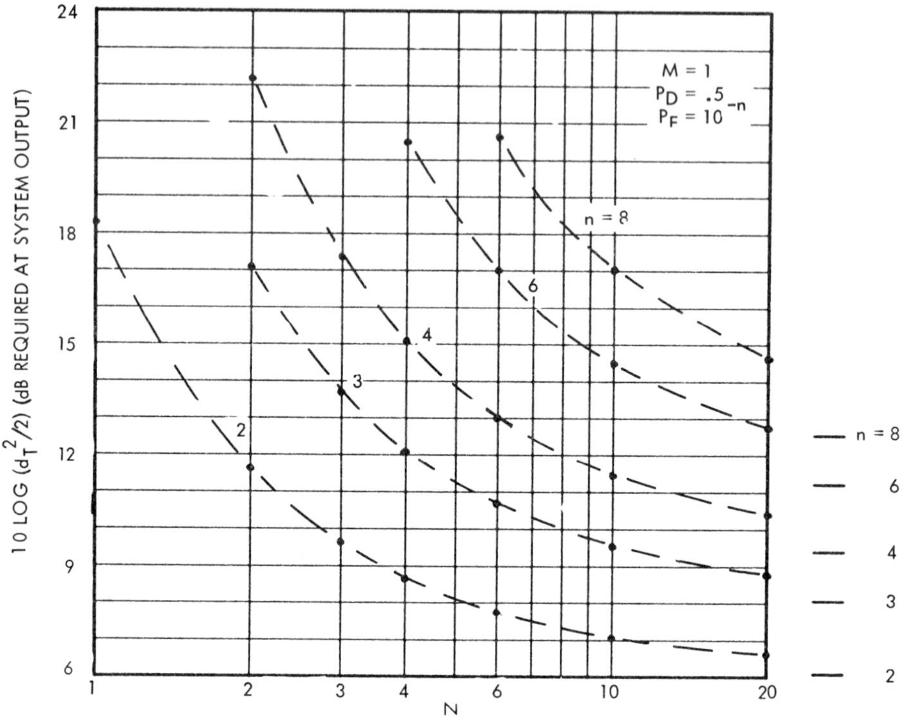

Figure 3. Required Signal-to-Noise Ratio for ML Processor

$$P_D = \int dR_T \; p(R_T) \; P_D(R_T) \; , \tag{24}$$

where $p(R_T)$ is the PDF of the total power SNR random variable R_T.

A general form of fading is attained by the PDF,

$$p(R_T) = \frac{R_T \exp(-R_T/\mu)}{\mu^{\nu+1} \; \Gamma(\nu+1)} , \qquad R_T > 0, \tag{25}$$

where $\nu > -1$,

$$\mu = \frac{\bar{R}_T}{\nu+1} , \tag{26}$$

and $\bar{R}_T$ is the average total power SNR on the M signal branches, averaged over the fading statistics. For example, the four fading cases considered by Swerling [8] are subsumed by (25) for particular choices of ν.

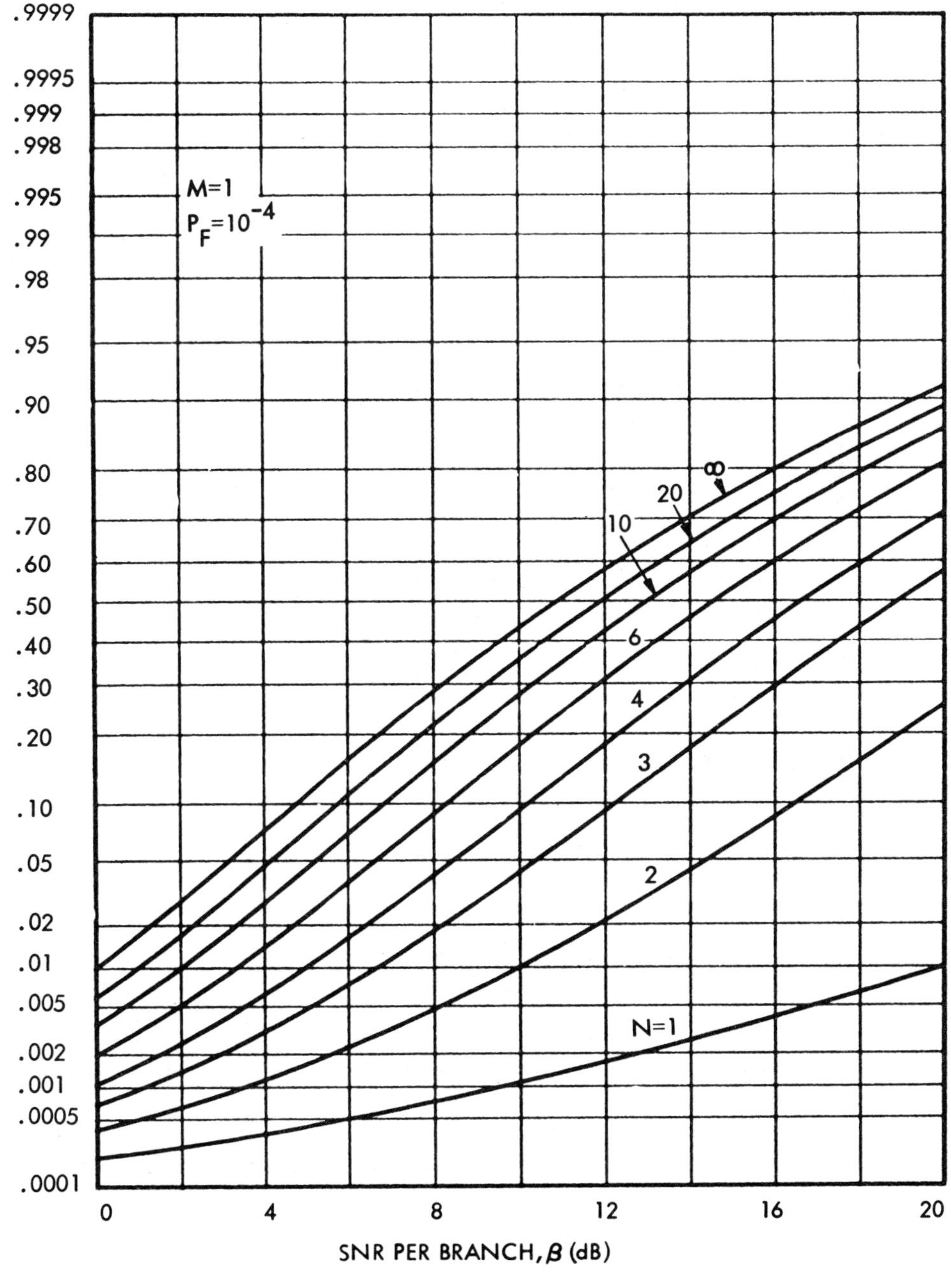

Figure 4. Detection Characteristics for ML Processor and Rayleigh Fading Signal

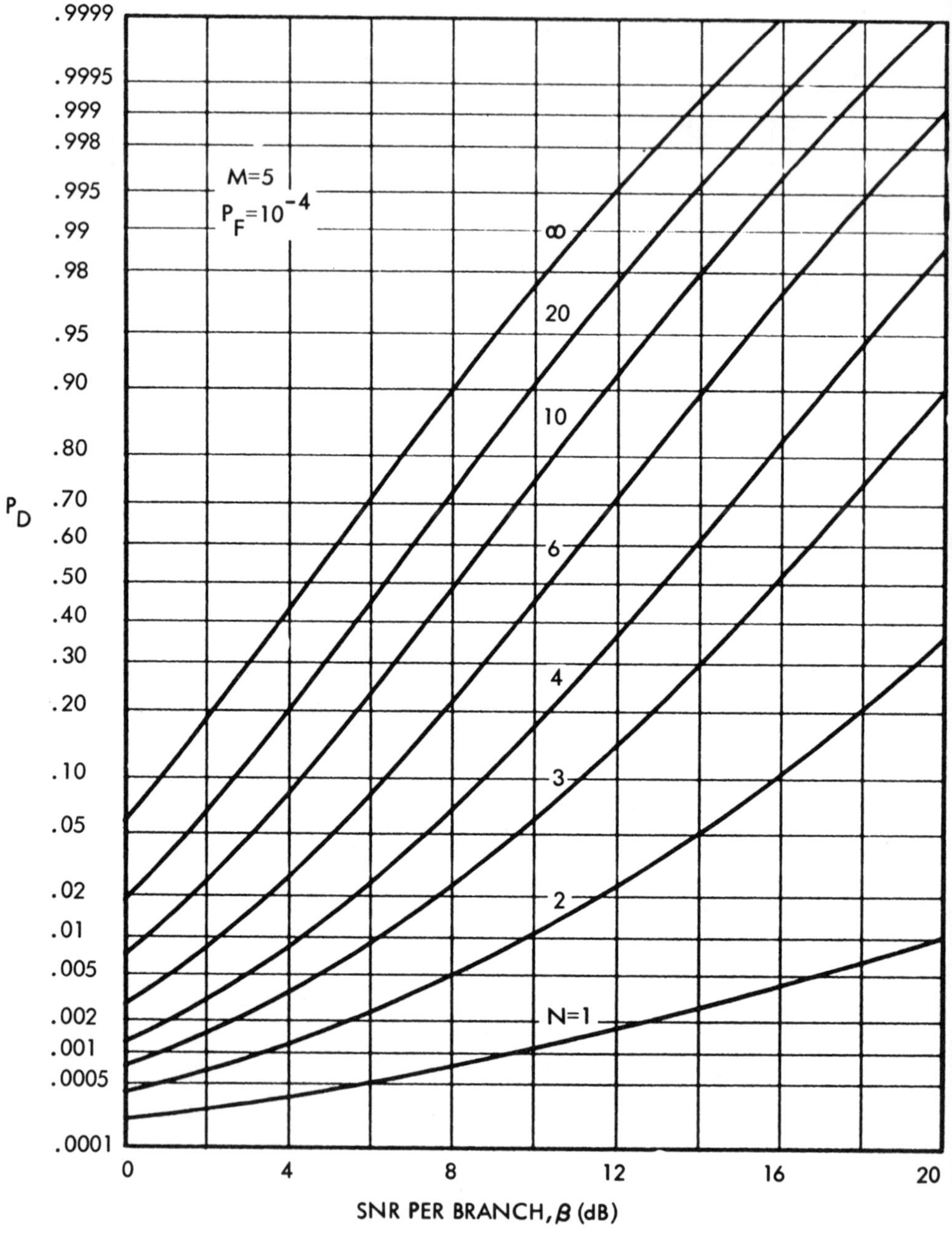

Figure 5. Detection Characteristics for ML Processor and Rayleigh Fading Signal

The average probability of detection is obtained by substituting (19) into (24) and employing equation 7.621 4 in [9]; the following form can then be obtained:

$$P_D = 1 - \left(\frac{\alpha}{1+\alpha}\right)^M \left(\frac{1+\alpha}{\mu+1+\alpha}\right)^{\nu+1} \sum_{k=0}^{N-1}$$

$$\cdot \binom{M-1+k}{k} (1+\alpha)^{-k} F\left(-k, \nu+1; M; \frac{-\alpha\mu}{\mu+1+\alpha}\right) , \quad (27)$$

where F is a hypergeometric function. This form is convenient for computer evaluation because the hypergeometric function terminates.

For ν=M-1, which corresponds to case 2 fading considered by Swerling [8], figures 4 and 5 show curves of detection probability P_D versus the SNR per branch, $\beta=\mu=\overline{R}_T/M$ in decibels. In figure 6, the required SNR with M=1 to achieve P_D=0.5, is shown versus N. The horizontal tic marks at the right edge of the figure show the required values of SNR for N=∞ and are, hence, the asymptotes of the curves.

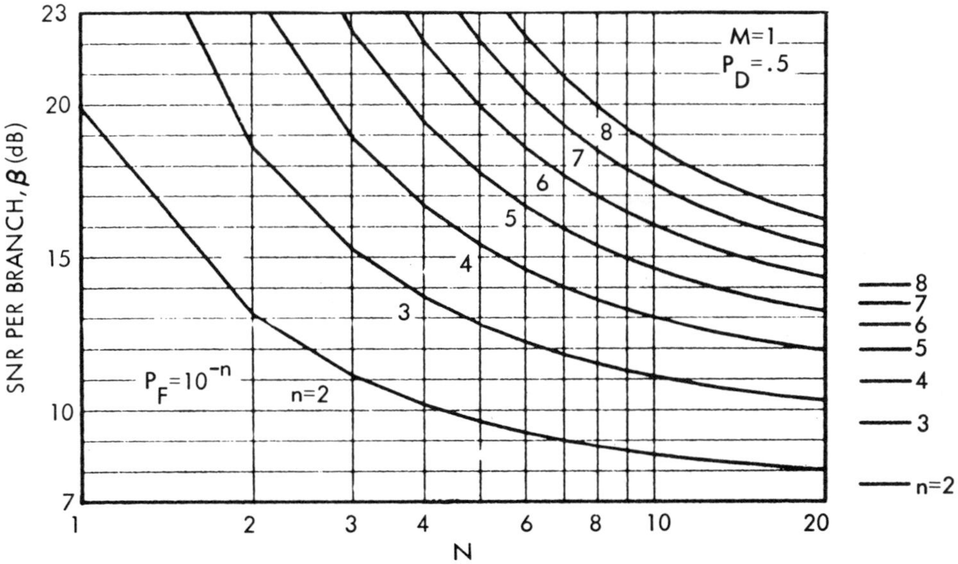

Figure 6. Required Signal-to-Noise Ratio for ML Processor and Rayleigh Fading Signal

Computation of the difference in the required SNR in decibels between the values required for finite N and infinite N for fluctuating signals (figure 6) shows that the results agree very well over a wide range of parameter values with the same calculation for the nonfluctuating signal case (figure 3). That is, the exact signal statistics, nonfluctuating or Swerling case 2, do not affect the additional SNR required to operate at finite N versus infinite N.

5. EFFECT OF LOCAL TRENDS IN NOISE LEVEL

The performance of the GLR test given by (18) will be affected if the actual variances of the $\{Z_k, Z_j\}$ are, contrary to assumption, unequal. In order to examine this, the case M=1 will be considered and it will be supposed that the variances of the N+1 observation variables are given by

$$\left(\frac{\sigma_0}{\sigma_m}\right)^2 = 1 + ms , \qquad 0 \leq m \leq N, \tag{28}$$

where, for specificity, it is assumed that the sample being tested for signal presence has an associated variance $\sigma_J^2 = \sigma_0^2/(1 + Js)$. $0 \leq J \leq N$. Provided $Ns \ll 1$, this model can be considered an approximation to a pth power range dependence of reverberation (or clutter) level by invoking the interpretation

$$s = p\Delta/R_0, \tag{29}$$

where Δ is the sonar (or radar) range resolution and R_0 is the range associated with the first observation.

In Appendix B the performance of the previously derived test, equation (18), is analyzed for the unequal noise level model (28) and a signal subject to Rayleigh envelope fading (i.e., the case ν=M-1 previously considered). The probability of detection for this example is found to be

$$P_D = \left(\frac{1+A+Js}{1+Js}\right) \prod_{m=0}^{N} \left(\frac{1+ms}{1+A+ms}\right) , \tag{30}$$

where $A \equiv \alpha(1+Js)/(1+\beta)$, and β is the power SNR in the signal branch. P_F is obtained by setting $\beta = 0$ in (30).

Because of the dependence of (30) on s, test (18) does not maintain a specified P_F independent of the degree of trend in the local noise level. In figure 7, the effect of s on P_F with J = 0 is shown with N as a parameter. Over the particular range of s values chosen, it is apparent that the adaptive procedure yields a P_F relatively immune to the degree of trend in noise level. However, as the number of samples N used in the ML estimate increases, the P_F achieved becomes more sensitive to the local changes in noise level. These considerations indicate that, in a practical situation where ML processing is contemplated, knowledge of the expected degree of noise stationarity would be a necessary ingredient in system design.

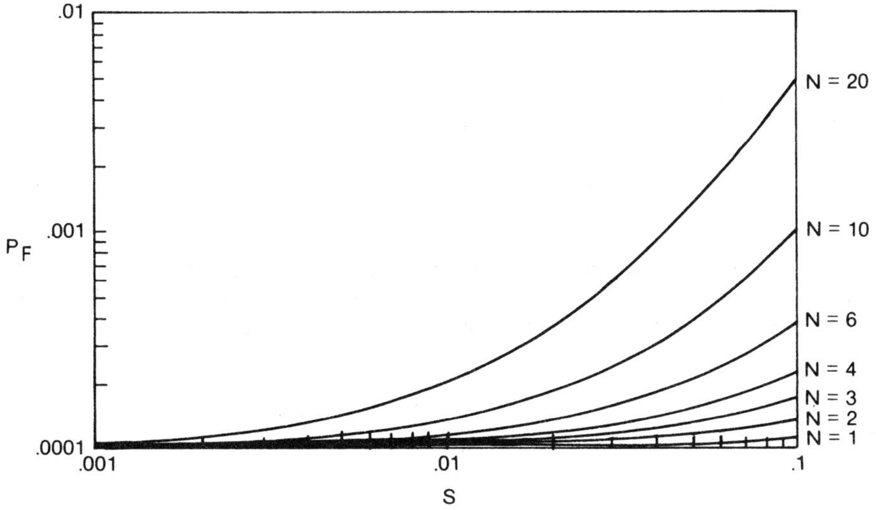

Figure 7. False Alarm Probability Versus Noise Level Trend Parameter S. $P_F = 10^{-4}$ when s = 0.

APPENDIX A. PERFORMANCE EVALUATION OF ML PROCESSOR

Let

$$Z_k = B_k \exp(i\theta_k) , \qquad 1 \le k \le M;$$

$$\tilde{Z}_j = \tilde{B}_j \exp(i\tilde{\theta}_j), \qquad 1 \le j \le N , \tag{A-1}$$

where $\{Z_k\}$ and $\{\tilde{Z}_j\}$ were defined in (5) and (6), and let

$$\underline{B}^T = \left[B_1 \ldots B_M \tilde{B}_1 \ldots \tilde{B}_N\right] . \tag{A-2}$$

Then

$$p(\underline{B}) = \prod_{k=1}^{M} \frac{B_k}{\sigma_0^2} \exp\left[-\frac{B_k^2 + A_{k0}^2}{2\sigma_0^2}\right] I_0\left(\frac{A_{k0}B_k}{\sigma_0^2}\right)$$

$$\cdot \prod_{j=1}^{N} \frac{\tilde{B}_j}{\sigma_0^2} \exp\left[-\frac{\tilde{B}_j^2}{2\sigma_0^2}\right],$$

$$B_k > 0, \; B_j > 0 , \tag{A-3}$$

Here A_{k0} is the actual amplitude of the complex sample of the signal output of the filter F(t); the parameter σ_0^2 in the PDF is proportional to the actual output noise variance of the filter:

$$\sigma_0^2 = \frac{1}{2} \iint dt \; du \; F(t)F^*(u)R_{11}(t-u) . \tag{A-4}$$

Let

$$U \equiv \sum_{k=1}^{M} \left(\frac{B_k}{\sigma_0}\right)^2 , \; V \equiv \sum_{j=1}^{N} \left(\frac{\tilde{B}_j}{\sigma_0}\right)^2 , \; \alpha \equiv \frac{rM}{N} ,$$

$$d_k \equiv \frac{A_{k0}}{\sigma_0} . \tag{A-5}$$

Then the GLR test (18) can be written as

$$U \gtrless \alpha \quad . \tag{A-6}$$

The random variables U and V are statistically independent of each other (see (A-3)). Therefore the characteristic function of U is

$$f_U(\S) = \overline{\exp(i\S U)} = \prod_{k=1}^{M} \int_0^{\infty} dx \, \exp(i\S x^2) x$$

$$\cdot \exp\left[-\frac{x^2+d_k^2}{2}\right] I_0(d_k x)$$

$$= (1 - i2\S)^{-M} \exp\left(\frac{id_T^2 \S}{1-i2\S}\right) , \tag{A-7}$$

where

$$d_T^2 \equiv \sum_{k=1}^{M} d_k^2 \ . \tag{A-8}$$

From (A-7), the PDF of U can be determined to be

$$P_U(x) = \frac{1}{2} \exp\left(-\frac{x+d_T^2}{2}\right) \left(\frac{x}{d_T^2}\right)^{\frac{M-1}{2}} I_{M-1}(d_T\sqrt{x}\,) ,$$

$$x > 0. \tag{A-9}$$

Similarly, the PDF of V can be shown to be

$$p_V(y) = \frac{1}{2} \exp\left(-\frac{y}{2}\right) \frac{(y/2)^{N-1}}{(N-1)!} , \qquad y > 0 \ . \tag{A-10}$$

From (A-6), (A-9), (A-10), there is obtained for the detection probability

$$P_D = \text{Prob}\ (U > \alpha V) = \int_0^{\infty} d\,x\ p_U(x) \int_0^{x/\alpha} dy\ p_V(y) \ . \tag{A-11}$$

Consider the inner integral in (A-11)

$$\int_0^{x/\alpha} dy\, p_V(y) = 1 - \int_{x/\alpha}^{\infty} dy \cdot \frac{1}{2} \exp\left(-\frac{y}{2}\right) \frac{(y/2)^{N-1}}{(N-1)!}$$

$$= 1 - \exp\left(-\frac{x}{2\alpha}\right) \sum_{k=0}^{N-1} \frac{1}{k!} \left(\frac{x}{2\alpha}\right)^k . \qquad \text{(A-12)}$$

Substitution of (A-9) and (A-12) in (A-11) and a change variable, $x = t^2$, yields

$$P_D = 1 - \frac{\exp(-d_T^2/2)}{d_T^{M-1}} \sum_{k=0}^{N-1} \frac{1}{k!} \left(\frac{1}{2\alpha}\right)^k \int_0^{\infty} dt\, t^{M+2k}$$

$$\cdot \exp\left[-\frac{t^2}{2}\left(1+\frac{1}{\alpha}\right)\right] I_{M-1}(d_T t) . \qquad \text{(A-13)}$$

From the identity

$$I_{M-1}(d_T t) = \frac{1}{i^{M-1}} J_{M-1}(i d_T t) , \qquad \text{(A-14)}$$

the integral in (A-13) can be evaluated using [9], 6.631 1 as

$$\int_0^{\infty} dt \frac{t^{M+2k}}{i^{M-1}} \exp\left[-\frac{t^2}{2}\left(1+\frac{1}{\alpha}\right)\right] J_{M-1}(i d_T t) \qquad \text{(A-15)}$$

$$= 2^k \left(\frac{\alpha}{\alpha+1}\right)^{M+k} d_T^{M-1} \frac{(M+k-1)!}{(M-1)!} \exp\left(\frac{d_T^2}{2} \frac{\alpha}{\alpha+1}\right)$$

$$\cdot {}_1F_1\left(-k; m; -\frac{d_T^2}{2} \frac{\alpha}{\alpha+1}\right) , \qquad \text{(A-16)}$$

where ${}_1F_1(a; b; x)$ denotes the confluent hypergeometric function. One form for the probability of detection is thus

$$P_D = 1 - \exp\left(-\frac{d^2}{2} \frac{1}{\alpha+1}\right) \left(\frac{\alpha}{\alpha+1}\right)^M \sum_{k=0}^{N-1} \binom{M-1+k}{k} \frac{1}{(\alpha+1)^k}$$

$$\cdot \; {}_1F_1 \left(-k;\, M;\, -\frac{d_T^{\,2}}{2}\,\frac{\alpha}{\alpha+1}\right) . \qquad \text{(A-17)}$$

An alternative expression for P_D can be obtained from

$$P_D = \text{Prob}\,(U > \alpha V) = \int_0^{\infty} dy\; p_V(y) \int_{\alpha y}^{\infty} dx\, p_U(x) \; . \qquad \text{(A-18)}$$

This form yields equation (23) for P_F when $d_T = 0$. Details of the calculation can be found in [2].

APPENDIX B: ML PROCESSOR PERFORMANCE IN THE PRESENCE OF A NOISE LEVEL TREND

For the case of unequal noise in each observation, Rayleigh fading of the signal and one signal branch, the actual PDFs of the sample envelopes are (cf Appendix A):

$$p(\underline{B}) = \frac{B_J}{2\alpha_J^2\,(1+\beta)} \exp\left[-\frac{B_J^2}{2\alpha_J^2\,(1+\beta)}\right]$$

$$\cdot \prod_{m=0}^{N}{}' \frac{\tilde{B}_m}{2\sigma_m^2} \exp\left[-\frac{\tilde{B}_m^2}{2\sigma_m^2}\right], \quad B_J > 0,\ \tilde{B}_m > 0, \qquad \text{(B-1)}$$

where β is the power SNR and the prime denotes that the index m = J is excluded from the product. The GLR test based on assumed equal noise levels is, from (18):

$$B_J^{\,2} \gtrless \frac{r}{N} \sum_{m=0}^{N}{}' \tilde{B}_m^2 \, , \qquad \text{(B-2)}$$

where the prime on the summation indicates exclusion of the m = J index.

Let

$$U \equiv \frac{1}{2}\left(\frac{B_J}{\sigma_0}\right)^2, \; V \equiv \frac{1}{2}\sum_{m=0}^{N}{}'\left(\frac{\tilde{B}_m}{\sigma_0}\right)^2, \; C_m \equiv \left(\frac{\sigma_0}{\sigma_m}\right)^2, \; \alpha \equiv \frac{r}{N} \, . \qquad \text{(B-3)}$$

Then test (B-2) can be written

$$U \gtrless \alpha V. \tag{B-4}$$

The PDF of U is, from (B-1),

$$P_U(x) = \frac{C_J}{1 + \beta} \exp\left(- \frac{C_J x}{1 + \beta}\right) . \tag{B-5}$$

The PDF of V can be obtained from the characteristic function as

$$p_V(y) = \prod_{m=0}^{N}{}' \int_{-\infty}^{\infty} \frac{d\xi}{2\pi} \frac{C_m}{C_m - i\xi} \exp(i\xi y)$$

$$= \sum_{k=0}^{N}{}' \frac{\prod_{m=0}^{N}{}' C_m}{\prod_{\substack{m=0 \\ m \neq k}}^{N}{}' (C_m - C_k)} \exp(-C_k y) . \tag{B-6}$$

From (B-4), (B-5) and (B-6), there is obtained for the probability of detection

$$P_D = \text{Prob } (U > \alpha V) = \int_0^{\infty} dy \; p_V(y) \int_{\alpha y}^{\infty} dx \; p_U(x)$$

$$= \int_0^{\infty} dy \sum_{k=0}^{N}{}' \frac{\prod_{m=0}^{N}{}' C_m}{\prod_{\substack{m=0 \\ m \neq k}}^{N}{}' (C_m - C_k)} \exp\left[- \left(C_k + \frac{\alpha C_J}{1 + \beta}\right) y\right]$$

$$= \int_0^{\infty} dy \sum_{k=0}^{N} \frac{\prod_{m=0}^{N} C_m (C_J - C_k)}{\prod_{\substack{m=0 \\ m \neq k}}^{N} C_J (C_m - C_k)} \exp\left[- \left(C_k + \frac{\alpha C_J}{1 + \beta}\right) y\right]. \tag{B-7}$$

When the model defined in (28),

$$C_M = (1 + ms), \tag{B-8}$$

is substituted in (B-7) there is obtained after some algebra:

$$P_D = \frac{\prod_{m=0}^{N} (1 + ms)}{(1 + Js)\, s^{N-1}} \int_0^\infty dy \exp[-(1 + A)y] \cdot \left\{ \sum_{k=0}^{N} \frac{sk(-)^{k-1}}{N!} \binom{N}{K} (J-k) \exp(-ksy) \right\}$$

$$= \frac{\prod_{m=0}^{N} (1 + ms)}{(1 + Js)s^{N-1}} \int_0^\infty dy \exp[-(1 + A)y] \cdot \left\{ \frac{(J - N)s}{(N - 2)!} \left[1 - e^{-sy}\right]^{N-1} e^{-sy} + \frac{s}{(N - 2)!} \left[1 - e^{-sy}\right]^{N-2} e^{-2sy} \right\}, \tag{B-9}$$

where $\binom{N}{k}$ is the binomial coefficient and $A \equiv \alpha(1 + Js)/(1 + \beta)$. A change of variable, $1 - \exp(-sy) = q$, permits the integral in (B-9) to be performed ([9], 3.251 1):

$$P_D = \frac{\prod_{m=0}^{N} (1 + ms)}{(1 + Js)s^N} \left[\frac{B(\frac{A + 1}{s} + 1, N-1)}{(N + 1)!} (J-N)\, s + \frac{B(\frac{A + 1}{s} + 2, N - 2)}{(N - 2)!} \right]$$

$$= \frac{\prod_{m=0}^{N} (1 + ms)}{(1 + Js)} \left[\frac{(J - N)s}{\prod_{m=0}^{N} (1 + A + ms)} + \frac{1}{\prod_{m=0}^{N} (1 + A + ms)} \right]$$

$$= \left(\frac{1 + A + Js}{1 + Js}\right) \prod_{m=0}^{N} \left(\frac{1 + ms}{1 + A + ms}\right) . \qquad \text{(B-10)}$$

REFERENCES

1. A. H. Nuttall and P. G. Cable, Operating Characteristics for Maximum Likelihood Detection of Signals in Gaussian Noise of Unknown Level. I. Coherent Signals of Unknown Level, NUSC Technical Report 4243, 27 March 1972.
2. P. G. Cable and A. H. Nuttall, Operating Characteristics for Maximum Likelihood Detection of Signals in Gaussian Noise of Unknown Level. II. Phase-Incoherent Signals of Unknown Level, NUSC Technical Report 4683, 22 April 1974.
3. A. H. Nuttall and P. G. Cable, Operating Characteristics for Maximum Likelihood Detection of Signals in Gaussian Noise of Unknown Level. III. Random Signals of Unknown Level, NUSC Technical Report 4783, 31 July 1974.
4. H. M. Finn, Proceedings of the National Electronics Conference, Vol. 22, 1966, pp. 562-567.
5. H. M. Finn and R. S. Johnson, RCA Review, Vol. 29, No. 3, September 1968, pp. 414-464.
6. S. S. Rappaport, Proceedings of the IEEE, Vol. 57, No. 8, August 1969, pp. 1420-1421.
7. A. D. Whalen, Detection of Signals in Noise, Academic Press, New York, 1971.
8. P. Swerling, IRE Transactions on Information Theory, Vol. IT-6, No. 2, April 1960, pp. 269-308.
9. I. S. Gradshteyn and I. M. Ryzhik, Tables of Integrals, Series and Products, Academic Press, New York, 1965.

THE ROLE OF COHERENCE IN TIME DELAY ESTIMATION

G. Clifford Carter

Naval Underwater Systems Center
New London, Connecticut 06320 U.S.A.

ABSTRACT. This paper investigates methods for passive estimation of the bearing to a slowly moving acoustically radiating source. The mathematics for the solution to such a problem is analogous to estimating the time delay (or group delay) between two time series. Since the estimation of time delay is intimately related to the coherence between two time series, a summary of the properties of coherence is presented

The maximum likelihood (ML) estimate of time delay (under jointly stationary Gaussian assumptions) is presented. The explicit dependence of time delay estimates on coherence is evident in the estimator realization in which the two time series are prefiltered (to accentuate frequency bands according to the strength of the coherence) and subsequently crosscorrelated. The hypothesized delay at which the generalized crosscorrelation (GCC) function peaks is the time delay estimate. The variance of the time delay estimate is presented and discussed.

INTRODUCTION. An acoustic source whose signal, $s(t)$, is transmitted through the ocean medium and received in the presence of additive noise can be characterized by

$$x_i(t) = s_i(t) + n_i(t) \qquad , \ i = 1,2 \tag{1}$$

For the main purposes of this paper $s_1(t)=s(t)$, $s_2(t)=\alpha s(t+D)$, and we desire to present an ML estimator for the time delay D. The delay parameter can be used, in a nondispersive medium with known speed of transmission, to estimate the bearing to an acoustic source (relative to the sensor baseline) or, more generally, to

G. Tacconi (ed.), Aspects of Signal Processing, Part 1, 251-256. *All Rights Reserved.*

estimate a hyperbolic "line" of position. Since the final result depends heavily on the coherence between x_1 and x_2, we precede the development with a concise review of the properties of the coherence function and of results that bear directly on the estimation of time delay.

THEORY OF COHERENCE. For any two jointly stationary random processes x_1 and x_2, the coefficient of coherency or the complex coherence has been defined by Weiner (1930) as the ratio

$$\frac{G_{x_1x_2}(f)}{\sqrt{G_{x_1x_1}(f)\, G_{x_2x_2}(f)}}$$

where $G_{x_1x_2}(f)$ is the cross power spectral density function between x_1 and x_2, and $G_{x_ix_i}(f)$, i=1,2 are the auto power spectral density functions at frequency, f.

The magnitude-squared coherence (MSC) or simply the coherence is defined by (see, for example, Carter, Knapp and Nuttall (1973))

$$C_{x_1x_2}(f) = \frac{\left|G_{x_1x_2}(f)\right|^2}{G_{x_1x_1}(f)\, G_{x_2x_2}(f)} \tag{2}$$

A useful property of the MSC is

$$0 \leq C_{x_1x_2}(f) \leq 1$$

provided the autospectra are positive (in particular non zero).

In order to attach some physical significance to what the coherence measures, consider that the ocean medium operators M_1 and M_2 are linear time-invariant filters. Thus $s_1(t)$ and $s_2(t)$ in equation (1) are the respective outputs of filters $M_1(f)$ and $M_2(f)$ when excited by source s(t). When the noise, $n_i(t)$, is uncorrelated with the signal, s(t), at the i-th sensor, the ratio of the received signal power at the output of the ocean channel to the corruptive noise power depends on the coherence between the source and the sensor. Specifically, from Carter, Knapp, and Nuttall (1973)

$$\frac{G_{s_i s_i}(f)}{G_{n_i n_i}(f)} = \frac{C_{sx_i}(f)}{1-C_{sx_i}(f)}, \quad i = 1,2 \tag{3}$$

That is, the received signal-to-noise ratio (SNR) at the i-th sensor depends on the coherence between the source and the received waveform. This result has been expressed by Carter and Knapp (1976) more compactly as

$$C_{x_1x_2}(f) = C_{sx_1}(f)\, C_{sx_2}(f) \tag{4}$$

These results apply only to the case where the medium can be accurately modeled by linear time-invariant filters corrupted by uncorrelated additive noise.

RESULTS. For the purpose of obtaining an ML estimate of delay, certain assumptions are required. In particular, for a signal emanating from a nearfield source and monitored in the presence of noise at two spatially separated sensors we require in equation (1) that $s_1(t) = s(t)$ and $s_2(t) = \alpha s(t+D)$. Further, we require that α is real and $s(t)$, $n_1(t)$, and $n_2(t)$ are real, jointly stationary, Gaussian random processes. Source $s(t)$ and noises, $n_1(t)$ and $n_2(t)$ are assumed to be mutually uncorrelated.

An estimated value of D is the hypothesized value τ that maximizes the generalized crosscorrelation (GCC) function defined by

$$\hat{R}(\tau) = \int_{-\infty}^{\infty} \hat{G}_{x_1x_2}(f) W(f) e^{j2\pi f\tau} df. \tag{5}$$

For $x_1(t)$ and $x_2(t)$ real, the ML estimator requires a particular weighting,

$$W(f) = H_1(f)H_2^*(f) = \frac{C_{x_1x_2}(f)}{|G_{x_1x_2}(f)| \left[1 - C_{x_1x_2}(f)\right]} \tag{6}$$

A complete derivation is given by Carter (1976).

Note from equation (9) that for the ML estimate of delay that $W(f)$ is real. The ML estimator is virtually equivalent to one proposed by Hannan and Thomson (1973). The ML estimator can be achieved by shaping $x_1(t)$ with filter $H_1(f)$ and $x_2(t)$ with filter $H_2(f)$ crosscorrelating the filter outputs, and observing what hypothesized value of delay achieves a maximum.

The estimator can also be achieved by other methods. For example, Hahn (1975), Carter and Knapp (1976) and Carter (1976) present a method of filtering and summing the outputs, squaring and averaging in order to estimate the delay D. The processor could also be realized as a number of "best" estimates of D for a variety of frequencies. The ML estimate is then achieved by performing a weighted average across frequency. For example, Clay, Hinich and Shaman (1973) develop ML estimates of bearing (analogous to delay) for each of a number of different frequencies. To obtain a single estimate of source bearing, these individual estimates should then be combined with weighting dependent upon the particular underlying signal and noise characteristics.

The role of coherence in the weighting used for ML estimation of D is specified in equation (6). Note that those values of coherence near unity are most important; conversely, in those frequency bands where there is no source signal power (hence, where the received waveforms are incoherent), the delay estimate, as would be expected, receives no weight. The ML estimator is actually a function of more fundamental spectral measurements than those specified in equation (6). However, expressing the processor in more fundamental but unnormalized quantities can make interpretation more difficult, though equally correct.

The ML weighting agrees with MacDonald and Schultheiss (1969), and Hahn (1975) under specific conditions (including when there are two sensors and no attenuation).

VARIANCE OF GENERAL TIME DELAY ESTIMATORS. The variance of the time delay estimate in the neighborhood of the true delay for general weighting function W(f) is given by

$$\mathrm{Var}\left[\hat{D}-D\right]=\frac{\int_{-\infty}^{\infty}|W(f)|^{2}(2\pi f)^{2}G_{x_1x_1}(f)G_{x_2x_2}(f)\left[1-C_{12}(f)\right]df}{P\left[\int_{-\infty}^{\infty}(2\pi f)^{2}\left|G_{x_1x_2}(f)\right|W(f)df\right]^{2}} \tag{7}$$

where P is the observation period (in seconds). From equations (6) and (7), the variance of the ML processor is

$$\mathrm{Var}^{ML}\left[\hat{D}-D\right]=\left[2P\int_{0}^{\infty}\frac{(2\pi f)^{2}\,C_{12}(f)}{1-C_{12}(f)}\,df\right]^{-1} \tag{8}$$

The ML processor achieves the Cramér-Rao lower bound (see Carter (1976)). Therefore, the ML processor achieves a variance less than or equal to that provided by other correlation processors.

These results for variance can be related to MacDonald and Schultheiss (1969) as follows. Define the bearing to an acoustic source, as in Nuttall, Carter and Montavon (1974)

$$\phi = \text{arc cos}\ \frac{\xi D}{d} \tag{9}$$

where ξ is the speed of sound in the nondispersive medium and d is the sensor separation. Consider the case where the estimated D equals the true delay plus a perturbation. By a Taylor series expansion it follows for the bearing error defined by the difference between the true bearing and the estimated bearing that the standard deviation of the bearing error is given by (Carter (1976)):

$$\left[\text{Var}\ (\hat{\phi} - \phi)\right]^{\frac{1}{2}} = \frac{\xi}{d \sin \phi} \left[\text{Var}\ (\hat{D} - D)\right]^{\frac{1}{2}} \tag{10}$$

The term d sin ϕ can be viewed as the effective array length (sensor separation) physically steered at the source. The combining of equations (8) and (10) suggests that, in order to reduce the variance of the bearing estimate, the observation period and the sensor separation should be made as large as possible. This agrees with one's intuition and the results of MacDonald and Schultheiss (1969). Further, the fact that equation (10) depends on the effective array length physically steered toward the source suggests the desirability of sensor mobility to maximize sin ϕ when d is limited.

REFERENCES.

1. G.C. Carter (1976), Time Delay Estimation, University of Connecticut, PH.D. Dissertation, Storrs, Connecticut (also NUSC Report 5335).

2. G.C. Carter and C.H. Knapp (1976),"Time Delay Estimation," Proc. IEEE 1976 International Conference on Acoustics, Speech and Signal Processing, pp. 357-360.

3. G.C. Carter, C.H. Knapp and A.H. Nuttall (1973), "Estimation of the Magnitude-Squared Coherence Function via Overlapped

Fast Fourier Transform Processing," IEEE Trans. on Audio Electroacoustics, Vol. AU-21, p. 337-344.

4. C.S. Clay, M.J. Hinich and P. Shaman (1973), "Error Analysis of Velocity and Direction Measurements of Plane Waves Using Thick Large-Aperture Arrays," J. Acoust. Soc. Amer., Vol. 53, No. 4, pp. 1161-1166.

5. W.R. Hahn (1975), "Optimum Signal Processing for Passive Sonar Range and Bearing Estimation," J. Acoust. Soc. Amer., Vol. 58, No. 1, pp. 201-207.

6. E.H. Hannan and P.J. Thomson (1973), "Estimating Group Delay," Biometrika, Vol. 60, pp. 241-253.

7. V.H. MacDonald and P.M. Schultheiss (1969), "Optimum Passive Bearing Estimation," J. Acoust. Soc. Amer., Vol. 46, pp. 37-43.

8. A.H. Nuttall, G.C. Carter, and E.M. Montavon (1974), "Estimation of the Two-Dimensional Spectrum of the Space-Time Noise Field for a Sparse Line Array," J. Acoust. Soc. Amer., Vol. 55, pp. 1034-1041.

9. N. Weiner (1930), "Generalized Harmonic Analysis," Acta Math, Vol. 55, pp. 117-258.

TARGET CLASSIFICATION AND HYPOTHESIS TESTING

L. Meier

NATO SACLANT ASW Research Centre
La Spezia, Italy

ABSTRACT. A simple target classification problem, illustrative of all target classification problems, is presented and its solution given. The problems created for classification by unknown target properties and their alleviation are discussed. Finally an extension of the ideas to dynamic situations is outlined.

1. INTRODUCTION

Figure 1 is a block diagram of the target classification problem. In a sonar system, for example, the measurement process might consist of transmission of a sonar ping, reception of the ping, detection, and measurement extraction. Closely related to target classification is target detection. In detection one seeks to separate targets from noise, while in classification one seeks to separate true targets from both noise and false targets. In a real system, classification may or may not be preceded by detection. In sonar systems, for example, the optimal classifier would operate on raw sonar returns — the use of detector and measurement extractor is purely to make the problem more tractable.

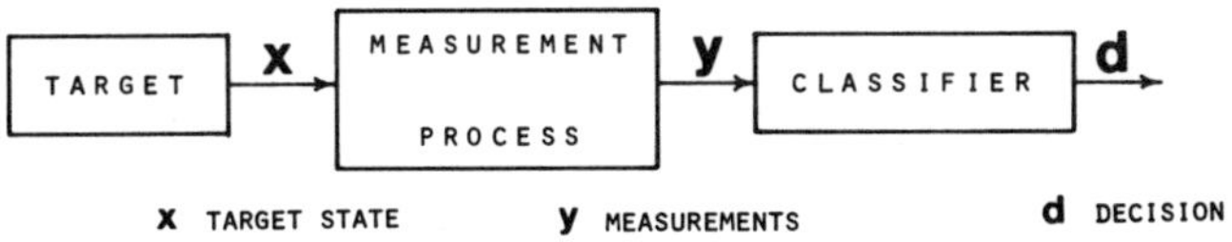

Fig. 1. Block diagram of a classification problem.

G. Tacconi (ed.), Aspects of Signal Processing, Part 1, 257-263.

The statistical theory of hypothesis testing has proved quite useful in detection theory and practice [1]. In this paper this application is extended to cover the target classification problem. One way in which classification differs from detection is that more than two hypotheses may be involved in classification problems. Work on two hypotheses predominates the hypothesis-testing literature, but the many-hypotheses problem has not gone unnoticed [2]. In this paper these results are applied to the classification problem. More importantly, in the target-classification problem, parameters may not be very well specified for some of the hypotheses — in particular statistical parameters of some classes of false targets may be virtually unknown. In this paper, the use of the statistical concept known as the power function [3] is proposed as an approach to the difficulties caused by unknown statistical parameters. Finally, an example is presented to illustrate some of the ideas developed.

2. A CLASSIFICATION PROBLEM

In this section a simple classification problem is defined, its solution presented and difficulties caused by uncertainties about target properties discussed. More general classification problems may readily be formulated; however, lack of time and space prevents their discussion. Fortunately, extension of the results given herein is, in general, obvious.

2.1 Problem statement

Figure 1 is a block diagram of the situation. For the present, only the static situation will be considered, i.e. all measurements are made at the same time and are a function of the target state at this time. Three hypotheses exist about measurements, namely they are a result of: noise only (H_0), a false target (H_1), or a true target (H_2). The classifier must decide whether or not H_2 is true.

Three error probabilities exist: the probability of missed classification (saying H_2 is not true when it is), the probability of false alarm for noise only (saying H_2 is true when H_0 is true), and the probability of false alarm for false targets (saying H_2 is true when H_1 is true). These may be combined into a performance criterion in two ways: in the Neyman-Pearson approach two probabilities are constrained while the third is minimized; in the Bayes approach a linear combination of the probabilities is minimized.

2.2 Problem solution

Figure 2 is a block diagram of the classifier that is optimum for both of the performance criteria. The parameters a and b must be chosen to optimize the criterion used. The proof of this result is both standard [2] and simple, so it is not repeated here. To compute the likelihoods $p(y/H_i)$, $i = 0, 1, 2,$ it is convenient if the target models take the form of probability densities $p(x/H_i)$, $i = 1, 2$ and if the measurement process model takes the form of the probability densities $p(y/H_0)$ and $p(y/x, \bar{H}_0)$ (where $\bar{H}_0$ means not H_0). Then $p(y/H_i)$, $i = 1, 2$ is given by

$$p(y/H_i) = \int_x p(y/x, \bar{H}_0)\, p(x/H_i)dx \tag{2.1}$$

2.3 Unknown target properties

In general, $p(y/H_0)$ and $p(y/x, \bar{H}_0)$ are determined more or less readily by the physics of the measurement process. On the other hand, determining $p(x/H_i)$ is not nearly so easy. For true targets, such as submarines and some classes of false targets such as decoys, the form of $p(x/H_i)$ is not too hard to determine and appropriate parameters may be selected with some degree of confidence. But what of unknown false targets such as schools of fish? A reasonable assumption is that the form of $p(x/H_1)$ for such false targets is similar to $p(x/H_2)$ (else they would not be readily confused with true targets) i.e. $p(x/H_i) = p(x/\alpha_i)$, $i = 1, 2$ where α is an appropriate parameter vector. Picking the statistical parameter values α_i is a much more complicated task.

In statistics it is not uncommon that the null hypothesis is better defined than the alternative hypothesis. In this case the power function is quite useful [3]. In our problem, the power plot is a plot of the probability that the decision is that a true target is present, given $x = \xi$ versus ξ. For a given classifier such a plot is a function of $p(y/x, \bar{H}_0)$ only and thus is more dependable than values of the error probabilities as a measure of system performances. The use of the power plot is better described using an example.

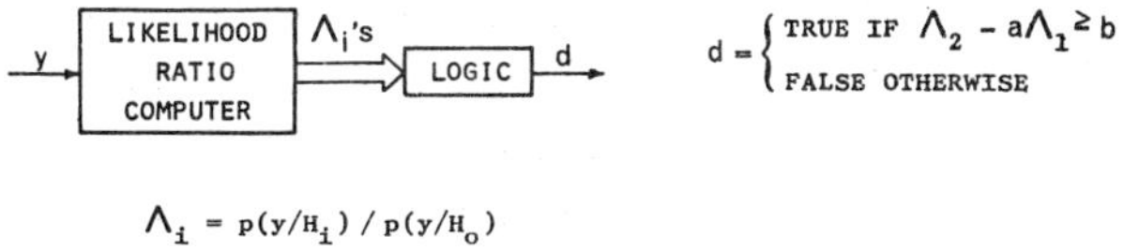

Fig. 2. Block diagram of the optimal classifier.

3. AN EXAMPLE

To illustrate these ideas consider the situation in which an LCW ping is sent out by a sonar to detect and classify a submarine.

3.1 Statement

With an LCW ping, range rate $\dot{r}$ and a nominal echo strength $\widetilde{ES}$ may be measured. (The former from the frequency shift of the return, the latter by dividing the measured S/N ratio by a nominal zero-target-strength S/N ratio.) Target-state variables pertaining to these measurements are: target speed v and average target strength $\overline{TS}$. The following assumptions are made:

(1) Range rate measurement error is negligible.
(2) All target aspects are equally likely.
(3) Measured echo strength is exponentially distributed and independent of aspect given the target strength.
(4) Target speeds are uniform between v_{MIN} and v_{MAX}.
(5) Average target strengths are uniformly distributed between $\overline{TS}_{MIN}$ and $\overline{TS}_{MAX}$.
(6) Detection thresholds are $|\dot{r}| > 1.5$ and $\widetilde{TS} > 1$.
(7) S/N is sufficiently high that only targets are detected.

Assumption (2) implies that $\dot{r}$ may be replaced by $|\dot{r}|$. Clearly for the example:

$$x = \begin{bmatrix} v \\ \overline{TS} \end{bmatrix} \qquad y = \begin{bmatrix} |\dot{r}| \\ \widetilde{ES} \end{bmatrix} \quad \text{and} \quad \alpha = \begin{bmatrix} v_{MIN} \\ v_{MAX} \\ \overline{TS}_{MIN} \\ \overline{TS}_{MAX} \end{bmatrix}$$

Furthermore, assumptions (1) to (3) define $p(y/x, \bar{H}_o)$ and assumptions (4) and (5) define $p(x/\alpha)$.

3.2 Solution

Because of assumption (7), $p(y/H_o) = 0$; therefore the decision rule of Fig. 1 may be replaced by

$$\lambda'_{FT} \triangleq \ell n \;\; p(y/H_1)/p(y/H_2) \;\; \leq \lambda_{TH} \Rightarrow \begin{cases} \text{a submarine} \\ \text{is present} \end{cases} \qquad (3.2)$$

Exact calculation of the likelihood $p(y/H_i)$, $i = 1, 2$ is rather complicated; therefore the following simplifying assumptions were made: $|\dot{r}|$ is normally distributed and $\widetilde{ES}$ is exponentially distributed. Given these assumptions:

$$\lambda'_{FT}(|\dot{r}|, \widetilde{ES}) = C_r(|\dot{r}| - |r|)^2 - C_{ES}\,\widetilde{ES} \tag{3.3}$$

3.3 Simulation results

A Monte Carlo simulation of the example was made. It was assumed that it was known that $v^{SUB}_{MIN} = 6$, $v^{SUB}_{MAX} = 10$, $\overline{TS}^{SUB}_{MIN} = 12.6$, $\overline{TS}^{SUB}_{MAX} = 200$. The corresponding values for a false target were not known; therefore it was assumed that the variation of false target speed was as large as possible and that false target strengths were less than submarine target strengths: $v^{FT}_{MIN} = 0$, $v^{FT}_{MAX} = 16$, $TS^{FT}_{MIN} = 0.3$, $TS^{FT}_{MAX} = 63$. Finally the following likelihood parameters were used: $\overline{|\dot{r}|} = 7$, $C_{\dot{r}} = 0.02$, $C_{ES} = 0.03$.

Note that while the classifier was designed on the basis of simplification, the actual distributions were used to generate the classifier inputs. The results are summarized in Figs. 3 and 4. Figure 3 is the familiar ROC: $P_C = 0.866$ is the highest probability of classification, which is obtained when all returns that are detected are classified as submarines. The power function for $\lambda_{TH} = 0.326$ shown in Fig. 4 depends only on the measurement process model and the classifier design and not on the target model. On the other hand the ROC does depend upon the target model in addition to the other factors and hence is less reliable.

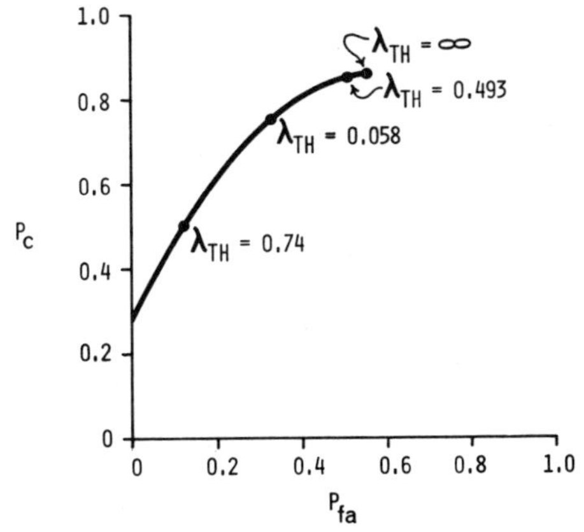

Fig. 3. ROC for the example.

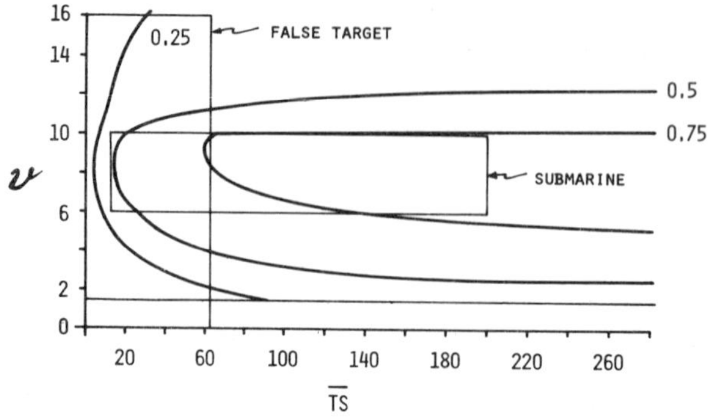

Fig. 4. Contour plot of probability that d=true for $\lambda_{TH} = 0.326$.

Now consider the question of how a power function such as that given in Fig. 4 may be used in classifier design. The classifier in the example has four parameters: $C_{\dot{r}}$, $|\bar{\dot{r}}|$, C_{TS} and λ_{TH}. It is reasonable to demand that, no matter what the submarine state, the probability of classification as a submarine be, say, 0.75. Then the four parameters must be adjusted until the contour P(d=true) = 0.75 surrounds the box labelled true target in Fig. 4. There may be many parameters sets for which this is true; one would then want to adjust the parameters so that the P(d=true) = 0.75 contour lies as close as possible to the true target boundary. There is still likely to be a degree of arbitrariness in classifier parameter selection and it may be both necessary and desirable to modify the parameters in actual operation. On the other hand, it may not be possible to find any parameters such that the constraint on probability of classifying submarines is met or, more commonly, that sets of parameters that do not meet the constraint do not eliminate many false targets. If so it must be concluded that the signal is not adequate for classification and a different signal must be tried.

4. AN EXTENSION

In the above it was assumed that the measurements are all made at a single time. Frequently, the measurements may be made at a series of times (for example, if a series of sonar pings is sent out) and a dynamic model of the target may then be required. From differential equations describing the dynamic behaviour of the model, difference equations for the state as a function of measurement time may be obtained. Equation (1.1) still applies but may be very difficult to evaluate from the state equation and equations describing the measurement process. In addition, there is the question of whether track detection is to be performed as part of the detection process or of the classification process. For a given track, (1.1) may be evaluated recursively by using a slight extension of the equations for Bayesian estimation [4]. In general, even these recursive equations are hopeless to solve. In many situations, however, they may be approximated by the extended Kalman filter [5].

In the dynamic case it is convenient to factor all the uncertainty about the target model into uncertainty about the initial target state. In this situation the power plot becomes a plot of probability of classification versus the value of the initial state.

5. CONCLUSIONS

Use of the statistical theory of hypothesis testing in target classification appears to be as fruitful as its application to target detection. The relative lack of knowledge of false-target statistical parameters, however, makes the classification problem far less straightforward. This difficulty is mitigated by consideration of the power function, which gives the probability of classifying a target as a true target as a function of target state (initial state in dynamic situations). Much future work remains to be done. To mention only a few possibilities: more realistic examples need to be treated; sensitivity to changes of classifier and target parameters needs to be investigated (preliminary indications are that sensitivity to classifier parameters is low); and ways of adaptively optimizing classifier parameters in operation should be explored.

REFERENCES

1. Van Trees, H.L. Detection, Estimation, and Modulation Theory, Pt 1. New York, Wiley, 1968: pp 239-418.
2. ibid: pp 46-52.
3. Kendall, M.G. and Stuart, A. The Advanced Theory of Statistics, Vol. 2: Inference and Relationship. New York, Hafner, 1967: pp 179-184.
4. Ho, Y.C. and Lee, R.C.K. A Bayesian approach to problems in stochastic estimation and control. IEEE Trans. on Automatic Control AC-9, 1966: 333-339.
5. Cox, H. On the estimation of state variables and parameters for noisy dynamic systems. IEEE Trans. on Automatic Control AC-9, 1964: 5-12.

AN ADAPTIVE KALMAN FILTER TRACKER FOR MULTI-MODE RANGE/DOPPLER SONAR

Dr. C. Nicholas Pryor

Technical Director
Naval Underwater Systems Center
Newport, Rhode Island, USA

In most modern sonar systems, each active ping produces a combination of information on both the range and the range rate of the target. This information is then used to form the basis of a target track. In this paper, I propose to discuss the use of Kalman filter methods to make optimum use of the incoming range and doppler information in forming the track. The first part of the discussion will show the fairly routine application of linear Kalman filter mathematics, and will expand the matrix equations to show the actual form of the filter and some of its characteristics. Two methods will then be discussed for making the filter adaptive to target maneuvers by sensing the statistical properties of the measurement residuals. Finally a means will be described for using the uncertainty matrix of the tracker to select the optimum waveform for the next ping, where the desire is to minimize the range error at some future time.

Before proceeding further, I would like to acknowledge the contribution of Mr. Ed Hug of the SacLant Center to this work. During discussions at the time of the last Institute in Loughborough, we found we were both interested in this topic. Certain of the ideas presented, and particularly in the area of adaption, were contributed by Mr. Hug during these discussions.

The steps required in defining a Kalman filter problem involve first describing the state vector X to be estimated, modeling the dynamic evolution and the random processes influencing this state vector, and then describing the relationship between this state vector and the measurements being made on it along with the uncertainty in these measurements. For the problem at hand, we will estimate the range and the range rate of the target. Thus,

G. Tacconi (ed.), Aspects of Signal Processing, Part 1, 265-278. *All Rights Reserved.*

the state vector X becomes

$$X = \begin{bmatrix} x_1 \\ x_2 \end{bmatrix} = \begin{matrix} \text{range in meters} \\ \text{range rate in meters/second} \end{matrix} \quad .$$

In describing the evolution of this state vector, we wish to show that the range is the integral of the range rate and that the range rate is influenced randomly by target maneuvers. This can be done by writing

$$\dot{x}_1 = x_2$$

$$\dot{x}_2 = (w - x_2)/\tau$$

where the differential equation for x_2 causes it to behave as a low-pass process with relaxation time τ with w as a white noise excitation. Thus, the range rate will evolve from its current value toward some new unknown value described only by its variance, in a time on the order of τ seconds. Putting this into the matrix notation of continuous Kalman filters gives

$$\dot{X} = F\,X + G\,w$$

where

$$F = \begin{bmatrix} 0 & 1 \\ 0 & -1/\tau \end{bmatrix} \quad \text{and} \quad G = \begin{bmatrix} 0 \\ 1/\tau \end{bmatrix} \quad .$$

Now given this information on the dynamics of the process, it can be shown that the uncertainty matrix P on the estimate of the state evolves according to the differential equation

$$\dot{P} = F\,P + PF^t + G\,Q\,G^t$$

where Q is the power density of the noise process w. Expanding this matrix equation for the three components of P gives

$$\dot{P}_{11} = 2\,P_{12}$$

$$\dot{P}_{12} = P_{22} - P_{12}/\tau$$

$$\dot{P}_{22} = -2\,P_{22}/\tau + Q/\tau^2 \quad .$$

The above differential equations are simple enough to be integrated to show the evolution in both the best state estimate and in the uncertainty matrix from a time t_o to a new time t.

These results are

$$x_1(t) = x_1(t_o) + \alpha\,\tau\,x_2(t_o)$$

$$x_2(t) = (1-\alpha)x_2(t_o)$$

$$P_{11}(t) = P_{11}(t_o) + 2\alpha\tau\,P_{12}(t_o) + \alpha^2\tau^2 P_{22}(t_o) + \left[2(t-t_o)/\tau - 2\alpha - \alpha^2\right]\tau^2\,V^2$$

$$P_{12}(t) = (1-\alpha)\,P_{12}(t_o) + \alpha(1-\alpha)\tau\,P_{22}(t_o) + \alpha^2\tau\,V^2$$

$$P_{22}(t) = (1-\alpha)^2 P_{22}(t_o) + \alpha(2-\alpha)\,V^2$$

where

$$\alpha = 1 - \exp(-(t-t_o)/\tau)$$

and $V^2 = Q/2\tau$ has been introduced as the value toward which the mean squared range rate uncertainty evolves.

It is interesting to look at these equations both for prediction times short compared to τ, where $\alpha \simeq (t-t_o)/\tau$, and for long prediction times where α approaches unity. In the short time limit, these equations may be reduced by dropping higher order terms in $t-t_o$ to those which are obtained if a discrete model of the process were used. In the long time limit the range rate uncertainty P_{22} evolves from $P_{22}(t_o)$toward the a priori uncertainty in velocity V^2. The covariance term P_{12} evolves toward the constant τV^2, which indicates that the range error and range rate error in the estimates will eventually be positively correlated. The range uncertainty P_{11} has perhaps the most interesting characteristic. This uncertainty contains a term contributed by each of the initial components of the uncertainty matrix, and each of these terms approaches a constant. In addition, there is a term which eventually increases linearly with time and is proportional to the variance of the target velocity and to its relaxation time.

It is now necessary to describe the process by which new measurement data are inserted into the filter. This is a discrete process and is described by a measurement matrix H giving the relationship between the measured parameters and state vector of the filter. Since the measured information from each ping is also range and range rate, it would seem that the relation between the state vector and the measurement is trivial. However, this is not so because of the acoustic propagation time delay involved. The measurement is taken at the time the signal reflects from the target, but is not available at the tracker input until the signal returns to the sonar. If an attempt is made to maintain the track

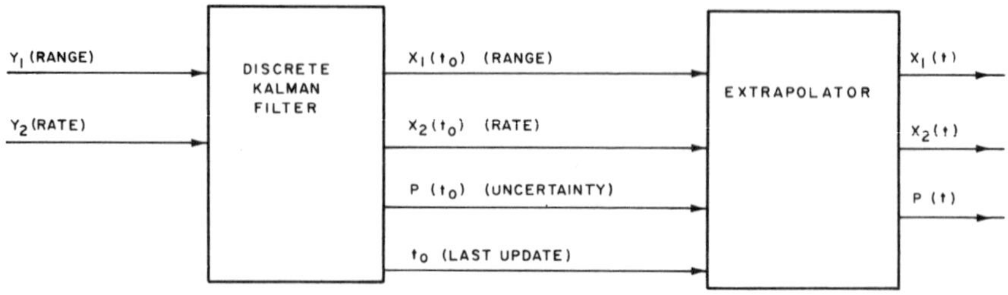

Fig. 1. Configuration of the Kalman filter.

directly in current time, a term of the form $x_1 - x_1x_2/c$ appears in the measurement equations and causes the tracker to be nonlinear. This complication can be eliminated by developing the tracker in the form shown in Figure 1. The first part is the actual Kalman filter, and it is updated once per ping cycle with a new state vector X, a new uncertainty matrix P, and a time t_0 representing the time of reflection of the last echo. This time is computed by subtracting the one-way travel time at the estimated range from the actual time of receipt. While this technique does not completely eliminate the nonlinearity, because of the uncertainty in the range estimate, it does reduce it to second order.

In most cases a range estimate is desired more frequently than once per ping cycle. Thus, a second element is shown in Figure 1, which uses the most recent information in the Kalman filter to provide a current (or future) estimate for X, and if desired for the estimated uncertainty P. With this technique the measured parameters become identical to the state parameters, so the measurement matrix H is simply a unit matrix. The error matrix for the measurement is a 2 by 2 matrix R.

With this definition of the problem, the process for updating the Kalman filter estimate involves first computing the new update time t by subtracting the one-way travel time from the time of echo reception (or adding it to the time of transmission). The state estimates and the P matrix are then updated to that time. The process of inserting the new data involves first computing a Kalman gain matrix of the form

$$K = P\,H^t \left[H\,P\,H^t + R\right]^{-1} \quad ,$$

which is multiplied by the vector of measurement residuals Y - H X to map them into corrections to the state vector in the form

$$X = X + K\left[Y - H\,X\right] \quad .$$

The K coefficients thus represent the relative weights attached to the new measurement as compared to the previous estimate. Finally the P matrix is adjusted according to the equation

$$P = P - K\,H\,P$$

to represent the reduction in uncertainty due to the introduction of the new data.

Since the H matrix has been found to be a unit matrix, we can expand these equations in terms of the P matrix at the time of the echo and the R matrix associated with the ping waveform to give

$$K_{11} = \left[P_{11}(P_{22} + R_{22})-P_{12}(P_{12} + R_{12})\right] / \left[(P_{11} + R_{11})(P_{22} + R_{22})-(P_{12} + R_{12})^2\right]$$

$$K_{12} = (P_{12}R_{11}-P_{11}R_{12}) / \left[(P_{11} + R_{11})(P_{22} + R_{22})-(P_{12} + R_{12})^2\right]$$

$$K_{21} = (P_{12}R_{22}-P_{22}R_{12}) / \left[(P_{11} + R_{11})(P_{22} + R_{22})-(P_{12} + R_{12})^2\right]$$

$$K_{22} = \left[P_{22}(P_{11} + R_{11})-P_{12}(P_{12} + R_{12})\right] / \left[(P_{11} + R_{11})(P_{22} + R_{22})-(P_{12} + R_{12})^2\right]$$

for the Kalman gain equations,

$$x_1 = x_1 + K_{11}(y_1-x_1) + K_{12}(y_2-x_2)$$

$$x_2 = x_2 + K_{21}(y_1-x_1) + K_{22}(y_2-x_2)$$

for the state vector corrections, and

$$P_{11} = (1-K_{11})\,P_{11} - K_{12}P_{12}$$

$$P_{12} = (1-K_{11})\,P_{12} - K_{12}P_{22}$$

$$P_{22} = (1-K_{22})\,P_{22} - K_{21}P_{12}$$

for the adjustments to the P matrix. Note that the differences between the observed and predicted values of both range and range rate are used in correcting both state estimates, as desired. The Kalman gains tend to be ratios of variances. In a simple example, if both P_{12} and R_{12} were zero, K_{11} becomes $P_{11}/(P_{11} + R_{11})$ or the variance of the range estimate divided by the variance of the measurement residual, as might have been expected. Note that for any cross-coupling to occur between the range and range rate estimates, either P_{12} or R_{12} must be non-zero. However, as we have seen earlier, the P_{12} covariance term will always evolve toward a positive non-zero value.

As shown before, the terms of the P matrix tend to increase between measurements due to the random process assumed in the model, while it is seen here that the measurement process tends to decrease the P matrix. Normally, if measurements are taken on a regular basis these two effects balance to give stable values to the P matrix. These effects point up the need for the random perturbations in the target model. If they were not included, the P matrix would tend to zero, causing all the Kalman gain terms to go to zero. The filter would then no longer respond to new measurements, being satisfied that it knew all there was to know about the target. This is the malfunction known as divergence in Kalman filter literature.

A simulated example of the performance of this Kalman filter is based on the target track shown in Figure 2a. The target proceeds inbound at 10 meters/ second (about 20 knots) from a range of 10 kilometers to a range of 5 kilometers. It then makes a 90-degree turn at 3 degrees per second and again proceeds in a straight line. This track, translated into a trajectory in range/range rate space, is shown in Figure 2b. Sonar data is assumed to come from a sonar operating at 5 kiloHertz with a 100 millisecond CW pulse. A pulse is transmitted immediately after each echo is received, giving an interpulse interval of 6.5 to 13 seconds, depending on range. While the measurement uncertainty in an echo strictly depends on signal-to-noise ratio, this dependency is ignored here and the standard deviations in the echo measurement are assumed to be 100 milliseconds in time and 10 Hertz in doppler. This is equivalent to 75 meters in range and 1.5 meters/second in range rate, and these two errors are uncorrelated.

The most interesting part of the track is the shaded area of Figure 2b, extending from just before the turn until the target range and range rate again begin to change significantly. This region is shown enlarged in Figure 3a, with the actual target track shown by the solid curve and the measured data shown by the open circles connected by the dotted lines. Since the measurement error is large compared to the target motion between pings, smoothing of the data is required in order to form a sensible target track. Actual rms error of range samples in the 20 pings before the maneuver was 67.4 meters, while the actual rms error in range rate samples was 1.45 meters/second. Performance of various forms of the Kalman filter will be compared in terms of the amount of smoothing of these data points during steady-state conditions and by the general nature of the response to the maneuver.

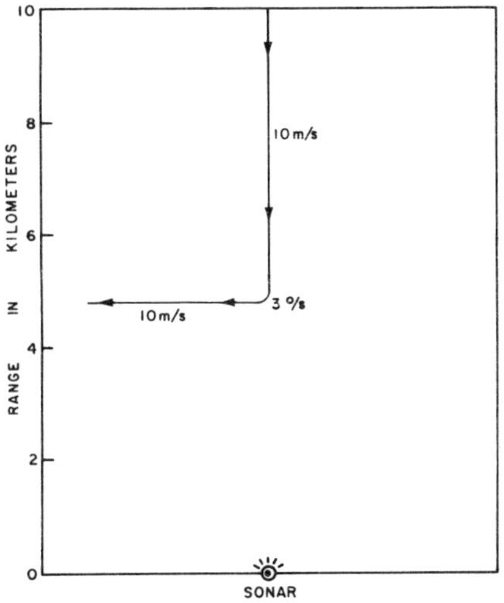

Fig. 2a. Geometry of simulated run.

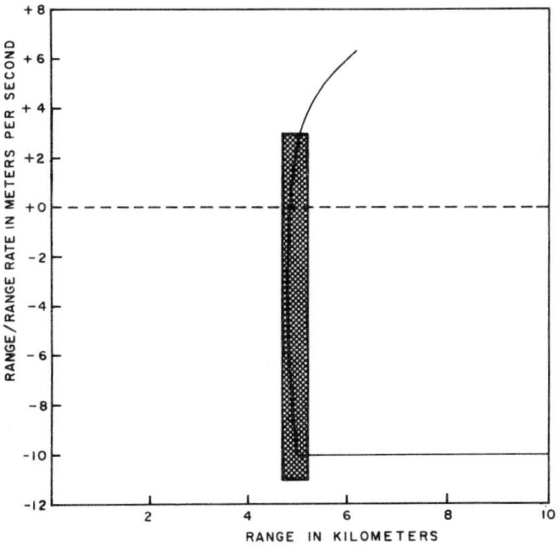

Fig. 2b. Range/range rate trajectory of simulated run.

Figure 3b shows the data as tracked by the Kalman filter, using an arbitrarily selected relaxation time τ of 300 seconds. Each circled dot represents the output of the filter immediately after a ping is received, while the straight line is the prediction from that point until the next ping, as would be performed by the second block in Figure 1. During the steady-state portion of the track, rms range error was reduced to 19.65

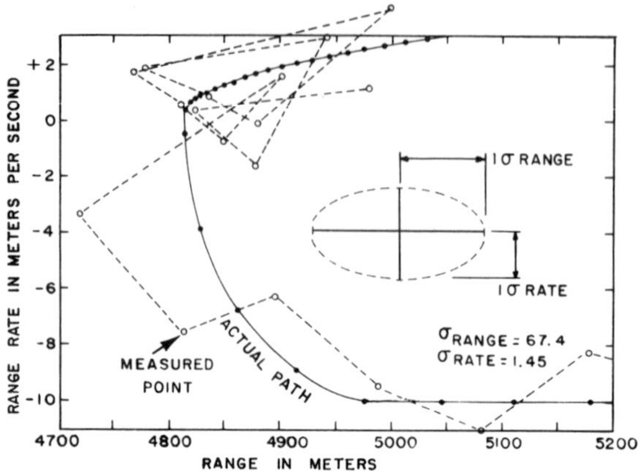

Fig. 3a. Raw measurement data points.

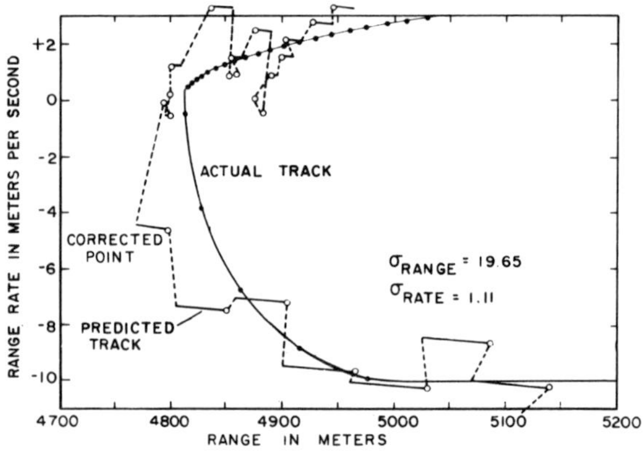

Fig. 3b. Kalman filter tracker output (τ = 300 sec).

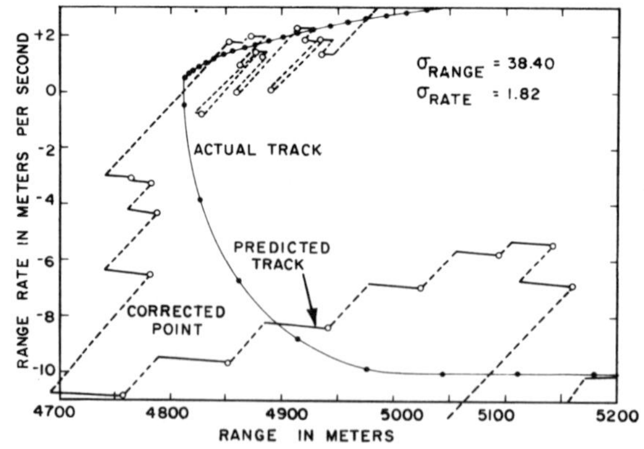

Fig. 3c. Kalman filter output with range data only (τ = 300 sec).

meters, or a factor of about 3.5. Rms range rate error was reduced to 1.11 meters/second, or only slightly below that of the input data. Response during the maneuver was good, primarily as a result of the limited smoothing of range rate data. Since the relaxation time τ determines the steady-state values of the P matrix, selection of this parameter determines the effective filter time constants and thus the balance between smoothing and responsiveness. Significantly larger τ would improve the smoothing, but at the expense of responsiveness to the maneuver. Note that the rate estimate stays nearly constant between pings, and there is no tendency for it to predict the actual range rate changes during the maneuver. This is due to the two-state model used; any prediction of range rate would require addition of a third acceleration variable to the state vector.

It is interesting to compare this track with one obtained using only the range data from each ping and ignoring the range rate data obtained from echo doppler. This track is shown in Figure 3c. The rms range error during the straight-line portion of the track was 38.40 meters, or about double the error obtained when the rate data was included. The rms range rate error, derived entirely from the range samples, was 1.82 meters/second, which is inferior to the samples of the input rate data. This form of the Kalman filter, estimating range and rate from range data only, is equivalent to the "α,β tracker" often used in radar systems.

As mentioned previously, the smoothing performance of the Kalman filter tracker shown can be improved by increasing the assumed value of τ, but the response to maneuvers eventually becomes unacceptable. This conflict can be overcome by attempting to sense target maneuvers and adapting the parameters of the filter to them. This is rather easy to do with a Kalman filter, based on the fact that the scalar function

$$z = \left[Y-HX\right]^t \left[HPH^t+R\right]^{-1} \left[Y-HX\right]$$

is a normalized chi-squared random variable with degrees of freedom equal to the size of the measurement vector, when the target and measurement data are properly modeled. If this statistic is formed for each data sample, it may be tested against a threshold to determine whether the input data residuals are reasonably explainable as random errors. If an abnormally large value of z occurs, it may be used as a signal that the target has maneuvered and that we wish to increase the responsiveness of the filter. Two simple ways of implementing this will be illustrated: the first using samples of z to adapt the parameter τ to the target behavior, and the second operating directly on the P matrix.

Adaptation in τ is performed by continually increasing τ (decreasing the assumed maneuvers of the target) as long as z does not exceed a threshold, and decreasing τ whenever a large value of z suggests that the target may be maneuvering faster than the model assumed. This can be done in rather crude ways, since changes in τ only affect the rate of growth of the P matrix which, in turn, determines the Kalman gains. Figures 3d and 3e show the performance of two trackers using adaption in τ. In the first case, a threshold z_0 was set at the median value 1.386 of the chi-squared variate, and τ was doubled whenever z fell below z_0 and cut in half when z exceeded z_0. With this approach, one would expect τ to stabilize around a value consistent with the random maneuvers of the target. During the straight-line portion of the example run, τ did increase to very large values consistent with the non-maneuvering target. This was reflected in additional smoothing of the input data, particularly in range rate. The rms range rate error at the tracker output was 0.37 meters/second, or about a third that of the non-adaptive tracker, while the range error was not significantly different from the non-adaptive case. Once the maneuver started, it was detected immediately by the z statistic test. However the filter output did lag significantly in both range and rate while the τ values were being reduced enough to affect the P matrix, and therefore, the input data weightings.

Figure 3e shows the somewhat different performance obtained when the threshold z_0 was changed to the 80 percentile point 3.219 on the chi-square distribution. In this case τ was increased by only $\sqrt[4]{2}$ or about 20 percent when z was below z_0, but still was cut in half when z exceeded the threshold. This still causes τ to seek the proper long-term value but to increase less rapidly during non-maneuvering periods. This led to a more responsive track when the maneuver occured in the example, but at the expense of an increase in the rate tracking error during the straight-line segment.

The second method of adaptation has less to do with the physical model of the target but is also quite effective. This technique simply increases the value of the P matrix (essentially saying I must know less about the state of the target than I thought I did) whenever z exceeds a threshold. Since each new measurement tends to decrease the P matrix anyway, it is not necessary to reduce the P matrix when z falls below the threshold. It is possible to make selective increases in the P matrix whenever a simple relation exists between the state variables and the measured data, for example, increasing both the row and column corresponding to a given variable whenever its residual exceeds a threshold. However in the example shown in Figure 3f, the entire P matrix is simply doubled whenever z exceeds the 95 percentile point 5.991. Since the P adaptation provides a means of

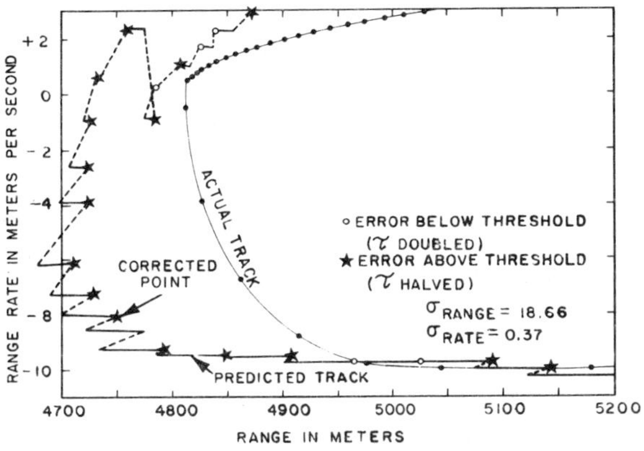

Fig. 3d. τ -adaptive Kalman filter output.

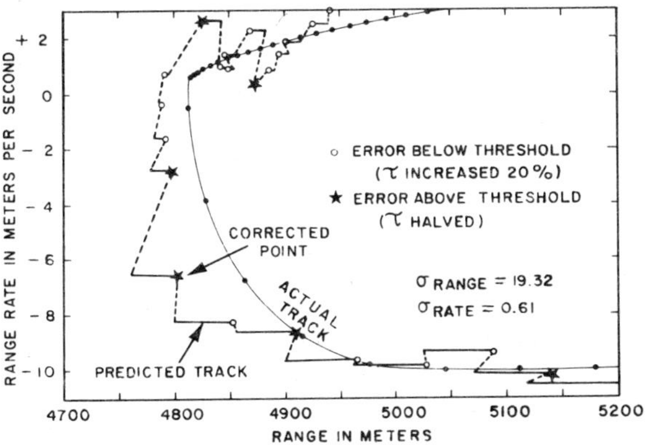

Fig. 3e. τ -adaptive Kalman filter output.

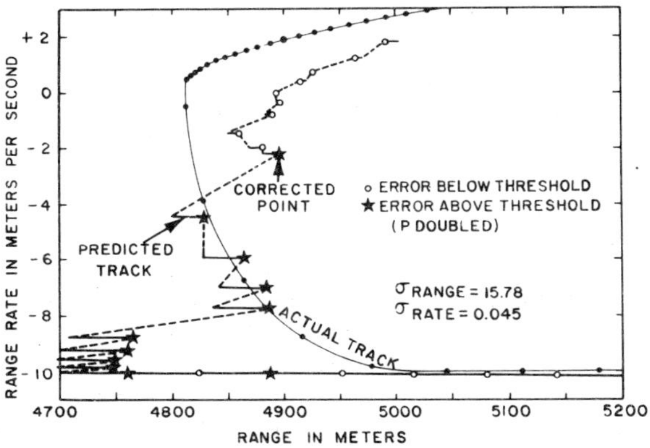

Fig. 3f. P-adaptive Kalman filter output.

increasing the P matrix, the random excitation in the target model was removed. This made the model correspond to a non-maneuvering target and left the adaptive mechanism as the only means of avoiding divergence in the filter. Note that during the straight-line portion of the track the rms range rate error was only 0.045 meters/second, reduced about a factor of 30 from the error of the input data. However because the P_{22} term of the P matrix gets quite small, the filter is quite slow to respond to changes in the range rate even after the maneuver is detected. This could be avoided by keeping a small randomness in the target model to limit the smoothing of the range rate data.

While neither of the methods of adaption discussed here completely reaches the goal of an optimum track for both maneuvering and non-maneuvering targets, both are quite simple to implement and provide some performance improvement. In a practical application, one or the other (or perhaps a combination of the two) could provide useful improvements over non-adaptive trackers.

A second interesting use can be made of the P matrix information in the Kalman filter. As we have seen earlier, the P matrix can be projected into the future to show the state of target uncertainty at some future time, assuming no more information is gained from the sonar. This in itself is tactically useful. We can also estimate the effect of an additional sonar ping by projecting the P matrix to the predicted time of the echo, adjusting it to account for the information expected from that echo, and then continuing the projection on into the future. This can be used as an algorithm for selecting ping waveforms in a multimode sonar.

Suppose there is some time in the future, such as the expected time of arrival of a weapon, at which the target range should be known as accurately as possible. By carrying out the above process for each of the available ping waveforms, the expected future range uncertainties can be compared and the waveform giving the lowest value for P_{11} selected. The form of the previously shown equation for $P_{11}(t)$ shows the intuitive result of this process. The portion of this expression which can be affected by the ping waveform selected is $P_{11}(t_o) + 2(\alpha\tau)P_{12}(t_o) + (\alpha\tau)^2P_{22}(t_o$. As long as the projection is far into the future so that $\alpha\tau$ is large, it is important to minimize $P_{22}(t_o)$ and waveforms with good doppler resolution will tend to be preferred. Once the projection time becomes short, however, the $P_{11}(t_o)$ term becomes the most important and waveforms with good range resolution will tend to be selected. It is also clearly an advantage to make $P_{12}(t_o)$ as negative as possible, which leads to a preference for waveforms with large negative R_{12}.

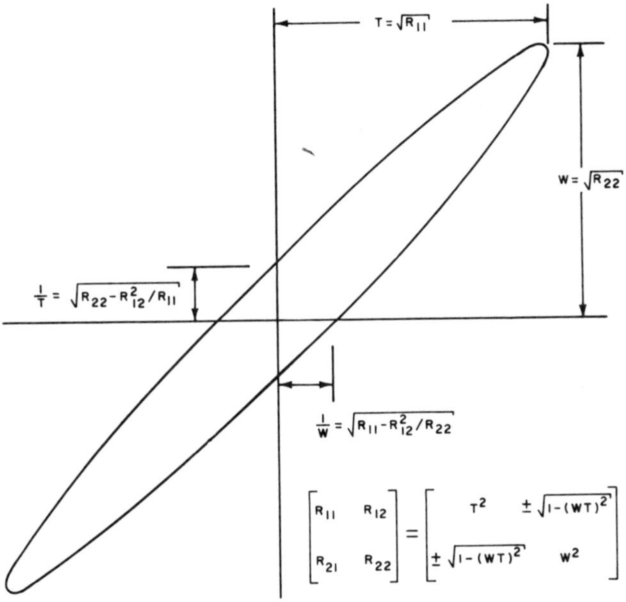

Fig. 4. Ambiguity matrix for linear FM signal.

Consider, as an example, a sonar with the option of either continuous wave (CW) or upward or downward linear FM transmissions. As before, we will assume for a ping duration of T seconds that the time (range) measurement variance of the CW signal is T^2 and the variance of the frequency (range rate) measurement is $1/T^2$ with no correlation between them and independent of signal-to-noise ratio. Figure 4 shows how the R matrix is derived for the linear FM signals, using the familiar ambiguity function for such signals. The mean square range error when doppler is unknown is still T^2, but this value reduces to $1/W^2$ when the target doppler is known. Similarly, the variance in frequency when range is unknown is W^2, while it becomes $1/T^2$ when the range is known. Since the diagonal terms of the R covariance matrix are the variances of the marginal error distributions, and since the conditional variances are smaller than the marginal variances by the factor $1 - R_{12}{}^2/R_{11}R_{22}$, we find that for a linear FM signal the R matrix is of the form

$$\begin{bmatrix} R_{11} & R_{12} \\ R_{21} & R_{22} \end{bmatrix} = \begin{bmatrix} T^2 & \pm\sqrt{W^2T^2-1} \\ \pm\sqrt{W^2T^2-1} & W^2 \end{bmatrix}$$

where the plus sign applies for upward FM signals while the minus sign applies for downward FM signals.

Limited simulation experiments with such a sonar confirm that when the objective time, at which the range error is to be minimized, is well into the future the CW mode is invariably selected because of its better range rate accuracy. As the objective time approaches, the sonar finally switches to a down-slope FM signal to take advantage of its conditional range accuracy. The down slope is preferred to the up-slope FM because the range rate error tends to be in a direction to reduce the range error at a future time. If all parameters of the problem were known in advance, it appears that one could pre-compute the optimum time to switch from CW to FM transmissions. However, if this technique is to be combined with adaption or with compensation for observed signal-to-noise ratios, it may be an advantage to do the waveform selection computations in real time on each ping.

In summary, this discussion has shown the application of Kalman filter techniques to range and doppler tracking in active sonars. The Kalman filter inherently provides an uncertainty matrix indicating the accuracy of the solution. Uses of this uncertainty matrix formation to provide adaptive operation and to aid in waveform selection in a multi-mode sonar were also discussed.

STRUCTURE ESTIMATION FROM ACOUSTIC REFLECTION MEASUREMENTS*

Kenneth B. Theriault† and Arthur B. Baggeroer††

†Graduate student, Department of Electrical Engineering, MIT, Cambridge, Massachusetts, USA
††Associate Professor, Departments of Ocean and Electrical Engineering, MIT, Cambridge, Mass., USA

ABSTRACT. A preliminary study of a unique approach to the determination of the structure of a medium from noisy normal-incidence acoustic reflection data is carried out. This approach employs a simple equal travel-time layer model for the medium, implemented by a state variable formalism. Structure determination is achieved by maximum-likelihood estimation of the parameters which define the system model, the reflection coefficients. The Cramer-Rao bound on estimation accuracy is derived for the model, and computed for two cases of practical interest. It is shown that physically significant conclusions may be drawn from the behavior of the bound. Monte Carlo tests are used to compare the noise sensitivity of this estimator with that of the prediction error filter estimator. This latter estimator, although exact in the absence of noise, is found to be unstable when noise is present, and has higher variances when it remains stable.

1. INTRODUCTION

The use of normal-incidence acoustic reflection data to infer the structure of a medium is a classical method in many fields, including geophysics. A definitive solution of the inverse problem, quantitative structural determination given such reflection data, has, however, not been found for realistic earth models and data. This is not surprising in view of the difficulty of the

---

* This research was supported, in part, by the Office of Naval Research, under Contract N00014-67-0204-0086.

G. Tacconi (ed.), Aspects of Signal Processing, Part 1, 279-295.

general problem. Indeed, to date, the structural information extracted from normal-incidence acoustic reflection data has been almost exclusively qualitative. The medium under investigation is characterized by the locations of the strong components of its reflection response, which are identified qualitatively as layer interfaces.

Despite the successes of qualitative analysis, quantitative results are certainly preferable. In some situations, such as studies of acoustic propagation paths involving ducting along the sea floor, quantitative structural information is a necessity. In geophysical exploration, of course, the more quantitative information available, the better. Several quantitative methods have been developed, at least in theory, but all are deficient in some aspects: none represents a generally useful method for the quantitative determination of the structure of the medium from a noisy acoustic reflection response.

One notable method, termed "bright spot," is used extensively in petroleum seismic prospecting [1,2]. Here, the response is subjected to a time-varying gain, tailored to compensate for geometrical spreading, earth attenuation, reflection losses, and deconvolution-related distortion. The time-varying gain ideally recovers the true relative amplitudes of returns from discrete reflectors, enabling the visual detection of large reflection coefficients and phase shifts, both indicative of earth/petroleum interfaces. In the hands of a skilled interpreter, this simple and ad hoc method has apparently enjoyed some success. Recent studies [3] have indicated, however, that this method's results are quite sensitive to the choice of gain, and can be very misleading.

Another approach is the characterization of a single discrete reflector within the medium, e.g., the ocean bottom, by a "reflectivity" parameter. Such a simple specification has proven to be of value in studies of seafloor propagation [4,5], and in the remote identification of seafloor sediments [6]. Baggeroer [7] has presented a reflectivity estimation method, and has successfully used it in the design of dereverberation filters for marine seismic data. Such a reflectivity characterization is quantitative, and can be made noise tolerant. It is very limited in scope, however, since it cannot be used to determine structure for the entire medium. Moreover, the "reflectivity" parameter includes the effects of geometrical spreading, and attenuation, as well as the reflection properties of the target interface. A major difficulty, even with relatively sophisticated reflectivity estimation techniques [7], is the variation of measured reflectivity over reflection environments which are, apparently, constant [8].

Fitting the received data to a parametric model has been suggested by several authors. Middleton and Whittlesey [9] mention, but do not develop or apply, maximum likelihood estimation of discrete reflector delay and reflection coefficients for dereverberation filter design, given noisy reflection observations. This method was proposed only for a partial characterization of a medium (estimation of parameters for a few interfaces). Schell, et al. [10], on the other hand, develop a state-variable model for an entire medium and approach structure determination as a problem of identification of the system parameters. They propose an exact algebraic processing scheme by which all the reflection coefficients of the medium may be extracted from a noiseless received signal. Parametric methods have not been developed for the estimation of complete medium structure in the presence of noise corruption of the observations.

The most sophisticated and commonly employed solution method formulates the structure-identification problem in terms of a Wiener-Hopf integral equation, in which the deconvolved received signal is the kernel. The reflection coefficient density of the medium (or, in a medium consisting of equal travel time layers, the sequence of reflection coefficients) is simply related to the solution of the integral equation. This approach was developed by Wakita [11], and Gopinath and Sondhi [12], in the context of vocal tract estimation, and by Pusey [13], and (implicitly) Claerbout [14] in the seismic context. This is the so-called prediction-error filter (PEF) method. The name arises from one of the common situations in which a Wiener-Hopf equation occurs, computation of a prediction error filter. We emphasize, however, that this method is <u>not</u> directly related to stochastic filtering theory, save that the equation involved is of the same form. A more extensive discussion indicating the parallel structure is described by Ljung and Kailath [19]. Indeed, the method is exact, and stable, only when there is <u>no</u> observation noise. This restriction is a severe limitation on the usefulness of the method, as is the requirement that the received signal be deconvolved. It appears, however, that Bolt, Beranek and Newman's Acousticore system utilizes the PEF approach, at least in an approximate sense [15]. The PEF solution seems to be the most useful of the developed methods, and will be evaluated in greater detail below.

This paper presents a preliminary study of a unique approach to the problem of <u>quantitatively</u> determining the structure of a medium from <u>noisy</u> acoustic reflection data. No existing method successfully achieves this goal. This approach essentially combines those of Middleton and Whittlesey [9], and Schell, et al. [10] in that it treats the structure determination problem as one of maximum-likelihood identification of a system modelled by a state-variable formalism.

Our objective here is not to develop a processor which has an immediate practical application, but rather to use a very simple model and statistical approach to delineate the salient points of the general problem. Specifically, using this simple model, we are able to bound the performance of any estimator applied to identify the system, to study the sensitivity of the estimator's performance to the structure of the medium, and to demonstrate the robustness of the ML estimator in the presence of noise (and the instability of the PEF method).

2. EARTH MODEL

The first step in solving the structure determination problem is to create a mathematical model for the medium. Care must be taken to ensure both that physical reality is accurately represented, and also that the model is mathematically tractable. The modelling assumptions listed below are largely conventional (see, for instance, Trorey [16]). They are, in many ways, simplistic and inaccurate, but do allow us to identify important issues without obscuring the development early on.

(1) The medium is an infinite, isotropic, linearly elastic half-space, in contact with the atmosphere at its one plane boundary, a free surface.

(2) The medium consists of a finite number of homogeneous, parallel, plane layers with equal travel time, overlying a homogeneous half-space. Thus λ, the elastic parameter, and ρ, the density, are simple functions of depth.

(3) All propagation is normal to the layer interfaces; the waves are infinite, longitudinal, plane displacement waves.

(4) The medium is lossless (no attenuation).

This model, although crude, contains the essentials of the problem. It is a fairly common model, and merits consideration in its own right. It is particularly useful for our purposes, since the PEF a-proach may be easily and exactly applied to this model, using the well-known Wiener-Levinson algorithm. In the continuous limit of vanishingly small discrete layer thickness, this model converges to a distributed-parameter system, modelled by the familiar wave equation. We thus expect that the properties of an estimator designed for the discrete model should, in some sense, also carry over in the limit. Finally, the estimator's performance for this model should give physical insight into the general problem.

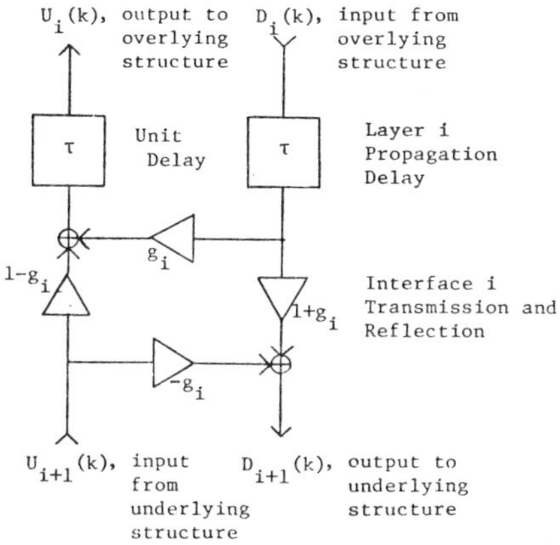

Fig. 1. Single layer-interface pair system model.

The wavefield within each homogeneous layer is separated into upward- and downward-travelling wave components, which are coupled to the adjacent layers by transmission and reflection at the layer interfaces. A single layer-interface pair is modelled by the system diagram shown in Figure 1. $U_i(k)$ and $D_i(k)$ are the upward- and downward-travelling waves at a point just below the (i-1)st interface, and g_i is the reflection coefficient for a wave travelling from layer i to layer i+1, given

$$g_i = \frac{\rho_i c_i - \rho_{i+1} c_{i+1}}{\rho_i c_i + \rho_{i+1} c_{i+1}} \tag{2.1}$$

where c_i is the propagation velocity in layer i, and ρ_i is the density. The unit delay indicated in Figure 1 corresponds to the one-way travel time through each layer. We have, for simplicity, assumed a discrete-time system model, in which the sampling interval is equal to this one-way travel time. This layer-interface model can be used iteratively to represent an entire layer structure, as illustrated in Figure 2 for a system with three interfaces.

Such a model may be implemented in several ways, including Laplace or z-transforms (see Theriault [17] for a Laplace transform example), or discrete-time state variables (Schell, et al. [10]). The state variable implementation was chosen because of its simple, compact canonic form. With the exception of $x_1(k)$, the (2n+1)st state variable is the level of the upward-travelling wave just below the nth interface. The (2n)th state

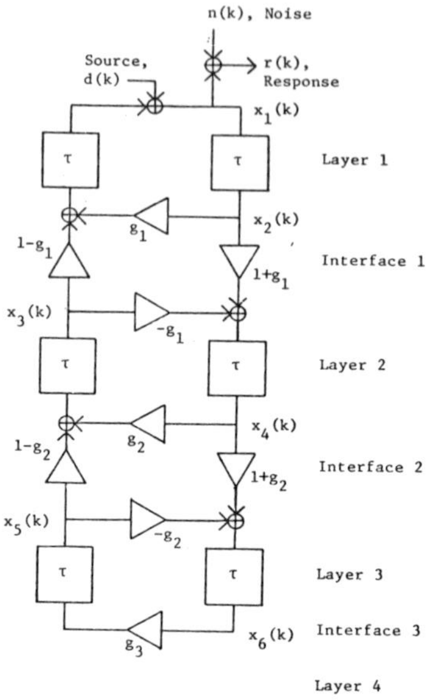

Fig. 2. Three-interface (four-layer) earth model, indicating the location of the state variables, $x_i(k)$.

variable is the level of the downward-travelling wave just above the nth interface. The state variables are indicated in Figure 2. This choice of state variables leads to dynamics equations which are considerably simpler than that employed by Schell, et al. [10]. The system dynamics for this model are expressed by two difference equations:

$$m \geq 0 \qquad x_{2m+1} = g_{m+1} x_{2m+2}(k) + (1-g_{m+1}) x_{2m+3}(k) + \delta_{mo} d(k+1) \tag{2.2a}$$

$$m > 1 \qquad x_{2m}(k+1) = (1+g_{m-1}) x_{2m-2}(k) - g_{m-1} x_{2m-1}(k) \tag{2.2b}$$

where d(k) is the source waveshape. Note that a system with L interfaces has 2L states. Also included in the dynamics are the boundary conditions at the surface:

$$x_2(k+1) = x_1(k) \tag{2.2c}$$

A boundary condition at infinity (no upcoming energy from the underlying half-space) is implicit in the model. The initial condition is

$$\underline{x}(0)=(x_1(0),\ldots,x_M(0))^T \tag{2.3}$$

The received signal, $r(k)$, is $x_1(k)$, corrupted by observation noise:

$$r(k)=x_1(k)+n(k) \tag{2.4}$$

or

$$r(k)=\underline{\Gamma}^T\underline{x}(k)+n(k) \tag{2.5}$$

where $\underline{\Gamma}=(1,0,\ldots,0)^T$, and $n(k)$ is zero-mean white Gaussian noise with variance $\sigma^2(k)$.

We have specified the earth model to be used for structure determination, and now develop a corresponding estimator.

3. MAXIMUM LIKELIHOOD ESTIMATOR

Determination of the structure of a medium modelled by Equations (2.2)-(2.5) is equivalent to identification of the parameters which specify the system behavior, the vector of reflection coefficients, $\underline{g}$. We will use a maximum likelihood (ML) estimator for $\underline{g}$. This estimate is formed by choosing that $\underline{g}$ which maximizes the joint probability density of the set of observations, $\underline{r}=(r(1),r(2),\ldots,r(N))^T$, where N is the length of the observation. The joint density of the observations, $p(\underline{r}|\underline{g})$, is

$$p(\underline{r}|\underline{g})=(2\pi)^{N/2}|\det\underline{\Sigma}|^{-1/2}\exp\left(-\frac{1}{2}(\underline{r}-\underline{X}(\underline{g}))^T\underline{\Sigma}^{-1}(\underline{r}-\underline{X}(\underline{g}))\right) \tag{3.1}$$

where

$$\underline{X}(\underline{g})=(\underline{\Gamma}^T\underline{x}(1|g),\underline{\Gamma}^T x(2|g),\ldots,\underline{\Gamma}^T\underline{x}(N|\underline{g}))^T$$

$$(\underline{\Sigma})_{jk}=\delta_{jk}\sigma^2(k) \qquad 1\leq j,\quad k\leq N$$

and $\underline{x}(k|\underline{g})$ is the state of the system defined by parameter $\underline{g}$ at time k. Maximization of $p(\underline{r}|\underline{g})$ as given by (6) is equivalent to the minimization

$$\hat{\underline{g}}:\quad \min_{\substack{\underline{g}\\|\underline{g}|\leq 1}} J(\underline{g})=\min_{\substack{\underline{g}\\|\underline{g}|\leq 1}}\sum_{k=1}^{N}(r(k)-\underline{\Gamma}^T\underline{x}(k|\underline{g}))^2/\sigma^2(k) \tag{3.2}$$

This is a constrained minimization, since $|g_i| \leq 1, \forall_i$, so $|g| \leq 1$. Deconvolution of r(k) is <u>not</u> required in (3.2); all that is needed is knowledge of the source waveform, d(k). This estimation scheme therefore avoids problems associated with deconvolution of the received signal, which may arise even if the waveform d(k) is exactly known (d(k) may not have a causal, stable inverse).

4. CRAMER-RAO BOUND

We now want to specify the performance of the maximum-likelihood estimator (3.2). A suitable performance measure is the mean-square error for each parameter (the estimate variance, for an unbiased estimator), $E[(g_i-\hat{g}_i)^2]$. Since the estimator does not have a closed form, however, the computation of such values is not possible, and we are forced instead to bound the variances. A lower bound on the estimate variance, which has found extensive application in communication theory, is provided by the Cramer-Rao bound (see, for instance, [18]), $E[(g_i-\hat{g}_i)^2] \geq \sigma^2_{CRB}(\underline{g}, i)$.

The numerical validity of the Cramer-Rao bound is, unfortunately, based on the assumption that the estimator is unbiased. For the same reason that the estimate variance cannot be calculated (lack of a closed form for the estimate), it is impossible to determine if the estimator is biased. Despite possible numerical inaccuracy, however, the Cramer-Rao bound provides qualitative indications of estimator performance as a function of the parameters of the problem. In particular, it gives physical insight into the effects of various earth structures on estimation accuracy. Moreover, the bound is purely a function of the model and noise structure, and does <u>not</u> depend upon the estimator; hence, conclusions we draw from the Cramer-Rao bound are valid for <u>any</u> estimator, not just the ML estimator developed in Section 3. (If one can demonstrate asymptotic efficiency, the maximum likelihood and the Cramer-Rao bound coincide asymptotically.)

The Cramer-Rao bound is given by

$$E[(g_i-\hat{g}_i)^2] \geq \sigma^2_{CRB}(g,i) = \left\{ E\left[\left(\frac{\partial \ \ell n \ p(\underline{r}|\underline{g})}{\partial \underline{g}} \right) \left(\frac{\partial \ \ell n \ p(\underline{r}|\underline{g})}{\partial \underline{g}} \right)^T \right] \right\}^{-1}_{ii} \tag{4.1}$$

Using (3.1), (4.1) becomes

$$E[(g_i-\hat{g}_i)^2]_{N}\geq\sigma^2_{CRB}(\underline{g},i,N)=$$

$$\left[\sum_{k=1}^{N}\left(\frac{\partial}{\partial \underline{g}}\,\underline{\Gamma}^T\underline{x}(k|\underline{g})\right)\left(\frac{\partial}{\partial \underline{g}}\,\underline{\Gamma}^T\underline{x}(k|\underline{g})\right)^T/\sigma^2(k)\right]^{-1}_{ii} \tag{4.2}$$

which may be computed in a straightforward manner. Note that the bound value for g_i depends on both the observation length, N, and on the complete set of reflection coefficients, $\underline{g}$.

4.1 Single interface model

We first evaluate (4.2) for the simple case of a single reflecting interface, which is a very crude model for seafloor reflectivity estimation. In this case, assuming stationary noise ($\sigma^2(k)=\sigma^2$), and a unit-amplitude impulsive source ($d(k)=\delta_{k0}$), a simple, closed-form expression for $\sigma^2_{CRB}(g,N)$ can be obtained:

$$\sigma^2_{CRB}(g,N)=$$

$$\sigma^2\left(\frac{4-(N+1)^2g^{N-1}}{4(1-g^2)}+\frac{3g^2-(N+2)g^{N+1}}{(1-g^2)^2}+\frac{2g^4-2g^{N+3}}{(1-g^2)^3}\right)^{-1} \tag{4.3}$$

$N=2m+1,\ m\geq 1$

(The value of the bound at N-2m is equal to the value at N=2m-1, since the source is an impulse.) The asymptotic value of the bound (as $N\to\infty$) is

$$\sigma^2_{CRB}(g,\infty)=\sigma^2\left(\frac{1+g^2}{(1-g^2)^3}\right)^{-1} \tag{4.4}$$

In Figure 3, $\sigma^2_{CRB}(g,N)/\sigma^2$ is plotted for various values of $|g|$ and N (here, $\sigma^2_{CRB}(g,N)=\sigma^2_{CRB}(-g,N)$).* The plot yields two conclusions: (1) performance improves monotonely with N to asymptotic value; (2) performance improves monotonely with $|g|$. Those conclusions are physically quite reasonable: the more <u>total</u> energy reflected from the interface, the better the estimate will be. The total received energy is, of course,

---

*I.e., if $\sigma^2=0.01$, the bound values would be 0.01 of those indicated in the plot. The bound values are linear in the noise variance, σ^2.

monotonely increasing in both $|g|$ and N. Note that the model includes, and the estimator is assumed to take advantage of, all multiples, although the bound has approached its asymptotic limit at small N, i.e. after a small number of observations, for low values of g.

For systems of two or more interfaces, the bounds may be computed by a general numerical method. The two-interface case, presented below, is the most complicated for which a general study has been made, if only because the results for three or more interfaces are difficult to present or visualize.

4.2 Two-interface model

The two-interface model is a crude model for an earth structure with a strong sub-bottom reflector in addition to the seafloor or for an isolated section of a seismogram which has three strong reflectors. Contour plots of $\sigma^2_{CRB}(\underline{g},i,N)/\sigma^{2*}$ for this case, with stationary noise, and a unit-amplitude impulsive source, are shown in Figure 4, for $N\to\infty$ (that is, the plot shows the asymptotic bound values for $N\to\infty$). The bound values are symmetric about $g_1=0$.

In contrast to the single-interface case, monotone improvement of the bound with $|g_i|$ is no longer generally true, because of interaction between the interfaces (we emphasize again that the model, and by assumption the estimator, includes <u>all</u> orders of internal multiples). Results for finite N, not presented here, show that the bound values do, as before, improve monotonely with N. Two significant observations for Figure 4 are: (1) as $|g_1|$ increases, the bound for g_2 gets worse; (2) as $|g_2|$ increases (about the axis of symmetry), the bounds for both g_1 and g_2 improve. These are attributed, respectively, to the physical phenomena: (1) the larger $|g_1|$ is, the less energy penetrates to illuminate the lower interface; (2) the larger $|g_2|$, the more energy is trapped in the layer structure as a whole, and the stronger is the return from the lower interface. In practice, then, one expects to see relatively better results above strongly reflecting regions, and worse results below.

5. PEF ESTIMATOR FOR THE DISCRETE SYSTEM

This system model is the natural one for the discrete PEF estimator, implemented via the solution of the normal equations. Here the normal equations are the exact specification of the PEF

---

* See preceding footnote.

estimator, rather than an approximation to the Wiener-Hopf integral equation (see Wakita [11], or Pusey [13]). The normal equations have the form

$$\begin{bmatrix} r(0) & r(1) & \cdots & r(m-1) \\ r(1) & r(0) & \cdots & r(m-2) \\ \cdot & \cdot & & \cdot \\ \cdot & & \cdot & \cdot \\ r(m-1) & \cdots & & r(0) \end{bmatrix} \begin{bmatrix} h_m(1) \\ \cdot \\ \cdot \\ \cdot \\ h_m(m) \end{bmatrix} = - \begin{bmatrix} r(1) \\ \cdot \\ \cdot \\ \cdot \\ r(m) \end{bmatrix} \tag{5.1}$$

The reader should take careful note that this equation involves the raw observations, $r(k)$, and <u>not</u> a correlation function. When the Wiener-Levinson algorithm is applied to solve (12), $h_{m+1}(k)$ is iteratively computed from $h_m(k)$. The PEF estimate of g_i is then

$$\hat{g}_i = -h_i(k) \tag{5.2}$$

When there is no observation noise, (5.2) is exact, that is, g_i is precisely $-h_i(i)$. The presence of additive noise in $r(k)$, however, as well as introducing error into the estimate (5.2), also destroys the positive-definiteness of $r(k)$ (proven by Pusey [13]) and, in turn, the positive-definiteness of the matrix of $r(k)$ values in (5.1). Hence, the solution of (5.1), by whatever means, while probably well-conditioned (i.e. the norm of the matrix is bounded away from zero) may lead to non-physical results (reflection coefficients of magnitude greater than 1) when $r(k)$ contains noise.

This difficulty is in addition to the requirement that $r(k)$ be the reflection response of the medium to an impulsive source. For a non-impulsive source, deconvolution of $r(k)$ is a necessity (and may not be possible even if $d(k)$ is known exactly). This is in contrast to the ML estimator, which does not require deconvolution.

It is interesting that the PEF estimator, (5.2), when implemented by the Wiener-Levinson algorithm, is easily seen to be an equivalent to a conditional maximum likelihood estimator. The received signal at time $2k$, $r(2k)$, contains the first return from interface k, and has the form

$$r(2k) = g_k \prod_{i=1}^{k-1} (1-g_i^2) + F_{k-1}(2k) + N(2k) \tag{5.3}$$

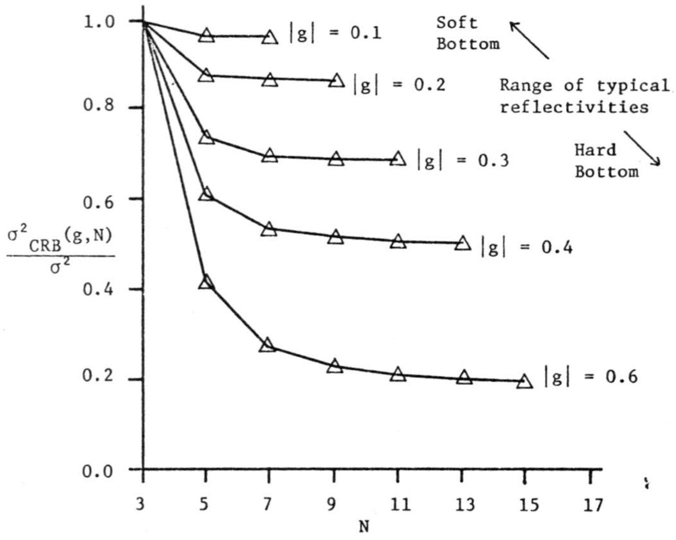

Fig. 3. Cramer-Rao bound values, as a fraction of σ^2, for a single-interface system.

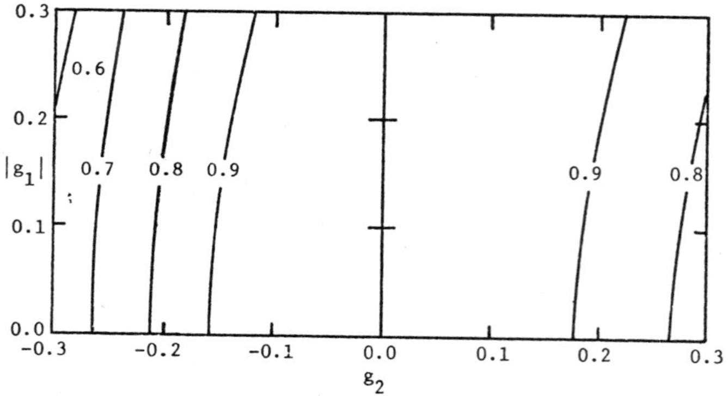

Fig. 4a. Two-interface Cramer-Rao bound contour plot of $\sigma^2_{CRB}(\underline{g},1,\infty)/\sigma^2$.

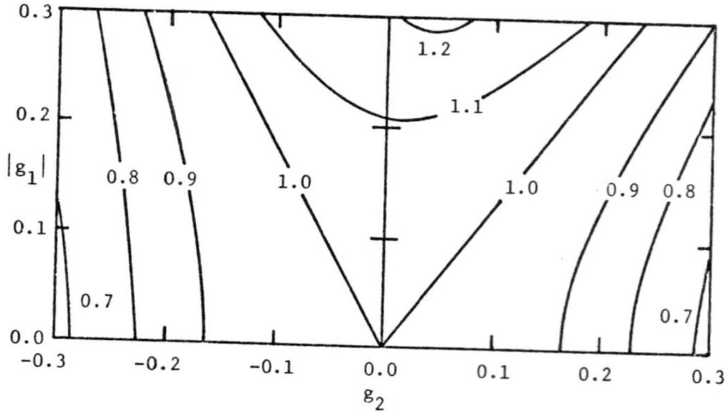

Fig. 4b. Two-interface Cramer-Rao bound contour plot of $\sigma^2_{CRB}(\underline{g},2,\infty)/\sigma^2$.

where $F_{k-1}(2k)$ is the response at time 2k given $g_1,\ldots g_{k-1}$. The ML estimate of g_k, conditioned on $g_1,\ldots g_k$ is, simply,

$$\hat{g}_k = \frac{r(2k)-F_{k-1}(2k)}{\prod_{i=1}^{k-1}(1-g_i^2)} \tag{5.4}$$

But this is the exact expression evaluated by the Wiener-Levinson algorithm to obtain $-h_k(k)$, and our contention is proven. Observe that the algorithm's computation, however, is conditioned on the previously obtained estimates, $\hat{g}_1,\ldots\hat{g}_k$, rather than the true reflection coefficients. In this procedure of sequential, conditional ML estimation, each successive estimate is affected by the errors of all the previous estimates, and we expect that the PEF estimator's performance will become worse as k increases. In contrast, the ML estimator described in Section 3 estimates the entire parameter vector, $\underline{g}$, simultaneously, and thus this degradation should not take place.

6. MONTE CARLO RESULTS

Some limited Monte Carlo test results for both ML and PEF estimators are presented in Tables 1 and 2 for $\underline{g}=(.1,.1)^T$ and $(0.,.1)^T$, respectively, with N=23, stationary noise, and a unit impulse source. The statistics are based on about 175 simulations for each set of noise and reflection coefficient parameters. As $\sigma \to 0$, the ML and PEF methods are virtually identical, both in their gross statistics, and sample function by sample function. As σ increases, however, the results diverge. In particular, the PEF estimate exhibits non-physical estimates (predicted in Section 5) for large σ (0.2). We have observed, in other experiments, that breakdown occurs in the PEF estimator at much larger signal-to-noise levels if the length of the filter (the number of interfaces) is large.

With the exception of the PEF method for high noise levels, it is interesting that the experimental variances closely approximate the noise variance (which is not significantly different from the Cramer-Rao bound in these cases). It appears that both methods may well be almost efficient (almost satisfy the bound with equality) below threshold, but more extensive Monte Carlo tests are needed to confirm this.

Table 1. Summary of Monte Carlo comparison of ML and PEF estimators for various noise levels, with $\underline{g}=(0.1,0.1)$. These results are based on about 175 runs for each estimator at the specified noise level.

	ML Results				PEF Results			
	$\hat{g}_1$		$\hat{g}_2$		$\hat{g}_1$		$\hat{g}_2$	
σ	mean	s.d.	mean	s.d.	mean	s.d.	mean	s.d.
0.01	0.1006	0.0103	0.0995	0.0091	0.1005	0.0103	0.0995	0.0098
0.05	0.1035	0.0512	0.0942	0.0451	0.1013	0.0515	0.0968	0.0503
0.10	0.1075	0.1026	0.0805	0.0916	0.0981	0.1048	0.0914	0.1097
0.20	0.1125	0.2053	0.0365	0.1907	0.0552	0.2616	0.1967	1.3060

Table 2. Summary of Monte Carlo comparison of ML and PEF estimators for various noise levels, with $\underline{g}=(0.0,0.1)$. These results are based on about 175 runs for each estimator at the specified noise levels.

	ML Results				PEF Results			
	$\hat{g}_1$		$\hat{g}_2$		$\hat{g}_1$		$\hat{g}_2$	
σ	mean	s.d.	mean	s.d.	mean	s.d.	mean	s.d.
0.01	0.0006	0.0103	0.0996	0.0091	0.0005	0.0103	0.0996	0.0960
0.05	0.0032	0.0513	0.0950	0.0458	0.0002	0.0515	0.0967	0.0497
0.10	0.0063	0.1026	0.0822	0.0932	-.0062	0.1056	0.0887	0.1098
0.20	0.0127	0.2053	0.0408	0.1869	-.0598	0.2783	-.0458	1.398

7. RESULTS AND CONCLUSIONS

The primary result of this paper is the specification and initial examination of a unique method of determining the complete structure of a medium from noisy observations of its normal-incidence acoustic reflection response. We have developed a very simple earth model, specified the structure determination procedure, and obtained the Cramer-Rao bound on estimator performance (indeed, the bound on the performance of <u>any</u> structure based on the proposed model). We have examined the performance bound in two cases of practical interest (albeit crudely represented by our model), and have demonstrated that physically incisive conclusions may be drawn from the behavior of the bound, as we had hoped. The proposed estimator, and the PEF estimator, have been compared by Monte Carlo experiments, which confirmed the predicted instability of the PEF estimator, and the comparative robustness of the ML estimator. Moreover, the Cramer-Rao bound appears to be a very tight bound on estimator performance for both estimators, at least for moderate noise levels.

The new estimator we have presented has, however, one severe drawback, its dimensionality. A priori, one reflection coefficient must be provided for every two data points. As a practical matter, if the observation contains only a few hundred samples, the optimization (3.2) must be carried out over more than a hundred independent g_i, a procedure sure to be numerically ill-conditioned. The difficulty is that we have allowed too many degrees of freedom in the model. While the acoustic impedance as a function of depth can be approximated by a simple function, and while such an approximation leads to a straightforward response synthesis method, it need not be true that the acoustic impedance should be estimated in terms of that simple function.

A more appropriate model, which we will develop in future work, is the representation of acoustic impedance as a more restricted function, e.g. a weighted sum of polynomials. The objective of the estimator would then be to estimate the weights, rather than the reflection coefficients.

Other topics which we will consider in future work are the expression of the estimation/identification problem in the continuous-time and space domain (using a distributed parameter system model), and the consideration of additional sensitivity questions, e.g., the effects of source waveshape on estimation accuracy, and of choice of sampling interval.

PRINCIPAL SYMBOLS USED*

$U_i(k)$:	Upward travelling wave at ith interface at time k
$D_i(k)$:	Downward travelling wave at ith interface at time k
g_i:	Reflectivity of ith layer
$x_i(k)$:	State variable representation of interface waves
$r(k)$:	Noisy observation of medium impulse response
σ^2:	Variance of observation noise
N:	Observation interval length
$\sigma^2_{CRB}(\underline{g},i,N)$:	Cramer-Rao bound on an estimate of the ith layer reflectivity with an N length observation interval

REFERENCES

1. R. Sheriff, "Seismic Detection of Hydrocarbons--The Underlying Physical Principles," Preprints of the 1974 Offshore Technology Conference, Vol. 1, pp. 637-649.
2. C. Savit, "Exploration Changes Radically," Oil and Gas Journal (May 21, 1973), p. 156.
3. B. May and F. Hron, "Synthetic Seismic Sections of Typical Petroleum Traps," paper presented at the 45th Annual International Meeting of the Society of Exploration Geophysicists, Denver (1975).
4. R. Urick, "Underwater Sound Transmission Through the Ocean Floor," in Physics of Sound in Marine Sediments, ed. L. Hampton (Plenum Press, New York, 1974).
5. H. Bucker, "Sound Propagation Calculations Using Bottom Reflection Functions," in Physics of Sound in Marine Sediments, ed. L. Hampton (Plenum Press, New York, 1974).
6. D. Bell et al., "Progress in the Use of Acoustics to Classify Marine Sediments," Ocean '73 Record, pp. 354-359.
7. A. Baggeroer, "Trapped Delay Line Models for Dereverberation of Deep Water Multiples," IEEE Transactions on Geoscience Electronics, Vol. GE-12, no. 2, pp. 33-54 (April 1974).
8. R. Goodman and A. Robinson, "Measurements Reflectivity by Explosive Signals," in Physics of Sound in Marine Sediments, ed. L. Hampton (Plenum Press, New York, 1974).
9. D. Middleton and Whittlesey, "Seismic Models and Deterministic Operators for Marine Reverberation," Geophysics, 33, 4, pp. 557-583 (August 1968).
10. J. Schell et al., "Dereverberation by Linear System Techniques," IEEE Transactions on Geoscience Electronics, Vol. GE-9, no. 1, pp. 38-24 (January 1971).
11. H. Wakita, "Direct Estimation of the Vocal Tract Shape by Inverse Filtering of Acoustic Speech Waveforms," IEEE Transactions on Audio and Electroacoustics, Vol. AU-21, no. 5,

pp. 417-427 (October 1973).
12. B. Gopinath and M. Sondhi, "Determination of the Shape of the Human Vocal Tract from Acoustical Measurements," Bell System Technical Journal, 49, 6, pp. 1195-1214 (July-August 1970).
13. L. Pusey, "An Innovations Approach to Spectral Estimation and Wave Propagation," Ph.D. Thesis, MIT, 1975.
14. J. Claerbout, "Synthesis of a Layered Medium from its Acoustic Transmission Response," Geophysics, 33, 2, pp. 264-269 (April 1968).
15. H. Wright and P. Miles, "'Acousticore', A New Concept in Acoustic Prospecting for Minerals and Preconstruction Surveying," Ocean '73 Record, p. 349.
16. A. Trorey, "Theoretical Seismograms with Frequency and Depth Dependent Absorption," Geophysics, 27, 6, pp. 766-785 (December 1962).
17. K. Theriault, "Optimum Arrival-Time Estimation in Exploration Seismology," M.S. Thesis, MIT, 1974.
18. H. Van Trees, Detection, Estimation, and Modulation Theory, Part I (Wiley, New York, 1968).
19. L. Ljung, T. Kailath, and B. Friedlander, "Scattering Theory and Least Squares Estimation--Part I: Continuous Time Problems," IEEE Proceedings, 64, 1, pp. 131-139.

S U B J E C T 3

TIME-SPACE PROCESSING, ADAPTIVE PROCESSING AND NORMALIZATION, QUANTIZATION METHODS

AN INTRODUCTION TO ADAPTIVE PROCESSING IN A PASSIVE SONAR SYSTEM

J.W.R. Griffiths and J.E. Hudson

Department of Electronic and Electrical Engineering,
University of Technology, Loughborough, U.K.

In a sonar system the carrier frequencies and system bandwidths are such that, taken together with the rapid reduction in size and cost of computer components, complicated signal processing in real time becomes viable.

Basically in a passive sonar system one is trying to estimate a two dimensional spectrum, i.e. to display the frequency spectrum of the signals being received from the various directions (Fig.1). It should be noted that the 'signals' in a passive sonar system are usually noise-like in character. In non-adaptive systems the limited resolution causes leakage from large signals and in particular from near-field noise.

For simplicity the case of a linear array of transducers is considered in this paper although the methods are easily extendable to other configurations and with more complexity to two dimensional arrays. The basic directional pattern of a linear array has large sidelobes and frequently the output of the elements of the array are weighted in some fashion before summation to reduce the sidelobes to an acceptable level. More sophisticated weighting can arrange for a null or nulls to appear at designated angles and in some cases these nulls can be 'steered' by an operator. In a sense an adaptive system is an arrangement whereby these nulls are steered automatically, although this is not how the system is defined. There is an extensive literature on the subject of adaptive processing but in view of the limitations of space no attempt has been made in this paper either to list these works or to carry out a critical review.

G. Tacconi (ed.), Aspects of Signal Processing, Part 1, 299-308.

The basic processor system to be described operates conveniently as a narrow band system and thus the practical system will be realised in the form shown in Fig. 2. The inputs from the sensors are separated into the various frequency bands by an F.F.T. processor and then fed to the adaptive processor which deals with each frequency band seaprately. The basic ideas can be applied to other configurations provided the process is linear but the one shown in Fig. 2 proves most convenient for this case. With a multi-element array of, say, K elements we can fix K constraints. Referring to Fig. 3 we see that the output is given by

$$y = w_1 x_1 + w_2 x_2 \,.....$$

$$= \sum_{r=1}^{K} w_r x_r$$

At each element the voltage comprises the wanted signal s_r plus the unwanted interference and noise n_r, i.e.

$$x_r = s_r + n_r$$

The basic constraint of unity gain in the signal direction is then

$$y = \sum_{r=1}^{K} w_r s_r = 1$$

Since we have K-1 degrees of freedom remaining we have K-1 other constraints which we could apply. We could use these to fix K-1 nulls at specified angles but the method used in this paper is to minimise the mean square output of the system i.e. $E(y^2)$ subject to the constraint that the gain is unity in the wanted signal direction. It is convenient to use matrix notation in formulating this problem and we define, referring to Fig. 3

$$S \equiv \begin{bmatrix} s_1 \\ s_2 \\ s_3 \\ \cdot \\ \cdot \\ \cdot \end{bmatrix} \qquad N \equiv \begin{bmatrix} n_1 \\ n_2 \\ n_3 \\ \cdot \\ \cdot \\ \cdot \end{bmatrix} \qquad X \equiv \begin{bmatrix} x_1 \\ x_2 \\ x_3 \\ \cdot \\ \cdot \\ \cdot \end{bmatrix} \qquad W \equiv \begin{bmatrix} w_1 \\ w_2 \\ w_3 \\ \cdot \\ \cdot \\ \cdot \end{bmatrix}$$

$$X = S + N$$

$$y = \sum_{r=1}^{K} w_r x_r = W^T X$$

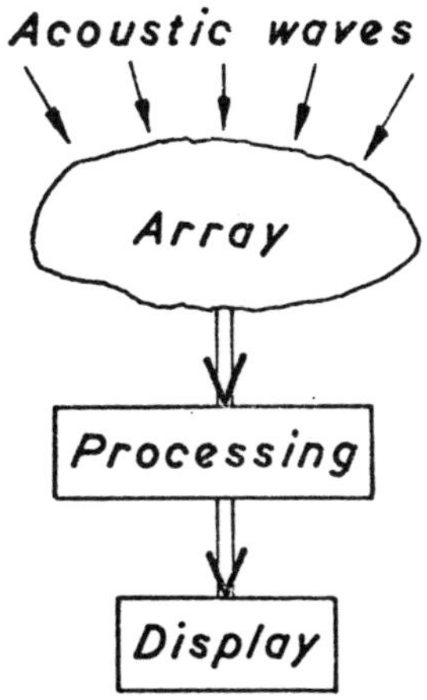

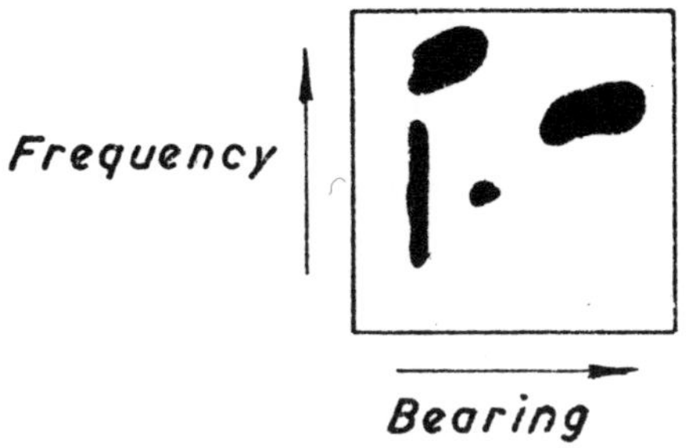

Fig. 1. Passive Sonar System.

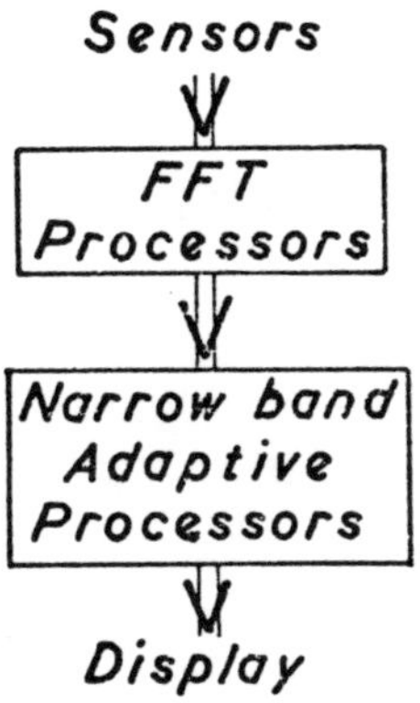

Fig. 2. Adaptive System.

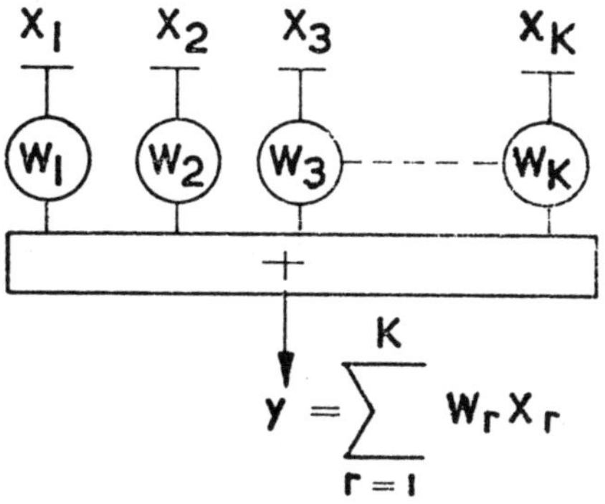

Fig. 3. Multi-element Array.

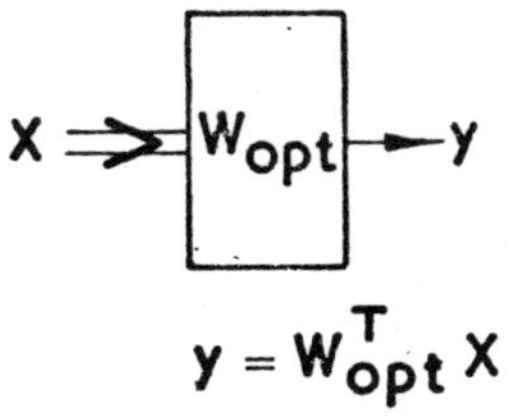

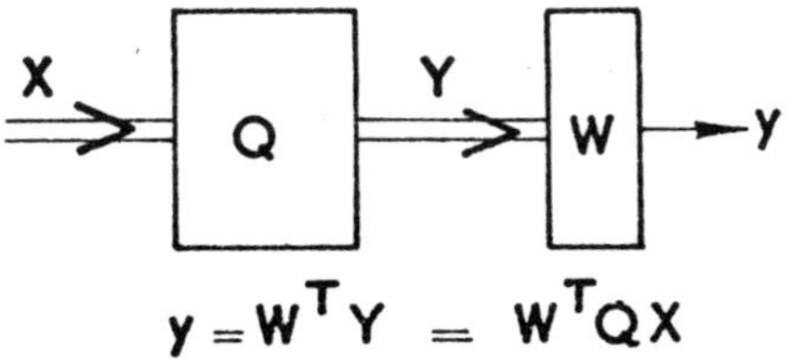

Fig. 4. Showing equivalence to pre-whitening and matching.

We wish to minimise the mean square output subject to the constraint of unity gain in·the signal direction, i.e. minimise $E\{y^2\}$ subject to the constraint $W^T S = 1$

$$E\{y^2\} = E\{W^T X\}^2 = E\{W^T X\ X^T W\} = W^T E\{X.X^T\} W$$

$$= W^T R_{xx} W.$$

R_{xx} is the covariance matrix for the vector X.

Using Lagrange Multipliers we write

$$H(W) = W^T R_{xx} W + \lambda(W^T S - 1)$$

and setting

$$\frac{\partial H(W)}{\partial W} = 0$$

we get after some manipulation

$$W_{opt} = -\frac{\lambda}{2} R_{xx}^{-1} S$$

but $S^T W_{opt} = 1$ and putting in this constraint gives

$$W_{opt} = \{S^T R_{xx}^{-1} S\}^{-1} R_{xx}^{-1} S \quad \ldots\ldots\ldots\ldots\ldots\ldots\ldots (1)$$

N.B. $$E\{y^2\} = E\{W^T (S + N)\}^2 = W^T E\{(S + N)^T\} W$$

$$= W^T R_{ss} W + N^T R_{nn} W$$

if the signal and noise are uncorrelated.

But $$W^T S = 1 = S^T W$$

$\therefore$ $$W^T S.S^T W = 1 = W^T R_{ss} W$$

$$E\{y^2\} = 1 + W^T R_{nn} W$$

Hence minimisation of $E\{y^2\}$ is the same as minimisation of $W^T R_{nn} W$.

Thus $$W_{opt} = \{S^T R_{nn}^{-1} S\}^{-1} R_{nn}^{-1} S \quad \ldots\ldots\ldots\ldots\ldots\ldots\ldots (2)$$

A way of viewing this result which might be more familiar is seen by considering Fig. 4. We break up the single operation W_{opt} into the two stages Q and W.

$$y = W^T Y = W^T Q X$$

$$E\{Y Y^T\} = E\{QXX^T Q^T\} = QR_{xx}Q^T$$

If the input is noise only then

$$E\{Y Y^T\} = Q R_{nn} Q^T$$

If we arrange the matrix Q such that $Q R_{nn} Q^T$ is a diagonal matrix (and conveniently the identity matrix) then this implies that the individual components of Y are uncorrelated.

$$\text{If } Q R_{nn} Q^T = 1 \text{ then } R_{nn}^{-1} = Q^T Q$$

Let the weight vector W be a constant factor times the signal vector after it has been through the matrix operation Q.

i.e. $\quad W = k S_y \quad$ where $\quad S_y = QS$

Applying the constraint $W^T S_y = 1$

i.e. $\quad k S_y^T S_y = 1$

$$k = \{S_y^T S_y\}^{-1} = \{S^T Q^T Q S\}^{-1}$$

$\therefore \quad k = \{S^T R_{nn}^{-1} S\}$

For the two systems to be equivalent

$$W_{opt}^T = W^T Q$$

i.e. $\quad W_{opt} = Q^T W$

$$Q^T W = k Q^T S_y = k Q^T Q S = k R_{nn}^{-1} S = W_{opt}$$

Basically then if we can estimate R_{xx} we can obtain an estimate of the optimum weighting vector. In fact since we require R_{xx}^{-1} certain problems arise and the remainder of this paper will be discussing the actual processes used in the adaptive procedure.

The covariance matrix R_{xx} is often ill-conditioned. This is especially true when there are only one or two strongly coherent sources illuminating the sensor array and the background incoherent noise due to other sources is relatively weak. In these circumstances the solution (1) becomes very sensitive to various unmeasurable errors in the medium/array/processor configuration. Such errors may be due to a summation of sensor and channel gain errors, errors in the field itself due to diffraction around the array mounting, mutual coupling, and propagation effects within the medium. The result of Eq. 1. is even more sensitive to errors in the bearing of an unknown source which it attempts to detect. Slight errors in the assumed bearing of the source cause the processor to regard it as interference and its output power is minimised.

Desirable system characteristics are (a) a target acceptance angle which is independent of input SNR and (b) insensitivity to errors of measurement of the free-field propagating waves. It should be able to tolerate errors of the order of 10%.

The use of the inverse matrix as in (1) is a very efficient way of forming an optimal digital processor since the same inverse may be used for any number of weight vectors for different look directions. It is not, however, possible to desensitise this processor to errors while retaining the computational advantages of matrix inversion and at the same time preserving the efficiency of the system. For example a measured quantity of white noise can be injected into the processor inputs (equivalent to adding an identity matrix to R) but to make a robust system the amount of noise required is high and it tends to drown out weak signals.

To reduce the sensitivity of the processor we have used a system of non-linear constraints[2] which are based on worst case weight vector and signal error combinations. These constraints allow the weight vector to vary only in a specified region. The processor must find the point within or on the boundary of the constraint region at which the processor output power is minimum. We need to distinguish now between the constraint vector C which represents the direction in which we are 'pointing' our system and the actual signal S which may not arrive exactly from this direction.

The conventional narrow-band beam forming system may be represented within our constraint as a weight vector W_c

where $$W_c = \frac{1}{N} C$$

Let the weight vector in the processor be defined as

$$W = W_c + \omega$$

Since $$C^T W = 1$$

and $$C^T W_c = 1$$

then $C^T \omega = 0$

Thus the vector ω is orthogonal to the constraint direction and the tip of the vector W lies in a plane to which C is a normal (see Fig. 5 which illustrates the problem in two dimensions). If an unwanted signal is present, say S_u, then W adjusts itself until it is orthogonal to S_u. If a wanted signal S is present but not exactly in the constrained direction, due to measurement or bearing error, W still tries to adjust itself until it is orthogonal to this direction and hence tries to null the wanted signal. However, if the wanted signal is close to the constrained direction the vector W has to adjust itself until it is almost normal to the constraint direction while at the same time remaining with its tip in the constraint plane. This results in very large values of the magnitude of W. It is thus possible to restrict the ability of the system to null signals close to the wanted direction by applying a bound to the maximum value of the magnitude of W.

Let the signal S = C + E and we assume E is orthogonal to C. This restricts S a little but since we are interested only in those signals close to the wanted direction C this restriction is not important.

$$y = W^TS = W^TC + W^TE$$

$$= 1 + (W_c + \omega)^TE$$

$$= 1 + \omega^TE$$

The Schwartz and triangle inequalities can be used to get the result

$$1 - (\omega^T\omega E^TE)^{\frac{1}{2}} < y < 1 + (\omega^T\omega E^TE)^{\frac{1}{2}}$$

If we set $\omega^T\omega E^TE < \delta^2$ then the output lies between $1 - \delta$ and $1 + \delta$. The constraint surface for W is then defined by

$$\omega^T\omega < \frac{\delta^2}{E^TE}$$

This constraint works well for signal bearing errors if there is only one signal within the conventional beam pattern. For two signals spaced equally either side of the wanted direction we find that in order to null these outside our defined acceptance angle limits it is necessary to increase the limit on the magnitude of the weight vector. This increase, however, would allow a single signal to be nulled inside the acceptance angle and thus it is necessary to introduce a further constraint to prevent this. Fortunately a simple constraint on the maximum slope of the adapted pattern at the centre is sufficient to

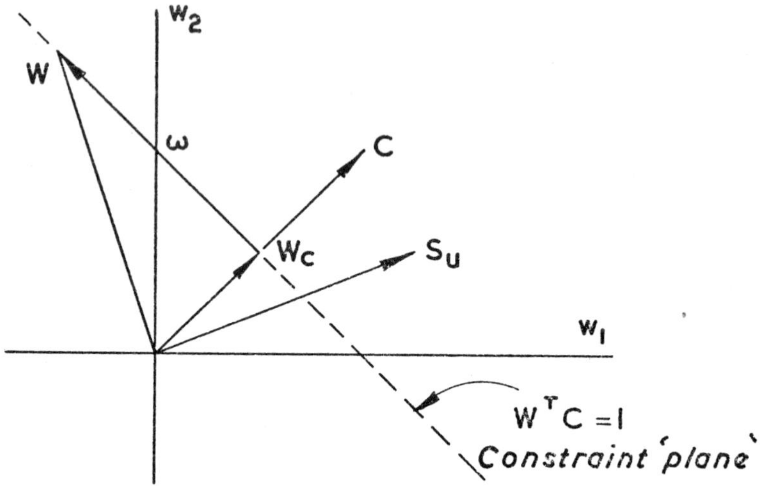

Fig. 5. Nulling of discrete source S_u by adding vector ω to W_c.

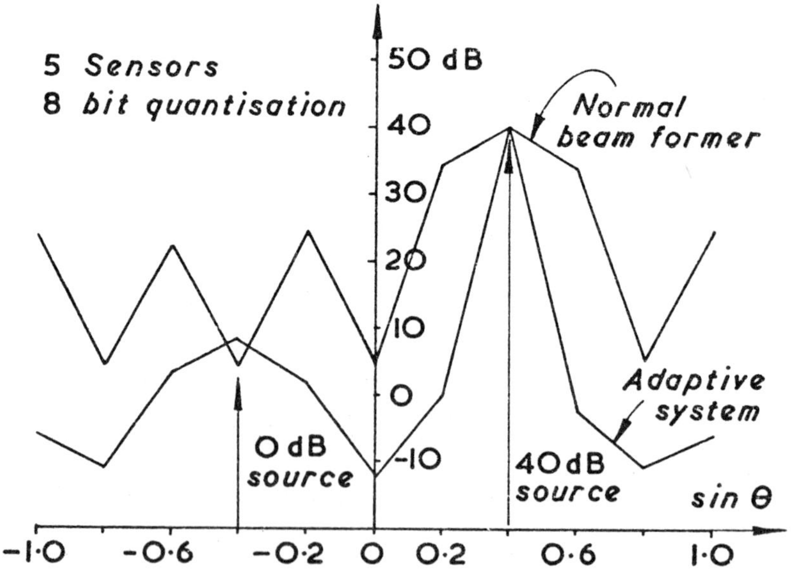

Fig. 6. Bearing display of 11 beam sonar.

prevent this nulling. Following this we can develop a generalised constraint surface.

$$\omega^T A \omega < \text{Bound}$$

where A is a positive-definite matrix which is usually very simple and easy to compute.

The final algorithm is thus of the form,

Minimise $W^T RW$ (output power)

with $C^T W = 1$ (look direction constraint) .... (3)

and $\omega^T A \omega <$ Bound (supergain constraint)

This approach has been found very successful on almost every type of signal field tested and turns the adaptive processor into a very robust system. It brings a penalty of reduced resolution, though not so low as the conventional system. This is because targets close in bearing can only be resolved with very super-directive weight vectors but the constraint system inhibits their development. If measurement errors can be reduced, then resolution can be increased accordingly.

Because the supergain constraint is an inequality constraint it is often inactive or 'slack'. Thus for many beams (look directions) the linearly constrained optimum Eq. (1) holds. The non-linear supergain constraint normally only becomes active in those beams in which there is a strong source in the neighbourhood of the look direction as defined by vector C.

Fig. 6 shows the results for a five sensor system. Eleven different constraint directions were used, i.e. effectively 11 different beam formers and the output of each beam former is indicated on the diagram at the appropriate bearing. In the illustration two targets are simulated one showing one of +40dB at about +20° and one weak one of 0dB at about -20°. As is seen on the diagram the weak source would not be resolved by the conventional system.

The minimum level of signal which can be resolved depends on the number of bits used for quantization and this matter will be discussed in a companion paper by one of the authors later in the Study Institute.*

Conclusions: A system has been described which can give considerable improvement in the performance of a passive sonar system.

*J.E. Hudson, 'Weight Coefficient Quantization in Adaptive Array Processors'

DISCUSSION

Comment : K.G. BEAUCHAMP

You have discussed adaptive arrays in terms of a single, linear array having a small number of elements. Have you carried out any work with other array forms, e.g. two linear arrays that are orthogonal to each other, as used extensively for seismic directivity work?

Reply : J.W.R. GRIFFITHS

No.

SOME LESSONS FROM ARRAY PROCESSING THEORY

Peter M. Schultheiss

Department of Engineering and Applied Science,
Yale University, New Haven, Connecticut, U.S.A.

1. INTRODUCTION

Practical problems arising in radar, sonar and seismology have stimulated extensive research in the area of array processing during the past two decades. This paper is not a review in the sense that it attempts to give a balanced presentation of the various elegant and highly mathematical analyses which have contributed to an understanding of arrays. It is the author's feeling that array processing as a field has now matured to the point where a number of fairly simple intuitive notions can be abstracted from the more formal studies. We will be concerned with these intuitive notions, with questions like: What have we learned and why are these conclusions plausible? For formal proofs the interested reader will be referred to the, by now, very extensive literature.

To begin with, we should perhaps briefly address the question: Why (or when) is the use of arrays important in signal processing? In the sonar case we are interested in determining the presence (and possibly location) of an object which radiates or reflects sound waves. These sound waves may have waveshapes about which a good deal is known (active sonar returns, radiated machinery noise) or they may be largely random (radiated broadband noise). In the former case, the knowledge can be exploited through coherent signal processing procedures. In the latter case, one is forced to reply primarily on statistical information.

Successful detection or localization can be accomplished if the properties (deterministic or statistical) of the signal are sufficiently different from those of the noise. Relevant statis-

G. Tacconi (ed.), Aspects of Signal Processing, Part 1, 309-331.

tical properties include both temporal features (e.g. power spectra) and spatial features (e.g. spatial correlation). If the signal is a pure sinusoid or a very narrowband random process, the temporal features may be most distinctive and temporal (frequency sensitive) filters will then be central to the processor structure. At the other extreme, broadband noise radiated by a ship may be nearly indistinguishable from sea noise on the basis of temporal characteristics. In such a case spatial features (the fact that the "signal" radiates from a localized source whereas the "noise" does not) form the only significant basis for distinguishing between signal and noise. To exploit spatial properties one must observe the sound pressure at more than one point in space and is therefore forced into the use of arrays.

The best possible system must clearly take advantage of both temporal and spatial features and most practical processors do, in fact, contain both types of filters. This paper is concerned with array processing. We therefore concentrate on problems in which spatial processing is central: Situations in which the signal is random with temporal properties similar to those of the noise. Specifically, we shall concentrate on signals and noises both of which are stationary Gaussian processes.

Simple array processors have been used in sonar applications for a very long time. The classical configuration is shown in Fig. 1.

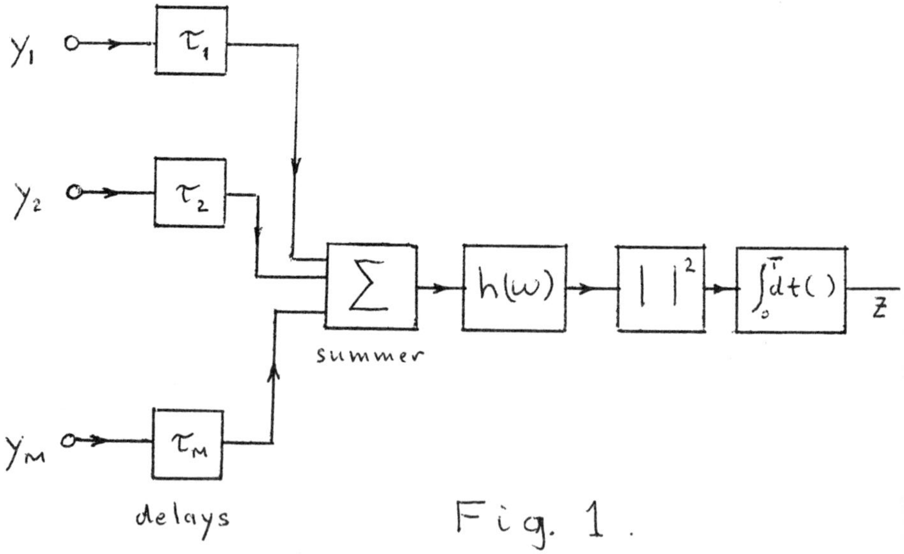

Fig. 1.

A series of the delays aligns the signal components at the outputs of the various sensors. In the summer the signals will add, whereas the noises, in general, will not. The subsequent power detector therefore yields a larger average output when the delays are properly matched to the signal wavefront than for all other situations. For a far field source, signal delay from sensor to sensor is readily converted to source bearing Θ. The processor, therefore, yields the average bearing response curve sketched in Fig. 2.

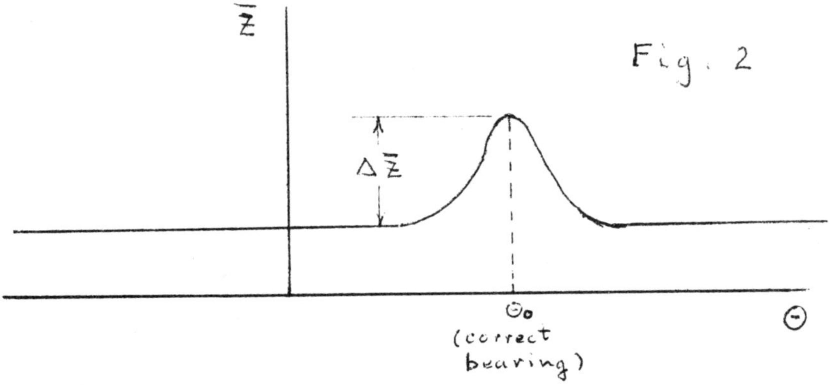

To assert that a signal is present one must be confident from a necessarily noisy measurement that a peak such as the one shown in Fig. 2 is indeed present. If the rms output fluctuation of Fig. 1 is D(z), the required condition is $\Delta\bar{z}/D(z) \gg 1$. The square of this ratio is often called the output signal to noise ratio d and is widely used as a figure of merit for detectors.

$$d = \frac{(\Delta\bar{z})^2}{D^2(z)} \tag{1}$$

In practice the delays in Fig. 1 are often implemented as digital shift registers. By forming sums for various combinations of delay one can then calculate and display $\bar{z}$ for many values of Θ in no more time than is required for a single such calculation (DIMUS system [1]).

The interest in array processing theory arises from the fact that Fig. 1, while intuitively appealing, does not guarantee optimal use of the available data. The suspicion that one might do substantially better led Bryn to carry out the first study of optimal detectors in 1961 [2]. His first results were very encouraging. Large gains were apparently realizable for very common types of noise fields (e.g. isotropic sea noise). In fact the potential gains were so large as to arouse suspicion. Pursuing this lead Vanderkulk demonstrated in 1962 that to realize the gains predicted by Bryn's theory for isotropic noise one would require enormously large numbers of sensors and an accuracy in their locations which is quite unreasonable [3]. This dampened enthusiasm for the use of optimal techniques considerably.

Interest in optimal (or near optimal) techniques revived only when it was discovered that the noise fields examined by Bryn and Vanderkulk were among the less favorable for practical improvement and that the achievement of important gains in other situations was by no means out of reach [4], [5]. One of the important theoretical questions is, therefore, the identification of situations in which major gains relative to Fig. 1 are possible.

2. BASIC THEORY

We shall need the following well-known results of detection theory:

a) If one wishes to maximize detection probability for a fixed false alarm rate, one must compute the likelihood ratio (L.R.) and compare it with a threshold determined by the desired false alarm rate. The likelihood ratio is given by the expression

$$\text{L.R.} = \frac{p(\underline{y}/S+N)}{p(\underline{y}/N)} \tag{2}$$

$p(\underline{y}/S+N)$ is the probability density of the received data vector $\underline{y}$ given that signal and noise are both present. $p(\underline{y}/N)$ is the probability density of $\underline{y}$ given that only noise is present. Instead of using L.R. one can use any monotone function of L.R. and compare with an appropriately adjusted threshold. For computational reasons it is often convenient to work with $z = \log$ (L.R.).

b) If the signal is a single known time function s(t) and the noise is white Gaussian with spectral level N_0, the best detector simply cross-correlates the received waveshape against a replica of the signal [Fig. 3]. The figure of merit [Eq. (1)] of this system is given by

$$\frac{(\Delta \bar{z})^2}{D^2(z)} = \frac{\text{signal energy}}{N_0} \tag{3}$$

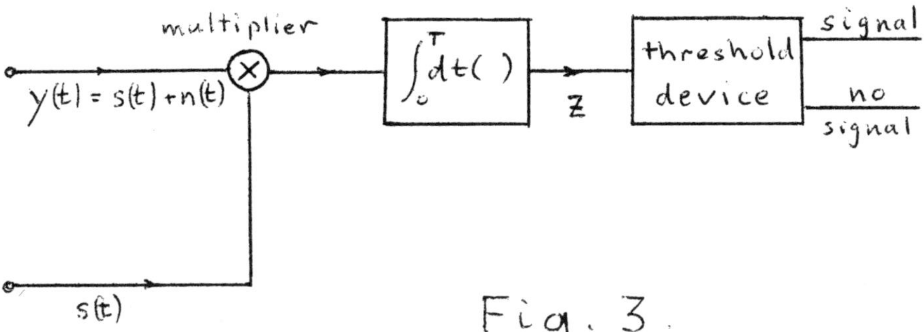

Fig. 3.

c) If the signal is a single known time function, the noise is Gaussian but has a non-white spectrum $N(\omega)$ and if the observation time T is large compared with the correlation time of the noise, the optimum detector assumes the form shown in Fig. 4. It

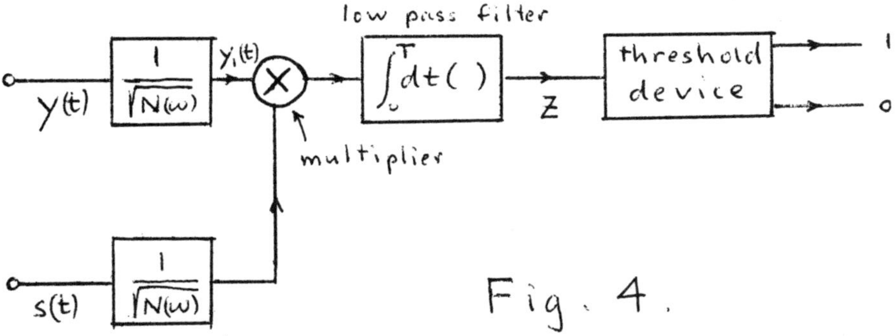

Fig. 4.

is identical with Fig. 3 except for a filter which pre-whitens the noise and a similar filter which insures that the signal replica entering the multiplier is of the same form as the signal component of $y_1(t)$. In particular, if the signal is a pure sinusoid at frequency ω_0, only those components of the noise which lie within approximately $2\pi/T$ rad/sec of ω_0 will generate inter-modulation products capable of passing through the low-pass filter. The form of the pre-whitening filter is therefore important only over the band $(2\pi/T)$ near ω_0 and if the noise spectrum is flat over this interval, Fig. 3 is very nearly optimal.

We are now ready to study the problem of detecting a Gaussian signal generated by a more or less localized acoustic source in an environment of Gaussian noise with arbitrary spatial properties. The components of the likelihood ratio [Eq. (2)] are given by

$$p(\underline{y}/S+N) = \frac{1}{(2\pi)^{r/2}\sqrt{\mathrm{Det}(P+Q)}} \exp\left\{-\frac{1}{2}\,\underline{y}^*\,(P+Q)^{-1}\underline{y}\right\} \qquad (4)$$

$$p(\underline{y}/N) = \frac{1}{(2\pi)^{r/2}\sqrt{\text{Det } Q}} \exp\left\{-\frac{1}{2}\underline{y}^* Q^{-1}\underline{y}\right\} \qquad (5)$$

$\underline{y}^*$ stands for the transpose (conjugated if complex) of vector $\underline{y}$. The data vector $\underline{y}$ can consist of any suitable representation of the received waveshapes. Thus, one might use n time samples of each of the M sensor outputs, n Fourier coefficients of each of the M sensor outputs or any number of other possibilities. P and Q are the covariance matrices of signal and noise in the chosen representation system. Because of the exponential form of Eqs. (4) and (5) it is convenient to use z = log L.R. as the detection statistic.

$$z = \log \text{L.R.} = \frac{1}{2}\underline{y}^* \left[Q^{-1} - (P+Q)^{-1}\right]\underline{y} \qquad (6)$$

Terms independent of the data vector $\underline{y}$ have been omitted. Eqs. (4) and (5) assume that $\underline{y}$ is real. If it is complex [e.g., if a complex Fourier representation is used] the factor of 1/2 disappears in the exponentials and hence in Eq. (6). This is an obviously trivial change.

In practical sonar problems, the receiving array often contains many sensors. Adequate representation of each received waveshape requires a large number of coefficients. The dimensionality of the matrices in Eq. (6) is therefore generally too large for convenient inversion. Careful choice of the representations becomes critical at this point. The following well-known result provides the necessary clue [6]:

If a stationary random process is observed for a time T large compared with its correlation time, its complex Fourier coefficients are uncorrelated.

Hence, if we regard the data vector $\underline{y}$ as a concatenation of vectors of Fourier coefficients $\underline{y}_k$ the matrices in Eq. (6) become block diagonal and one can write

$$z = \sum_{k=1}^{n} \underline{y}_k^* \left[(N_k Q_k)^{-1} - (S_k P_k + N_k Q_k)^{-1}\right]\underline{y}_k \qquad (7)$$

$\underline{y}_k$ is the vector of Fourier coefficients associated with frequency $\omega_k = 2\pi k/T$ Its dimension is M, the number of sensors. Q_k is the covariance matrix of noise Fourier coefficients at frequency ω_k normalized so that the trace of the matrix satisfies $\text{Tr}(Q_k) = M$.

N_k is simply the noise spectrum at frequency ω_k. The signal covariance matrix P_k is similarly normalized and S_k is the signal spectrum at ω_k. The factor of 1/2 has been omitted from the definition of z because of the complex form of the representation.

If the number of sensors is at all large, the matrix inversions required by Eq. (7) are still difficult. Relatively simple results can be obtained in two important special cases:

a) $S_k/N_k << 1$ (low input signal to noise ratio)

Here $$(S_k P_k + N_k Q_k)^{-1} \cong (N_k Q_k)^{-1} [I - S_k P_k (N_k Q_k)^{-1}] \quad (8)$$

It follows that

$$z = \sum_{k=1}^{n} \frac{S_k}{N_k^2} \underline{y}_k^* Q_k^{-1} P_k Q_k^{-1} \underline{y}_k \quad (9)$$

Since P_k is a covariance matrix, it is non-negative definite so that one can write

$$P_k = B_k B_k^* \quad (10)$$

Hence

$$z = \sum_{k=1}^{n} \frac{S_k}{N_k^2} ||B_k^* Q_k^{-1} \underline{y}_k||^2 \quad (11)$$

The symbol $||A||$ denotes the norm of matrix A. If P_k is a matrix of rank M, B_k is an MxM matrix and Eq. (11) requires the formation of M distinct linear combinations of the $(y_k)_i$ at each frequency. In terms of filters, the optimum processor contains M^2 frequency sensitive filters. For large M the realization problem is therefore formidable.

A much simpler instrumentation results when the signal comes from a point source and therefore produces a completely coherent wavefront.

b) Coherent signal wavefront.

Here P_k has rank one and can therefore be represented as an outer product of vector $\underline{V}_k$ with itself.

$$P_k = \underline{V}_k \underline{V}_k^* \tag{12}$$

In physical terms, $\underline{V}_k$ is simply the vector of relative phase shifts at various sensors.

$$(V_k)_i = \exp\{j\omega_k \tau_i\} \tag{13}$$

where τ_i is the relative delay at the ith sensor. The inversion required by Eq. (7) can now be performed without any assumption of low signal to noise ratio and one obtains

$$z = \sum_{k=1}^{n} \frac{\dfrac{S_k}{N_k^2}}{1 + \dfrac{S_k}{N_k} V_k^* Q_k^{-1} \underline{V}_k} \, |\underline{V}_k^* Q_k^{-1} \underline{y}_k|^2 \tag{14}$$

The only spatial operation is $|\underline{V}_k^* Q_k^{-1} \underline{y}_k|^2$. It differs from the operation $||B_k^* Q_k^{-1} \underline{y}_k||$ demanded by Eq. (11) in that $\underline{V}_k$ is a vector so that $\underline{V}_k^* Q_k^{-1} \underline{y}_k$ is a number, not a vector, and one only needs to compute its square. The optimum processor, therefore, requires only M frequency sensitive filters (rather than M^2), an enormous simplification. Its realization is, in fact, not at all complex. Using Parseval's theorem to convert the k sum into a time integral one arrives at the block diagram shown in Fig. 5.

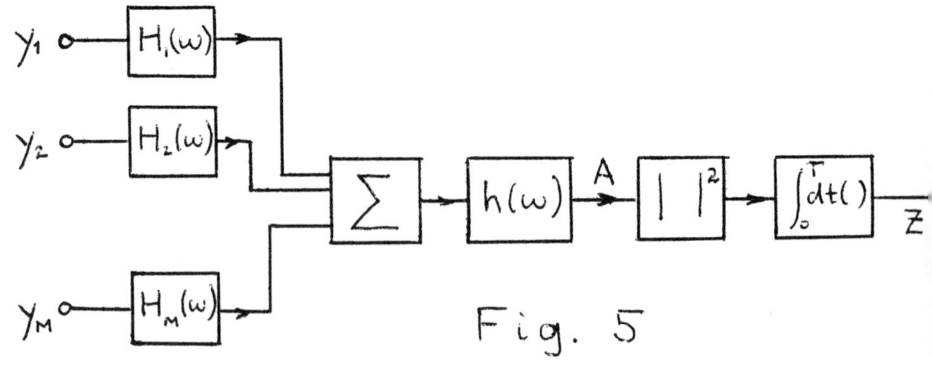

Fig. 5

The filter functions $H_i(\omega)$ are given by

$$H_i(\omega_k) = (\underline{V}_k^* Q_k^{-1})_i \qquad (15)$$

$h(\omega)$ is the common frequency filter

$$h(\omega_k) = \frac{\frac{S_k}{N_k 2}}{1 + \frac{S_k}{N_k} \underline{V}_k^* Q_k^{-1} \underline{V}_k} \qquad (16)$$

The quadratic form $\underline{V}_k^* Q_k^{-1} \underline{V}_k$ is called the array gain

$$G_k = \underline{V}_k^* Q_k^{-1} \underline{V}_k \qquad (17)$$

It may be interpreted as the signal to noise ratio at point A divided by the signal to noise ratio at each sensor. It is therefore a direct measure of the signal to noise ratio improvement gained through array processing.

Fig. 5 represents the optimal processor, exploiting both temporal and spatial features of signal and noise. A few comments about its operation and properties are therefore in order.

1) The primary spatial processing is done by the filters $H_i(\omega_k)$ specified by Eq. (15). The operation Q_k^{-1} to which they first subject the received waveshape may be regarded as spatial pre-whitening. The operator $\underline{V}_k^*$ simply aligns the signal components. The overall spatial operation may therefore be described as "prewhiten and match the signal." We shall see shortly that we may view this as a special case of Fig. 4.

2) In general, the frequency filter $h(\omega)$ does depend on noise spatial properties through the array gain G_k. However, in the practically very important case of low input signal to noise ratio it becomes independent of G_k.

$$h(\omega_k) \cong \sqrt{S_k}/N_k \qquad (18)$$

This is the well-known Eckart filter which exploits differences in the temporal spectra of signal and noise to enhance signal to noise ratio.

3) The likelihood ratio processor maximizes the detection probability for a given false alarm rate. If one is interested not so much in detection, but in extracting information from the signal component, one might prefer another criterion of performance. One might, for instance, choose to

a) maximize the signal to noise ratio at the processor output or

b) design the processor to yield the best rms estimate of the signal component of the input (Wiener filter).

It is interesting to observe that optimization under these criteria yields the same basic spatial operation. The resulting processors differ only in the common frequency functions $h(\omega)$ which now assume the form [7]

$$\text{case a):} \quad h(\omega_k) = \sqrt{S_k}/N_k \tag{19}$$

$$\text{case b):} \quad h(\omega_k) = \frac{S_k/N_k}{1+(S_k/N_k)G_k} \tag{20}$$

The fact that the Wiener filter performs the same basic operations as Fig. 5 suggests strongly that estimators for such signal parameters as sensor to sensor delay (hence bearing and range) will have much in common with Fig. 5.

4) The detection index {Eq. (1)} is readily calculated for the detector of Fig. 5. For low signal to noise ratios it assumes the simple form

$$d = \frac{T}{2\pi}\int_0^\infty \left[\frac{S(\omega)}{N(\omega)}\right]^2 G^2(\omega)\,d\omega \tag{22}$$

$S(\omega)$, $N(\omega)$ and $G(\omega)$ are simply the continuous versions of S_k, N_k and G_k. Thus the array gain completely characterizes the performance of the detector.

5) When the noise is spatially incoherent $Q_k = I$,

$$G_k = M \quad \text{and}$$

$$H_i(\omega_k) = (\underline{V}_k^*)_i = \exp(-j\omega_k\tau_i) \tag{23}$$

Fig. 5 now reduces to Fig. 1 and the conventional beamformer is optimal. In at least one very important situation, therefore, absolutely nothing can be gained by employing processing procedures more complex than conventional beamforming. The question whether optimal techniques are of practical or only academic interest is therefore a very central one.

6) Fig. 5 is very much simpler than the matrix processor implied by Eq. (11). Practical sources are never points but are very often confined to a small spatial region. How small must this region be before one can ignore the spatial distribution without incurring a major degradation in performance? Formal calculations (not confined to small signal to noise ratios) yield the very reasonable result that a source may be regarded as a point source if it subtends at the receiving array an angle small compared with the main lobe width of the beam pattern [8].

3. OPTIMAL VS. CONVENTIONAL PROCESSING

In the last section we left open the question whether there are practically important situations in which optimal processing yields meaningful improvements over conventional procedures. To gain some insight into this problem consider a single frequency component, the signal being radiated by a Gaussian point source in the far field. To keep the geometry simple we shall work with a linear receiving array as suggested by Fig. 6. If the (complex) signal component received at the origin is $s_o = A \exp\{j\omega_o t\}$, the corresponding signal component received at point x is

$$s(x) = s_o \exp\{j\omega_o \frac{x}{c} \sin\theta\} \equiv s_o \exp\{j\nu_s x\} \qquad (24)$$

Here $\nu_s \equiv (\omega_o/c)\sin\theta$ is the space frequency (wave number).

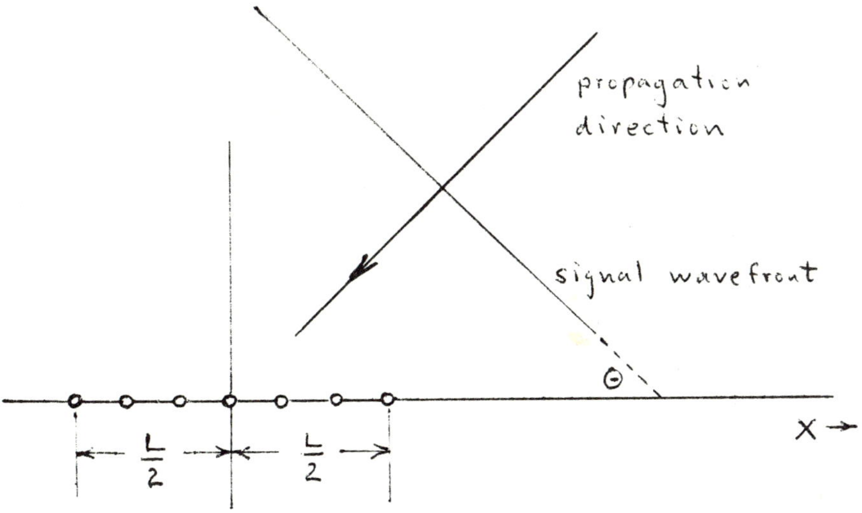

Fig. 6.

Considered as a function of x the signal is a simple sinusoid of frequency ν_s. When the source bearing θ is known, ν_s is fixed and we are therefore dealing with the detection of a known signal in Gaussian noise. If the noise is white (here spatially) we conclude immediately {Fig. (3)} that the best detector correlates the received data with a replica of the signal. The required mathematical operation is

$$z = s_o \int_{-L/2}^{-L/2} y(x)\exp\{-j\nu_s x\}dx \underset{\substack{\text{discrete}\\ \text{sensors}}}{\Longrightarrow} s_o \sum_{r=1}^{M} y(x_r)\exp\{-j\nu_s x_r\}$$

$$= s_o \sum_{r=1}^{M} y(x_r)\exp\{-j\omega_o \tau_r\} \tag{25}$$

where $\tau_r = (x_r/c) \sin \theta$, the sensor to sensor delay of the signal (see Eq. 13). Eq. (25) describes precisely the spatial operation performed by the conventional beamformer of Fig. 1. Thus we could have deduced the optimality of the conventional beamformer for spatially incoherent noise directly from the basic theory of coherent detection. Suppose now that the noise is not spatially incoherent but that its correlation distance is short compared with the length L of the array. We are then dealing with the spatial equivalent of Fig. 4 and our prescription is "pre-whiten and cross-correlate with an appropriately modified replica of the signal". The notion of pre-whitening has meaning only if the noise is stationary. Spatial stationarity means that the cross-correlation between the waveshapes received at points x_1 and x_2 depends only on (x_1-x_2). This condition is satisfied by noise coming from the far field, i.e., from a distance large compared with the array length L. Such noise can be regarded as a superposition of plane waves of the form of Eq. (24). Its power in any increment of space frequency depends on the number and strength of the noise sources located in the corresponding angular segment. Fig. 7 shows typical spatial spectra $N(\nu;\omega_o)$. Fig. 7a represents the spherically isotropic noise model used in much of the early work on optimal detection [2], [3]. Fig. 7b shows noise incident primarily from angles to the left of broadside.

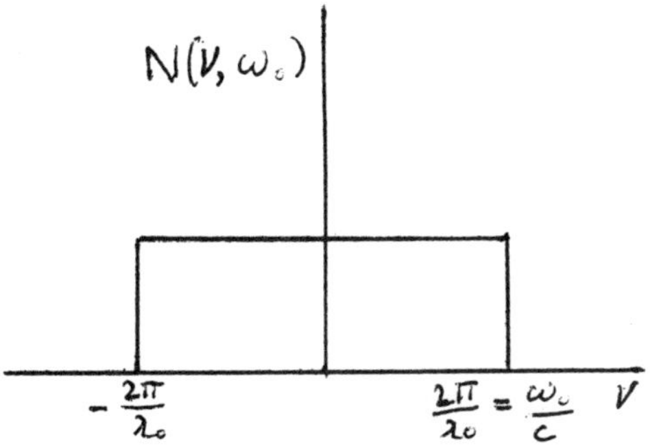

Fig. 7a

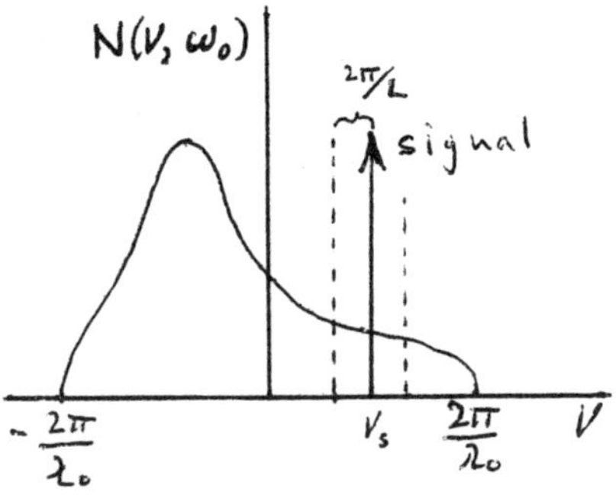

Fig. 7b

Also shown in Fig.7b is a plane wave signal which appears on this spectral plot as a delta function.

We note that all far - field spectra are confined to the interval $|\nu| \leq 2\pi/\lambda o$, the endpoints of which correspond to plane waves incident from the endfire direction. Observation at a discrete set of sensors spaced at regular intervals of d units produces the usual effect of sampling on spectra: The sampled spectra periodically repeat the un-sampled spectra with a period $2\pi/d$. To avoid spectral overlap sensor spacing must therefore satisfy

$$2\pi/d > 4\pi/\lambda_o \tag{26}$$

or

$$d < \lambda_o/2$$

where λ_o is the acoustic wavelength at frequency ω_o.

The analogy with Fig. 4 (detection of known signal in non-white Gaussian noise) is now complete and we conclude immediately:

If the correlation distance of the noise is short compared with the array length <u>and</u> if the spatial noise spectrum is essentially flat over a ν interval of the order $(2\pi/L)$ near $\nu = \nu_s$, then the conventional beamformer is nearly optimal.

In our example 7b, therefore, little could be gained by optimal space processing in spite of the macroscopically quite uneven spatial distributions of the noise.

To gain more insight into potential improvements when the noise correlation distance is not short or the noise spectrum is not locally flat near ν_s, consider the following point of view:

We ignore the spatial sampling problem and assume that the sound field is observed continuously over the interval $-L/2 \leq x \leq L/2$. The conventional beamformer computes the Fourier coefficient

$$C_s = \int_{-L/2}^{L/2} y(x)\exp\{-j\nu_s x\}dx \tag{27}$$

The received sound field could be reconstructed completely from the set of Fourier coefficients

$$C_n = \int_{-L/2}^{-L/2} y(x)\exp -j\{\frac{2\pi n}{L} x\}dx \qquad n = 0,\pm1,\pm2,\ldots \tag{28}$$

By proper choice of the space frequency origin one can always achieve $\nu_s = 2\pi s/L$, s an integer. In that case, the signal contributes only to the coefficient C_s. All of the other C_n depend on noise only. Note that all of the C_n are ordinary beamformer outputs (as they might be supplied by a DIMUS system) with one important reservation: When the index n becomes large, the space frequency $2\pi n/L$ does not correspond to a real arrival angle. To represent the received sound field fully one therefore needs not only the usual beams of the DIMUS system, but also the "pseudo-beams" obtained by introducing element to element delays larger than those encountered in the endfire direction.

Given that all of the signal will appear on a single beam, the other beams can aid the detection process only by providing information about the noise. If the noise on the signal beam is uncorrelated with that on all other beams, the conventional detector is optimal. In other cases the potential for improvement is measured by the cross-correlation between the noise component on the signal beam C_s and that on the various noise beams C_n. For spatially stationary noise a simple computation yields (for $s \neq n$)

$$E_N\{C_s C_n^*\} = \frac{L^2}{2\pi}\int_{-\infty}^{\infty} d\nu N(\nu;\omega_o)\;\frac{\sin(\frac{\nu L}{2} - s\pi)}{\frac{\nu L}{2} - s\pi}\;\frac{\sin(\frac{\nu L}{2} - n\pi)}{\frac{\nu L}{2} - n\pi} \tag{29}$$

If $N(\nu;\omega_o)$ is constant (spatially white noise) the orthogonality of the sinc functions insures lack of correlation. Thus once again we confirm the optimality of the conventional beamfomer for spatially incoherent noise.

Under what conditions will the correlation be high? An obvious candidate is a highly concentrated spatial noise spectrum such as

$$N(\nu;\omega_o) = I(\omega_o)\delta(\nu-\nu_I) \tag{30}$$

which corresponds to a plane wave interference from a direction corresponding to ν_I. Here

$$E_N\{C_s C_n^*\} = \frac{L^2}{2\pi} \frac{\sin\left(\frac{\nu_I L}{2} - s\pi\right)}{\frac{\nu_I L}{2} - s\pi} \frac{\sin\left(\frac{\nu_I L}{2} - n\pi\right)}{\frac{\nu_I L}{2} - n\pi} \tag{31}$$

Thus there is correlation unless $\nu_I L/2$ is an integral multiple of π, i.e. unless the interference happens to lie in a null of the conventional beam pattern. For a broadband interference this would certainly not happen simultaneously at all frequencies.

More informative than Eq. (31) is the normalized correlation coefficient which is easily computed

$$\frac{E_n\{C_s C_n^*\}}{\left[E_n\{|C_s|^2\}E_n\{|C_n|^2\}\right]^{1/2}} = 1, \quad \nu_I L/2\pi \text{ not an integer} \tag{32}$$

Thus the output of any non-signal beam completely characterizes the interference and can be used to eliminate it. When the total noise field contains components other than pure interference, the beam best suited for generating a replica of the interference is one aimed in the interference direction. This leads to the notion of the null steering detector discussed by Anderson and Rudnick [4], [5], a simple version of which is shown in Fig 8. One first beamforms on the interference, adjust the amplitude of the

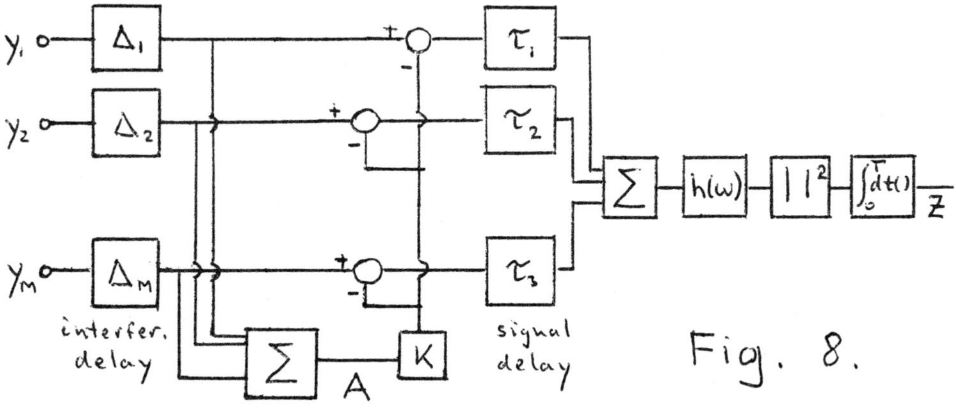

Fig. 8.

interference component at point A to match that at each sensor, subtracts to eliminate the interference completely and then beamforms on the signal in conventional manner. For interferences not too close to the signal in bearing this relatively simple scheme comes quite close to the optimum and can greatly out-perform the conventional detector if the interference is strong.

Rather than pursue the intrinsically interesting and important topic of interference rejection further we return to the main theme of this article: When do processing procedures other than conventional beamforming hold promise of substantial improvement? What makes interference rejection possible is that the interference has only one spatial degree of freedom - it can be characterized by a simple beamformer output. Far field noise is naturally characterized by beamformer outputs. Major improvements with instrumentations of tolerable complexity can therefore be expected if and only if a major fraction of the noise power is contained in spectral features spanning no more than a few beams of the form of Eq. (28).

When important noise components come from the near field (e.g. noise generated near the observing platform) conventional beams no longer provide the most natural characterization. A point source in the near field generates a spherical wavefront, not the plane wavefront matched by the conventional beamformer. Nevertheless it has only two degrees of freedom (bearing and range of the source). One can use that fact to characterize it and eliminate it by fairly simple procedures (e.g. use of a few sensors widely scattered over the receiving array). Some of the near optimal schemes suggested by Cox make use of this basic idea [9].

Before concluding this discussion we must return briefly to the problem which generated the initial controversy about optimal detection. We are now concerned with an isotropic noise field [Fig. 7a] and Eq. (29) assumes the form

$$E_N\{C_s C_n^*\} = \frac{L^2}{2\pi} \int_{-\frac{2\pi}{\lambda_o}}^{\frac{2\pi}{\lambda_o}} d\nu \; \frac{\sin(\frac{\nu L}{2} - s\pi)}{\frac{\nu L}{2} - s\pi} \; \frac{\sin(\frac{\nu L}{2} - n\pi)}{\frac{\nu L}{2} - n\pi} \tag{33}$$

If the array length L is large compared with the acoustic wavelength λ_o, the integral extends over many lobes of the sinc function. It differs from zero only because of end effects generated by the fact that the limits of integration are large but finite. The correlation coefficients are therefore small and each noise beam (or pseudo-beam) makes only a minor contribution to noise reduction. This does not say that by using enough beams one might not make substantial gains. In fact, formal computations show that one can reduce the noise level on the signal beam below any preassigned limits by using a sufficient number of additonal beams. This conclusion is not as startling as it may appear at first glance. We are dealing with a strictly bandlimited noise spectrum. According to Wiener's theory of extrapolation [10] a random process with such a spectrum is perfectly predictable from observation of any finite segment. In our terminology, after observing the sound field over $-L/2 \leq x \leq L/2$ we can calculate its value over the entire x axis. We therefore have the equivalent of an array of infinite length from which we would certainly expect perfect detection capability. The words "superdirectivity" and "supergain" have become associated with systems which seek to capitalize on the spatial band limitation of waveshapes received from the far field.

The practical obstacles to exploiting this feature are obviously formidable. Each noise beam makes only a minor contribution so that we are forced to use many such beams. Once the associated space frequencies exceed $(2\pi/\lambda o)$ (i.e. we are beginning to use pseudo-beams) we must increase the spatial sampling rate to avoid spectral overlap. The sensor spacing must now be reduced below a half wavelength and the required number of sensors increases rapidly. The problem is accentuated by the fact that there is invariably local (spatially incoherent) noise associated with each sensor. This spatially white noise quickly overwhelms the useful contribution provided by space frequencies well in excess of $2\pi/\lambda o$. The improvement available with realistic numbers of sensors is therefore sharply limited and generally not at all commensurate with the instrumentational complexity required to achieve it [3]. A possible exception to this statement concerns detection of endfire signals. Here the signal δ function is located at the edge of the noise spectrum, at $\nu = \pm 2\pi/\lambda o$, so that the end effects in Eq. (33) are maximized. In this case one finds a more rapid improvement in performance as sensor spacing is reduced below a half wavelength. It remains debatable, however, whether even this improvement justifies the necessary instrumentational complexity.

There is at least one situation of practical interest in which the super-directive approach does deserve serious consideration. Sometimes one is forced to work with arrays small compared with the acoustic wavelength [Bryn's original analysis was partly motivated by this problem]. Here the conventional beamformer is almost omnidirectional and its array gain is close to unity. On the other hand the spacing between sensors is inherently smaller than a half wavelength so that psendo-beams are automatically accessible. Hence some benefit of the super-directive variety can be gained with moderate numbers of sensors. As an example, consider the problem of placing an array into a sonobuoy whose dimensions are small compared with the acoustic wavelength. With only two sensors the optimum (super-directive) array gain is

$$G_o = 1 + 3\cos^2\theta \tag{34}$$

where θ is the angle relative to endfire. Thus one can achieve improvements of the order of 6 db for signals coming from the near-endfire direction. The optimum detector uses two beams, one a broadside beam forming the sum of the sensor outputs, the other a pseudo-beam forming their difference. In looking directions not too close to broadside the latter is dominant. One need therefore only compute the difference of the two sensor outputs (i.e. measure pressure gradient), a relatively easy task.

4. CLIPPING AND QUANTIZING

The processing procedures suggested in the previous section can become rather complex, so much so that analogue implementation may well become impractical. One is therefore forced into the use of digital techniques. An extreme version of the digital approach - and one which has found much favor because of its simplicity as well as of other features discussed later - is to hard-clip the output of each sensor before further processing. Fig. 9 shows a hard-clipped digital form of the conventional beamformer.

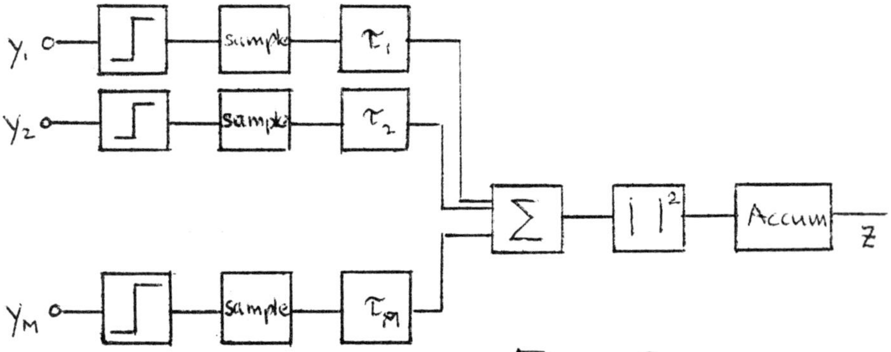

Fig. 9

The sampler outputs are binary numbers, indicating merely whether the received waveshape is positive or negative at the sampling instant. Clipping and sampling clearly destroy information and can therefore be expected to degrade detector performance.

The extent of the degradation is easily calculated for the case of spatially incoherent noise when signal and noise have white (or prewhitened) spectra of bandwidth W. With Nyquist rate sampling one finds that the digital system is inferior to the corresponding analogue system of Fig. 1 by about 2 db of equivalent input signal to noise ratio [11]. Not even all of this modest loss is due to clipping. Clipping spreads the spectrum so that sampling at the rate of (1/2W) samples per second is no longer adequate to represent the data. It turns out that one can recover about half of the loss by increasing the sampling rate, so that only about 1 db must be written off as irretrievable clipping loss [12]. The small cost of an operation which reduces complex waveshapes to a simple set of binary numbers may appear surprising at first glance. To understand what is happening we must keep in mind how the analogue detector identifies the presence of a signal. It observes that the received waveshape has a consistent pattern of delays from sensor to sensor. Delay is readily identified by the relative location of zero crossings at adjacent sensors, an item of information which is preserved in the clipping process. It is now also evident why it is beneficial to increase the sampling rate: The more frequent samples identify the location of zero crossings more accurately.

One should not infer from the above that clipping is never costly. In fact, our line of reasoning suggests strongly that there are situations in which the cost of clipping can be substantial. Consider, for example, a noise field dominated by a strong interfering plane wave. The sign of the waveshape is now determined almost entirely by the interference, except in those short time intervals when the interfering signal has an amplitude near zero. Fig.10 illustrates the mechanism at work. Only samples whose polarity is not determined by the interference contri-

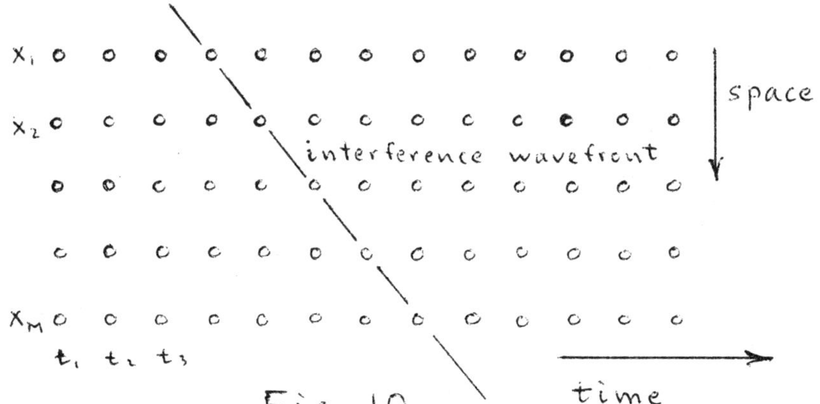

Fig. 10.

bute to the detection process. For any given interference to noise ratio it is a simple matter to compute the fraction of the total samples of which this is true. One therefore has an estimate of the inherent degradation for <u>any</u> processing system, an estimate which is confirmed by formal computations for the likelihood ratio processor [12]: Once the interference to noise ratio exceeds unity, one loses approximately 1/2 db for every 1 db increase in interference to noise ratio if one samples at Nyquist rate. About half of this loss is recoverable through faster sampling but the increase in sampling rate may have to be quite large if the interference to noise ratio is high [12].

If the clipping loss is serious one might turn to multilevel quantizing to improve performance. Aside from greater computer requirements, however, this approach sacrifices one of the most attractive features of hard clipping. The clipped outputs preserve only polarity information - they are independent of amplitude. The system is therefore unaffected by changes in power level. If one or a few sensors in a large array become unduly noisy the effect on array performance is minor [13]. Even changes in the probability distribution of signal and noise do not influence system performance as long as the correlation properties (both temporal and spatial) remain the same. The clipped system is therefore quite robust under variations of precisely those parameters of signal and noise statistics about which one is likely to have the least reliable a priori information.

5. ROBUST DETECTORS

In the previous section we have alluded to one of the major problems associated with optimal processing. The optimal detector exploits all features of the specified signal and noise statistics and is therefore apt to be very sensitive to deviations from the postulated form of these statistics. Efforts to deal with this difficulty have taken several directions:

a) One can design the detector for the nominal statistics and check how its performance varies when the actual statistics deviate from the postulated form. Cox has given expressions for the sensitivity of performance to such deviations [14]. He finds, in particular, that the sensitivity is likely to be high when the optimum processor is of the super-directive variety.

b) If the uncertainty about signal or noise statistics can be described by a single parameter (e.g. noise power or signal bearing) one can, in effect, test for all possible values of the parameter. The loss in performance relative to a system working with fully known statistics is usually small. This approach can be extended to more than one unknown parameter, but the resulting complexity quickly becomes prohibitive.

c) If there is considerable uncertainty about the relevant statistics one can turn to nonparametric procedures, designed to work with only the most general information. Not surprisingly, the cost in terms of performance is apt to be high.

d) One can measure the actual statistics and design a detector which adapts to the measurements. The cost here is measured primarily in terms of complexity, but the approach may well be the only feasible one if a priori knowledge of the relevant statistics is seriously deficient.

Since adaptive processing is the principal topic of other papers in this series we shall not pursue the subject here. Instead we confine ourselves to a few remarks about a non-adaptive procedure which is appropriate when the available a priori knowledge is less than in b) but considerably greater than in c). Specifically we shall assume that noise only statistics q_0 and signal plus noise statistics q_1 lie "near" nominal forms p_0 and p_1 respectively as suggested by Fig. 11.

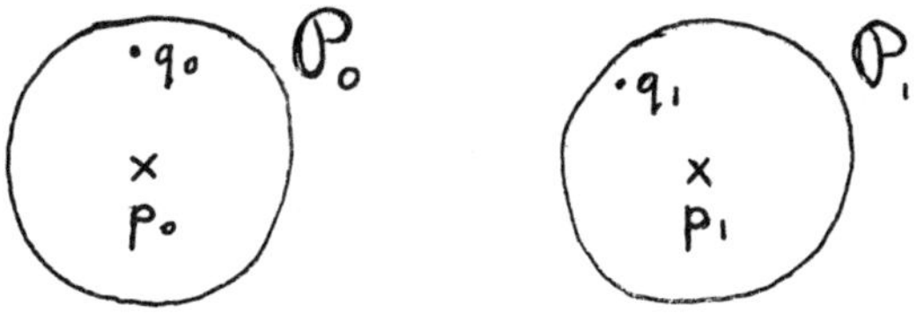

Fig. 11.

The regions $\mathcal{P}_0$ and $\mathcal{P}_1$ about p_0 and p_1 are neighborhoods within which the actual statistics are expected to fall. There are many possible ways of defining these neighborhoods. One rather general definition which turns out to be analytically tractable is

$$q_i = (1-\varepsilon_i)p_i+\varepsilon_i h_i \qquad i = 0,1 \tag{35}$$

h_i is a completely arbitrary probability density. Thus q_i is a linear combination of the nominal probability density and some arbitrary perturbing density function. ε_i measures the amount of perturbation: $\varepsilon_i = 0$ implies $q_i = p_i$ (complete a priori knowledge), $\varepsilon_i = 1$ implies total absence of a priori knowledge.

In the face of uncertain statistics it is no longer possible to demand a fixed false alarm rate or detection probability. A reasonable alternative might be to fix the maximum false alarm rate (as q_0 ranges over $\mathcal{P}_0$) and maximize the minimum detection probability (as q_1 ranges over $\mathcal{P}_1$). It is by no means obvious that such a minimax detector actual exists. One of the major theoretical contributions of recent years was Huber's demonstration that a minimax test not only exists for broad classes of perturbation models [including that of Eq. (35)], but that it is simply an ordinary likelihood ratio test between a "worst pair" of probability densities $q_o^* \epsilon \mathcal{P}_o$ and $q_1^* \epsilon \mathcal{P}_1$ [15]. The required likelihood ratio also has a pleasingly simple relation to the nominal likelihood ratio, illustrated in Fig. 12. The

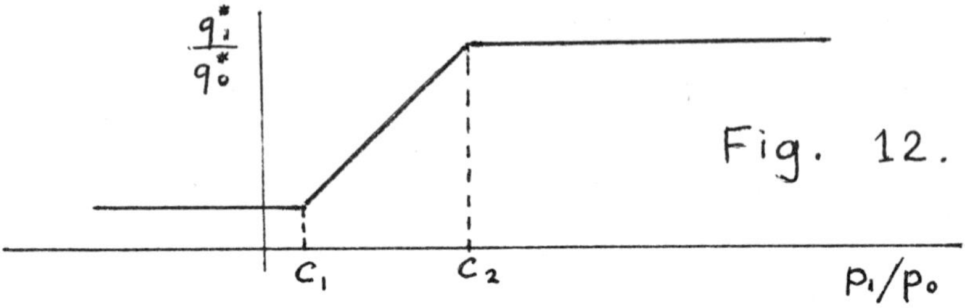

Fig. 12.

robust test is simply a clipped version of the nominal test. As $\epsilon_i \to o$, $c_1 \to o$, $c_2 \to \infty$ so that the robust test degenerates to the nominal test. The physical interpretation of Fig. 12 is obvious: One basically works with the nominal likelihood ratio. However, if any one data sample yields an extremely large or small value of p_1/p_o one discounts such an extreme piece of evidence because of the possibility that it may have been generated by unexpected statistical features. Such safeguards become particularly important when many samples are taken (low S/N) because the basically rare events of pathologically high or low p_1/p_o are then very likely to occur at some time during the observation interval. If one fails to include such protection, faulty decisions can become very likely under apparently minor deviations from the nominal distribution [16]. On the other hand, the protective clipper is obviously simple to implement and for small ϵ_i (good a priori knowledge) its cost in terms of system performance is small even if q_i happens to equal p_i. With good but not perfect knowledge of statistics one therefore has a simple and effective way to avoid potential disasters.

6. REFERENCES

1. Anderson, V. C. Digital Army Phasing JASA 32, 867, July 1960.

2. Bryn, F. Optimal Signal Processing of Three-dimensional Arrays Operating on Gaussian Signals and Noise JASA 34, 289, March 1962.

3. Vanderkulk, W. Optimum Processing for Acoustic Arrays. J. Brit. IRE Vol. 26, October 1963.

4. Anderson, V. C. DICANNE, A Realizable Adaptive Process. JASA 45, 398, February, 1969.

5. Anderson, V. C. and Rudnick, P. Rejection of Coherent Arrival at an Array JASA 45, 406, February, 1969.

6. Papoulis, A. Probability, Random Variables and Stochastic Processes, Section 13.1 McGraw - Hill 1965.

7. Edelblute, D. J., Fisk, J. M. and Kinnison, G. L. Criteria for Optimum Signal Detection Theory for Arrays, JASA 41, 199, January, 1967.

8. Bangs, W. J. Array Processing with Generalized Beamformers. Ph.D. Dissertation, Yale University, 1971 (also available as a technical report Naval Underwater Systems Center, New London, Conn. September, 1971)

9. Cox, H. Array Processing Against Interference. Symposium on Information Processing, Purdue University 1969. p. 453.

10. Wiener, N. Extrapolation, Interpolation, and Smoothing of Stationary Time Series, Wiley, 1949.

11. Kanefsky, M. Detection of Weak Signals with Polarity Coincidence Arrays. IEEE Trans. on Information Theory, IT-12, 260, April, 1966.

12. Kanefsky, M. Progress reports No. 26 (Suboptimal techniques employing hard limiting for detecting passive sonar targets in the presence of interference) and No. 28 (Optimal techniques employing quantized inputs for detecting passive sonar targets in the presence of interference), Yale University, Department of Engineering and Applied Science, November, 1965, April, 1966 (SUBIC program)

13. Schultheiss, P. M. Likelihood Ratio Detection of Gaussian Signals with Noise Varying from Element to Element of the Receiving Array. Progress report No. 11, Yale University, Department of Engineering and Applied Science, January, 1964 (SUBIC program)

14. Cox, H. Sensitivity Considerations in Adaptive Beam Forming. NATO Advanced Study Institute on Signal Processing, Loughborough 1972, P. 619.

15. Huber, P. J. A Robust Version of the Probability Ratio Test Ann. Math. Stat. 36, 1753, December, 1965.

16. Schultheiss, P. M. and Wolcin J. J. Robust Sequential Probability Ratio Detectors. EASCON 1975 Convention Record, p. 36.

NUMBER THEORETIC TRANSFORMS

P. J. W. Rayner

Department of Engineering, Cambridge University,
Cambridge, England.

ABSTRACT. The theory of Number Theoretic Transforms having circular convolution properties is developed from the definition of circular convolution. The application of these transforms to digital signal processing is discussed. The lectures include the following sections. 1. Circular convolution; 2. Circulant diagonalisation; 3. Finite arithmetic structures; 4. Application of Number Theoretic Transforms.

1. CIRCULAR CONVOLUTION

In digital signal processing one of the most important operations is that of the circular convolution of two data sequences. If the sequences $\{x\}$ and $\{h\}$ each contain N data points then the convolution product $\{y\}$ is defined by:

$$y_n = \sum_{m=0}^{N-1} x_m . h_{(n-m)} = \sum_{m=0}^{N-1} h_m . x_{(n-m)} \qquad (1)$$

where the subscript (n-m) is interpreted modulo N. The relationship of equation 1 may also be expressed as:

$$\underline{y} = \underline{C}\ \underline{x} \qquad (2)$$

where:

G. Tacconi (ed.), Aspects of Signal Processing, Part 1, 333-353.

$$\underline{y} = \begin{bmatrix} y_o \\ y_1 \\ \vdots \\ y_{N-1} \end{bmatrix} \quad \underline{x} = \begin{bmatrix} x_o \\ x_1 \\ \vdots \\ x_{N-1} \end{bmatrix} \quad \underline{C} = \begin{bmatrix} h_o & h_{N-1} & h_{N-2} & \cdots & h_1 \\ h_1 & h_o & h_{N-1} & \cdots & h_2 \\ \vdots & \vdots & \vdots & & \vdots \\ h_{N-1} & h_{N-2} & h_{N-3} & \cdots & h_o \end{bmatrix} \tag{3}$$

The matrix $\underline{C}$ is said to be a circulant.
Since the circular convolution is of such importance we are interested in efficient computational methods for evaluating the convolution product.

2. CIRCULANT DIAGONALISATION

The circulant matrix $\underline{C}$ is normal and is therefore unitarily similar to a diagonal matrix

$$\underline{C} = \underline{T} \;\underline{\Lambda}\; \underline{T}^{-1} \tag{4}$$

where: is a diagonal matrix with the eigenvalues of $\underline{C}$ on the leading diagonal.
$\underline{T}$ is an unitary matrix with columns equal to the eigenvectors of $\underline{C}$.

From equations 2 and 4 we can express the convolution relationship as:

$$\underline{y} = \underline{T} \;\underline{\Lambda}\; \underline{T}^{-1} \underline{x} \tag{5}$$

We shall see presently that if the eigenvalues and eigenvectors are derived by normal continuous mathematical methods then the matrix $\underline{T}$ is simply the Discrete Fourier Transform (D. F. T) and the diagonal matrix $\underline{\Lambda}$ contains the D. F. T. of the data vector $\underline{h}$.

Bellman [1] has presented an elegant method for determining the eigenvectors and eigenvalues of a circulant matrix. We shall now generalise this method in order to obtain Number Theoretic transforms having the circular convolution property.

Let W_1 be an Nth root of unity

and form:

$$\lambda_1 = h_o W_1^{o} + h_{N-1} W_1^{1} + h_{N-2} W_1^{2} \;----\; h_1 W_1^{N-1} \tag{6}$$

Note that the h_p in the above expression correspond to the top row of the circulant matrix given in equation 3. Multiply both sides of equation 6 by W_1 to give:

$$\lambda_1 W_1^{1} = h_1 W_1^{o} + h_o W_1^{1} + h_{N-1} W_1^{2} + \;----\; h_2 W_1^{N-1} \tag{7}$$

Note that the order of the h_p in the above expression corresponds to the second row of the circulant matrix of equation 3. Continuing this process gives the set of equations:

$$\lambda_1 W_1^{2} = h_2 W_1^{o} + h_1 W_1^{1} + h_o W_1^{2} \;----\; h_3 W^{N-1}$$

$$\lambda_1 W_1^{N-1} = h_{N-1} W_1^{o} + h_{N-2} W_1^{1} + h_{N-3} W_1^{2} + \;----\; h_o W_1^{N-1} \tag{8}$$

Equations 6, 7 and 8 may be combined and written in matrix form.

$$\underline{W}_1 \lambda_1 = \underline{C}\, \underline{W}_1 \tag{9}$$

where: $\underline{C}$ is defined by equation 3

$$\underline{W}_1 = \begin{bmatrix} W_1^{o} \\ W_1^{1} \\ \vdots \\ W_1^{N-1} \end{bmatrix} \tag{10}$$

Note that equation 9 has the form defining an eigenvalue λ_1 and eigenvector $\underline{W}_1$ of matrix $\underline{C}$. Equation 9 is defined in terms of W_1, an Nth root of unity. However there are N distinct N^{th} roots of unity. i. e. if $W_1{}^N = 1$

then W_1^{n} is an Nth root of unity for $n = 0, 1, 2, ---, N-1$

Thus we can write N equations of the form of equation 9. These N equations may be combined to form a single matrix equation

$$\underline{T}\, \underline{\Lambda} = \underline{C}\, \underline{T} \tag{11}$$

where:

$$= \begin{bmatrix} \lambda_0 & & & & \\ & \lambda_1 & & & \\ & & \lambda_2 & & \\ & & & \ddots & \\ & & & & \lambda_{N-1} \end{bmatrix} \tag{12}$$

$$\underline{T} = N^{-1} \begin{bmatrix} (W_1^0)^0 & (W_1^1)^0 & (W_1^2)^0 & ---- & (W_1^{N-1})^0 \\ (W_1^0)^1 & (W_1^1)^1 & (W_1^2)^1 & ---- & (W_1^{N-1})^1 \\ (W_1^0)^1 & (W_1^1)^2 & (W_1^2)^2 & ---- & (W_1^{N-1})^2 \\ & & & & \\ (W_1^0)^{N-1} & (W_1^1)^{N-1} & (W_1^2)^{N-1} & ---- & (W_1^{N-1})^{N-1} \end{bmatrix} \tag{13}$$

$$\lambda_k = h_0(W_1^k)^0 + h_{N-1}(W_1^k)^1 + h_{N-2}(W_1^k)^2 + ---- + h_1(W_1^k)^{N-1} \tag{14}$$

The constant N^{-1} which has been introduced in the definition of $\underline{T}$ does not affect equation 11 since both sides are multiplied by $\underline{T}$. The reason for introducing the constant will become clear later. Postmultiplying both sides of equation 11 by $\underline{T}^{-1}$ gives the required result:

$$\underline{C} = \underline{T}\,\underline{\Lambda}\,\underline{T}^{-1} \tag{15}$$

and the convolution relationship

$$\underline{y} = \underline{C}\,\underline{x} \tag{16}$$

may be written as:

$$\underline{y} = \underline{T}\,\underline{\Lambda}\,\underline{T}^{-1}\,\underline{x} \tag{17}$$

The results obtained so far are completely general in that they require that N^{th} roots of unity exist and that the matrix $\underline{T}$

has an inverse.

Let us consider first the continuous complex N^{th} roots of unity

$$\text{i.e.}\quad W_1 = e^{j\frac{2\pi}{N}} \tag{18}$$

If this value for W_1 is substituted in equation 13 then we see that the matrix $\underline{T}$ is simply the inverse D. F. T. matrix, and $\underline{T}^{-1}$ becomes the D. F. T. matrix. The diagonal matrix now contains the values of the D. F. T. of the data vector $\underline{h}$. Equation 17 may be interpreted as a frequency domain representation of time domain circular convolution in that $\underline{T}^{-1}\,\underline{x}$ is the discrete frequency spectrum of the data $\{x\}$ and each value of the discrete frequency spectrum is multiplied by the system discrete frequency transfer function. The resulting frequency spectrum is transferred back to the time domain by matrix $\underline{T}$.

Equations 16 and 17 are two alternative formulations of the circular convolution operation. The importance of the transform representation is that a very efficient computational algorithm, known as the Fast Fourier Transform for the case

$$W_1 = e^{j\frac{2\pi}{N}},$$

is available for computing the product of the transform matrix $\underline{T}^{-1}$ and a data vector. Use of this algorithm allows the circular convolution product to be evaluated more efficiently by equation 17.

In many realisations of convolution the arithmetic operations will be concerned with integer data in that the elements of the data vector $\underline{x}$ and the convolution matrix $\underline{C}$ are integer thus giving a vector $\underline{y}$ having integer elements. If however, the convolution is evaluated from equation 17 with

$$W_1 = e^{j\frac{2\pi}{N}}$$

then infinite precision arithmetic is required to completely represent the complex roots of unity. Thus the somewhat anomolous situation exists in which the efficient computation of an essentially bounded integer problem requires the use of infinite precision arithmetic. Of course, in a practical situation the arithmetic operations are not performed with infinite precision so that the computed data will not be completely accurate. The need for infinite arithmetic precision arises from the non-rational nature of the complex roots of unity. In an attempt to overcome this problem it is necessary to consider the existence of integer roots of unity. Such roots do not exist in continuous mathematical structures but they do exist in certain finite arithmetic structures.

It will be convenient for our investigation if the matrix expression for the transform $\underline{T}$ is written in summation form. The matrix-vector product:

$$\underline{u} = \underline{T}\ \underline{U}$$

may be expressed as:

$$u_n = N^{-1} \sum_{q=0}^{N-1} W_1^{nq}\ U_q \tag{18}$$

We require therefore, arithmetic structures in which N^{th} roots of unity exist and in which inverse transforms of equation 18 exist. If such arithmetic structures can be found then the resulting transforms will have a form identical to that of the Discrete Fourier Transform and the Fast Fourier type of algorithm may be used to compute the transforms.

3. FINITE ARITHMETIC STRUCTURES

3.1 Prime number moduli

Fermat's theorem [2, 3] states that for any prime number p and any integer b, prime to p, the following congruence relationship is satisfied:

$$b^{p-1} \equiv 1 \pmod{p} \tag{19}$$

This congruence relationship for integers is analogous to the expression for the continuous complex roots of unity:

$$\left(e^{j\frac{2\pi}{p-1}.k}\right)^{p-1} = 1$$

As an example consider the case of b = 2 and p = 5

$$\left.\begin{aligned} 2^1 &\equiv 2 \\ 2^2 &\equiv 4 \\ 2^3 &\equiv 3 \\ 2^4 &\equiv 1 \end{aligned}\right\} \pmod{5}$$

Thus the integer b = 2 may be considered as a fourth root of unity in a modulo 5 arithmetic system. Note that Fermat's theorem does not imply that (p-1) is necessarily the smallest exponent for which equation 19 is satisfied. For example, if p = 5 and b = 4 then:

$$4^4 \equiv 1 \pmod 5$$

but $4^2 \equiv 1 \pmod 5$

This is analogous to the continuous case where:

$$(-1)^4 = 1$$

but $(-1)^2 = 1$

Integers b which have the property that (p-1) is the smallest exponent for which b^{p-1} is congruent to unity are said to be primitive roots. More generally, if in the congruence relationship:

$$a^d \equiv 1 \pmod p$$

d is the smallest integer satisfying the expression then a is said to be of order d in a modulo p system.

Methods exist [2] for determining primitive roots and tables are available in the literature [4]. It may be shown that any prime p has $\emptyset(p-1)$ primitive roots. $\emptyset(k)$ is the Euler totient function and denotes the number of positive integers, 1 included, which are prime to k and not greater than k.

The set of integers $\{0, 1, ---, p-1\}$, where p is prime, constitute a field under multiplication and addition modulo p, so that each element of the field, except 0, has a multiplicative inverse. More general algebraic structures, termed rings, also have N^{th} roots of unity but discussion of these structures will be deferred.

We must now determine whether the roots of unity defined in a prime number field can be used in the transform [5, 6, 7] defined in Section 2 (equation 18). In particular we must prove the existence of the inverse transform. Substituting equation 19 in equation 18 defines a transform:

$$u_n \equiv (p-1)^{-1} \sum_{q=0}^{p-2} b^{nq} \cdot U_q \pmod p \qquad (20)$$

where: $N \triangleq (p-1)$ (21)

Appendix 1 shows that the following orthogonality relationship is valid for primitive roots b modulo p.

$$(p-1)^{-1} \sum_{q=0}^{p-2} b^{kq} \cdot b^{-lq} \equiv \begin{Bmatrix} 1, k=l \\ \\ 0, k \neq l \end{Bmatrix} \pmod p \qquad (22)$$

Multiply both sides of equation 20 by b^{-mn} and sum over n.

$$\sum_{n=0}^{p-2} u_n \cdot b^{-mn} \equiv (p-1)^{-1} \sum_{n=0}^{p-2} \left\{ \sum_{q=0}^{p-2} b^{nq} \cdot U_q \right\} b^{-mn} \pmod p$$

$$\therefore \sum_{n=0}^{p-2} u_n \cdot b^{-mn} \equiv \sum_{q=0}^{p-2} U_q \left\{ (p-1)^{-1} \sum_{n=0}^{p-2} b^{nq} \cdot b^{-mn} \right\} \pmod p$$

Applying the orthogonality condition, equation 20, to the above yields:

$$U_q \equiv \sum_{n=0}^{p-2} u_n \cdot b^{-mn} \pmod p \qquad (23)$$

Thus we have shown that an inverse transform exists so that the matrix $\underline{T}$ exists. Equations 20 and 23 are a Number Theoretic transform (N. T. T.) pair which have been derived from the circular convolution operation. Consideration of equation 14 for the elements λ_k of the diagonal matrix $\underline{\Delta}$ shows that the λ_k are simply the inverse transform of the data vector $\underline{h}$.

$$\lambda_k \triangleq H_k = \sum_{n=0}^{p-2} h_n \cdot b^{-nk} \pmod p \qquad (24)$$

It is perhaps worthwhile at this point to consider a numerical example in that it will demonstrate some of the shortcomings of this particular N. T. T. We wish to convolve the data sequence $\{x\} = \{0, 1, 2, 1\}$ with the sequence $\{h\} = \{1, -2, 1, 0\}$. The sequences contain N = 4 data points. Equation 21 relates N and prime p:

$$N \triangleq (p-1)$$

$$\therefore \quad 5 = p$$

Note that p must be prime so we are not able to choose N arbitrarily in a general convolution. We may write out the required convolution in matrix form as in equations 2 and 3.

$$\underline{y} = \underline{C} \; \underline{x}$$

or $$\begin{bmatrix} y_o \\ y_1 \\ y_2 \\ y_3 \end{bmatrix} = \begin{bmatrix} 1 & 0 & 1 & -2 \\ -2 & 1 & 0 & 1 \\ 1 & -2 & 1 & 0 \\ 0 & 1 & -2 & 1 \end{bmatrix} \begin{bmatrix} 0 \\ 1 \\ 2 \\ 1 \end{bmatrix} = \begin{bmatrix} 0 \\ 2 \\ 0 \\ -2 \end{bmatrix}$$

The transform formulation of the convolution is given by equation 17:

$$\underline{y} = \underline{T}\, \underline{\Lambda}\, \underline{T}^{-1}\, \underline{x}$$

where $\underline{T}^{-1}$ is given in summation form by equation 25

$$X_q \equiv \sum_{n=0}^{p-2} x_n \cdot b^{-qn} \pmod p \qquad (25)$$

The diagonal matrix elements $\lambda_q \triangleq H_q$ are given by

$$H_q \equiv \sum_{n=0}^{p-2} h_n \cdot b^{-qn} \pmod p \qquad (26)$$

Now b = 2 is a primitive root modulo 5, so that we may calculate the transforms $\{X\}$ and $\{H\}$ from equations 26 and 27.

$$\begin{bmatrix} X_0 \\ X_1 \\ X_2 \\ X_3 \end{bmatrix} \equiv \begin{bmatrix} (2^0)^{-0} & (2^1)^0 & (2^2)^0 & (2^3)^0 \\ (2^0)^{-1} & (2^1)^{-1} & (2^2)^{-1} & (2^3)^{-1} \\ (2^0)^{-2} & (2^1)^{-2} & (2^2)^{-2} & (2^3)^{-2} \\ (2^0) & (2^1)^{-3} & (2^2)^{-3} & (2^3)^{-3} \end{bmatrix} \begin{bmatrix} 0 \\ 1 \\ 2 \\ 1 \end{bmatrix} \equiv \begin{bmatrix} 1 & 1 & 1 & 1 \\ 1 & 3 & 4 & 2 \\ 1 & 4 & 1 & 4 \\ 1 & 2 & 4 & 3 \end{bmatrix} \begin{bmatrix} 0 \\ 1 \\ 2 \\ 1 \end{bmatrix} \equiv \begin{bmatrix} 4 \\ 3 \\ 0 \\ 3 \end{bmatrix} \pmod 5$$

$$\begin{bmatrix} H_0 \\ H_1 \\ H_2 \\ H_3 \end{bmatrix} \equiv \begin{bmatrix} 1 & 1 & 1 & 1 \\ 1 & 3 & 4 & 2 \\ 1 & 4 & 1 & 4 \\ 1 & 2 & 4 & 3 \end{bmatrix} \begin{bmatrix} 1 \\ -2 \\ 1 \\ 0 \end{bmatrix} \equiv \begin{bmatrix} 0 \\ 4 \\ 4 \\ 1 \end{bmatrix} \qquad \pmod 5$$

To evaluate the product $\underline{\Lambda}\, \underline{T}^{-1}\, \underline{x}$ we simply form the vector

$$\begin{bmatrix} X_0 H_0 \\ X_1 H_1 \\ X_2 H_2 \\ X_3 H_3 \end{bmatrix} \equiv \begin{bmatrix} 0 \\ 2 \\ 0 \\ 3 \end{bmatrix} \pmod 5$$

Finally we evaluate $\underline{y} = \underline{T}(\underline{\Lambda}\,\underline{T}^{-1}\underline{x})$ where $\underline{T}$ is given in summation form by equation 20.

$$\begin{bmatrix} y_0 \\ y_1 \\ y_2 \\ y_3 \end{bmatrix} \equiv 4^{-1} \begin{bmatrix} (2^0)^0 & (2^1)^0 & (2^2)^0 & (2^3)^0 \\ (2^0)^1 & (2^1)^1 & (2^2)^1 & (2^3)^1 \\ (2^0)^2 & (2^1)^2 & (2^2)^2 & (2^3)^2 \\ (2^0)^3 & (2^1)^3 & (2^2)^3 & (2^3)^3 \end{bmatrix} \begin{bmatrix} X_0 H_0 \\ X_1 H_1 \\ X_2 H_2 \\ X_3 H_3 \end{bmatrix}$$

$$4^{-1} \begin{bmatrix} 1 & 1 & 1 & 1 \\ 1 & 2 & 4 & 3 \\ 1 & 4 & 1 & 4 \\ 1 & 3 & 4 & 2 \end{bmatrix} \begin{bmatrix} 0 \\ 2 \\ 0 \\ 3 \end{bmatrix} \equiv \begin{bmatrix} 0 \\ 2 \\ 0 \\ 3 \end{bmatrix} \equiv \begin{bmatrix} 0 \\ 2 \\ 0 \\ -2 \end{bmatrix} \pmod 5$$

This result agrees with the one obtained by direct evaluation of the convolution relationship.

This example demonstrates a number of important points concerning N. T. T.

1) The length N of the data vectors cannot be chosen arbitrarily since $N+1 = p$ must be prime.

2) All the elements y_n of the vector $\underline{y} = \underline{C}\,\underline{x}$ must satisfy $|y_n| \leq \frac{p-1}{2}$ otherwise ambiguity will exist when the convolution is evaluated by the N. T. T. Any element y_n not satisfying the above condition cannot be represented since it will be interpreted modulo p. Moreover it is unlikely that the above inequality will be satisfied since each element y_n is the sum of $(p-1)$ numbers. Therefore if N. T. T. are to be of practical significance it is necessary that the prime modulus be rather greater than the sequence length.

3.2 Extension of prime modulus transform.

Fermat's theorem states that:

$$b^{p-1} \equiv 1 \pmod p$$

and we shall assume that b is a primitive root modulo p. Now (p-1) is non-prime for p an odd prime and may therefore be factorised as (p-1) = d.k where d and k are integer.

$$\therefore \quad (b^k)^d \equiv 1 \pmod p$$

Thus the integer (b^k) is of order d modulo p. i.e. (b^k) is a dth root of unity modulo p. We can therefore, satisfy the relationship

$$a^d \equiv 1 \pmod p \tag{27}$$

where $d|(p-1)$

It is shown in Appendix II that the following orthogonality expression exists:

$$d^{-1} \sum_{q=0}^{d-1} a^{kq} \cdot a^{-lq} \equiv \begin{cases} 1, & k=l \\ 0, & k \neq l \end{cases} \mod p \tag{28}$$

where: a is of order d mod p. $d|(p-1)$

The dth root of unity defined in equation 27 may be substituted in the transform definition (equation 18) and the orthogonality relationship applied to derive the inverse transform. Thus we have the transform pair:

$$x_n \equiv d^{-1} \sum_{q=0}^{d-1} X_q \cdot a^{nq} \pmod p \tag{29}$$

$$X_q \equiv \sum_{n=0}^{d-1} x_n \cdot a^{-nq} \pmod p \tag{30}$$

where: a is of order d mod p. $d|(p-1)$ (31)

This transform may therefore be used to evaluate circular convolution products of data sequences of length d where d is related to the prime modulus p by equation 31.

3.3 Composite moduli

We have seen that roots of unity exist in prime number

arithmetic systems and that we are able to define a N. T. T. and inverse transform based on these roots. Under the appropriate conditions we are able to extend these results to composite moduli.

The generalisation of Fermat's theorem is the Euler-Fermat theorem which states that for two mtually prime integers a and m [i. e. (a, m) = 1] then:

$$b^{\phi(m)} \equiv 1 \pmod{m} \qquad (32)$$

where ϕ is the Euler totient function defined in section 3. 1. Equation 32 does not imply that $\phi(m)$ is necessarily the smallest exponent for which the power of b is congruent to unity. In general $\phi(m)$ is composite and we have

$$b^{q.d} \equiv 1 \pmod{m} \quad \text{where } q.d = \phi(m)$$

$$\therefore \quad a^{d} \equiv 1 \pmod{m} \qquad (33)$$

$$d \mid \phi(m)$$

If d is the smallest exponent for which equation 33 is true a is of order d modulo m.

We can now consider defining a transform based on equation 33 but before we proceed it is necessary to introduce, without proof, two results from number theory [3] .

Theorem A. Let the composite modulus $m = p_1^{k_1} p_2^{k_2} - - - p_n^{k_n}$ where p_i are prime.

The solutions of the congruence $f(x) \equiv 0 \pmod{m}$ are equal to the solutions of the simultaneous congruences:

$$f(x) \equiv 0 \pmod{p_i^{k_i}} \qquad i = 1, 2, - - -, n$$

Theorem B. An element $\alpha \pmod{m}$ has a multiplicative inverse $\alpha^{-1} \pmod{m}$ if and only if α and m are mutually prime i. e. $(\alpha, m) = 1$.

For the existence of the transform and inverse transform the following conditions must be satisfied:

$$a^{d} \equiv 1 \pmod{m} \text{ where } a \text{ is of order } d \qquad (34)$$

Orthogonality: $$d^{-1} \sum_{q=0}^{d-1} a^{kq} \equiv \begin{Bmatrix} 0, & k=0 \\ 1, & k \neq 0 \end{Bmatrix} \pmod{m} \qquad (35)$$

Theorem A may be applied to equation 34.

If $a^d - 1 \equiv 0 \pmod{m}$

then $a^d - 1 \equiv 0 \pmod{p_i^{k_i}}$ $\quad 1 = 1, 2, ---, n$

or $$a^d \equiv 1 \pmod{p_i^{k_i}} \qquad (36)$$

The Euler-Fermat theorem implies that for equation 36 to be satisfied:

$d \mid \phi(p_i^{k_i})$

$$\therefore \quad d \mid p_i^{k_i-1}(p_i-1) \quad i = 1, 2, ---, n \qquad (37)$$

If the orthogonality relationship is to be satisfied then $d^{-1} \pmod{m}$ must exist. Applying theorem B:

$(d, m) = 1$

$$\therefore \quad d \nmid p_i \qquad (38)$$

The constraints imposed on d by equations 27 and 38 give the result that:

$$d \mid (p_i - 1) \quad i = 1, 2, ---, n \qquad (39)$$

The orthogonality condition equation 35, for k = 0 is obviously satisfied now that the existence of d has been established. Consider now the condition

$$\sum_{q=0}^{d-1} a^{kq} \equiv 0 \pmod{m}, \quad k = 1, 2, ---, d-1$$

Application of theorem A shows that the above equation can be satisfied if the simultaneous congruences

$$\sum_{q=0}^{d-1} a^{kq} \equiv 0 \pmod{p_i^{k_i}} \quad k = 1, 2, ---, d-1; i = 1, 2, ---, n \qquad (40)$$

are satisfied. Equation 40 is a geometric progression and may be written in closed form as:

$$\sum_{q=0}^{d-1} a^{kq} \equiv \frac{1 - a^{kd}}{1 - a^{k}} \pmod{p_i^{k_i}}$$

But a is of order d, so that $1 - a^{kd} \equiv 0 \pmod{p_i^{k_i}}$ provided that division by $(1 - a^k)$ is a valid operation mod $(p_i^{k_i})$. Theorem B allows the existence of $(1 - a^k)^{-1}$ if:

$$(1 - a^k, p_i^{k_i}) = 1$$

$$\therefore \quad 1 - a^k \not\equiv l p_i \pmod{p_i^{k_i}} \quad l = \text{integer},\ i = 1, 2, ---, n$$

$$\therefore \quad a^k \not\equiv 1 - l p_i \pmod{p_i^{k_i}} \quad k = 1, 2, ---, d-1$$

The above condition is satisfied if a is of order d mod p_i.

Thus we have the result that a Number Theoretic Transform with a composite modulus m can be defined for data length d such that:

$$d \mid (p_i - 1) \qquad i = 1, 2, ---, n \tag{41}$$

or $d = \text{g.c.d.}\ (p_1 - 1, p_2 - 1, ---, p_n - 1)$

where the composite modulus $m = p_1^{k_1}, p_2^{k_2} --- p_n^{k_n}$

The transform and inverse transform are:

$$x_n = d^{-1} \sum_{q=0}^{d-1} a^{qn} X_q \pmod{m} \tag{42}$$

$$X_q = \sum_{n=0}^{d-1} x_n \cdot a^{-nq} \pmod{m} \tag{43}$$

where $a^d \equiv 1 \pmod{m}$ and is of order d.

4. APPLICATION OF NUMBER THEORETIC TRANSFORMS

4.1 Prime modulus transforms

We showed in sections 3.1 and 3.2 that a Number Theoretic Transform with circular convolution properties can be defined over the field of integers modulo p, where p is prime. The transform and inverse transform is given in equation 29, 30 and 31 and are reproduced below for convenience.

$$x_n = d^{-1} \sum_{q=0}^{d-1} X_q \cdot a^{nq} \pmod{p} \qquad ((29))$$

$$X_q = \sum_{n=0}^{d-1} x_n \cdot a^{-nq} \pmod{p} \qquad ((30))$$

$$a^d \equiv 1 \pmod{p} \text{ and } a \text{ is of order } d \qquad ((31))$$

$$d \mid (p-1)$$

Three parameters, namely a, d and p, have to be chosen before the transform can be applied to a circular convolution problem and we must examine the implications of these parameters. If the transforms are to be computed efficiently by means of a Fast Fourier type of algorithm then the data vector length d must be a highly composite number, and preferably of the form $d = 2^k$. However equation 31 places an additional constraint on the value of d in that $d \mid (p-1)$ where p is prime. Table 1 gives a partial list of primes p with the property that (p-1) has a factor of the form 2^k. As an example suppose that we chose to perform a $2^5 = 32$ point transform with arithmetic modulo 1153. Now $(p-1) = 1152$ factors as $2^7 . 3^2$ and we have

$$b^{p-1} \equiv 1 \pmod{p}$$

Let b be a primitive root i. e. of order (p-1) then

$$\left(b^{2^2 . 3^2}\right)^{2^5} \equiv 1 \pmod{p}$$

and $a = \left(b^{2^2 . 3^2}\right)$ is of order 2^5 as required.

Table 1 shows that for a particular choice of d there are a number of choices for the prime modulus p. The data determines which of these possibilities is to be used since the choice of p determines the range of data values that can be unambiguously represented. In section 1 the circular convolution was defined as:

$$y_n = \sum_{m=0}^{N-1} x_m \cdot h_{(n-m)} = \sum_{m=0}^{N-1} h_m \cdot x_{(n-m)} \qquad ((1))$$

If the output values y_n are to be computed unambiguously by means

Table 1. List of primes p where (p-1) has a factor 2^n

p	Factors of p - 1	Transform length d
41	$2^3 . 5$	2^2
73	$2^3 . 3^2$	2^3
137	$2^3 .17$	2^3
233	$2^3 .29$	2^3
521	$2^3 . 5.13$	2^3
1033	$2^3 . 3.43$	2^3
97	$2^5 . 3$	2^4
113	$2^4 . 7$	2^4
241	$2^4 . 3 . 5$	2^4
577	$2^6 . 3^2$	2^4
1009	$2^4 . 3^2 . 7$	2^4
193	$2^6 . 3$	2^5
257	2^8	2^5
577	$2^6 . 3^2$	2^5
1153	$2^7 . 3^2$	2^5
193	$2^6 . 3$	2^6
257	2^8	2^6
577	$2^6 . 3^2$	2^6
1153	$2^7 . 3^2$	2^6
2113	$2^6 . 3 . 11$	2^6
257	2^8	2^7
641	$2^7 . 5$	2^7
1153	$2^7 . 3^2$	2^7
2689	$2^7 . 3 . 7$	2^7
3329	$2^8 .13$	2^8
7681	$2^9 . 3 . 5$	2^8
7937	$2^8 .31$	2^8
9473	$2^8 .37$	2^8

of a Number Theoretic Transform, then the prime p must be greater than the largest possible value of y_n, x_n and h_n. The impulse response h_n is usually known a priori and a useful bound [7] on the maximum value of y_n is given by:

$$|y_n| \leq |x_n|_{max} \sum_{n=0}^{N-1} |h_n| \qquad (44)$$

In many practical applications the maximum value of x_n can be estimated and equation 44 can be used to determine the required prime modulus p.

If Number Theoretic Transforms are to compare favourably with the Discrete Fourier Transform, methods for computing the arithmetic operations at high speed are required. Techniques for fast multiplication and addition are well-known but we have the additional requirements that these operations are to be performed modulo some prime number. This is not a particular problem with addition since the modulus will need to be added, or subtracted from, the conventional sum once, at most. However with multiplication it is necessary to divide the conventional product by the modulus and determine the remainder. Since division is a slow operation it is necessary to consider other methods of performing the multiplication operation for prime modulus.

In the field of numbers modulo some prime, each number, except zero, may be represented as the power of a primitive root. For example, if p=5 then 2 is a primitive root and powers of 2 generate all the numbers modulo 5.

$$\begin{aligned} 2^1 &\equiv 2 \\ 2^2 &\equiv 4 \\ 2^3 &\equiv 3 \\ 2^4 &\equiv 1 \end{aligned} \qquad \pmod 5$$

Thus multiplication of two numbers could be carried out in terms of the powers of the primitive roots corresponding to the numbers. The required operation being that of adding the indices modulo (p-2). This method is well-suited to look-up tables since two tables of length (p-2) are required rather than a single table of length $(p-1)^2$ that would be required for conventional multiplication.

4. 2 Composite modulus transforms.

Since there are so many possible transforms with composite modulus it is necessary to concentrate on those which have a particularly simple realisation. First we require that d be highly composite and preferably a power of 2. Multiplication by powers of the transform kernel a should be a simple operation which implies that a should have a binary representation of few bits; an ideal situation would be that a is some power of 2 so that multiplication by a corresponds to binary shifting. Finally the modulus should have few bits in its binary representation to facilitate the modulo interpretation of arithmetic results. Agarwal and Burrus [6, 7] consider the possibility of using moduli m of the form $(2^k + 1)$ and $(2^k - 1)$ which have simple binary representation. They show that for $m = 2^k + 1$ the maximum length of the transform will be governed by the maximum length possible when k is prime. Numbers of this form are known as Mersenne numbers and Rader [8] has discussed, in detail, transforms based on these numbers. It can be shown that transforms of length, at least, 2 P exist. However 2P is not highly composite so that transforms based on Mersenne numbers cannot be computed by means of the Fast Fourier type of algorithm.

For composite moduli $m = 2^k+1$ we need consider only the case of k even since if k is odd then 3 divides 2^k+1 and the maximum transform length would be d=2. In many computers the word length is a power of 2 so that a particularly important case for even k is when $k=2^t$. Therefore moduli of the form $m=2^{2^t}$ are of considerable interest. Such numbers are called Fermat numbers F_t, named after Fermat who conjectured that all such numbers are prime; in fact it appears that Fermat numbers are prime only for t=0, 1, 2, 3 and 4. For prime F_t the transform length d must divide $F_t - 1$ so that transform lengths of length 2^n where $n \leqslant 2^{2t}$ are available. However if the additional constraint that the transform kernel a = 2 is imposed then the possible transform lengths are reduced since a= 2 is not necessarily primitive root modulo $(2^{2t} + 1)$. For t 4 the Fermat numbers are not prime and Lucas [9] has shown that every prime factor of a composite F_t is of the form $K. 2^{t+2} + 1$. Therefore the transform length must divide 2^{t+2}. Table 2 shows the maximum sequence length d_{max} available for the first few Fermat numbers. Also shown is the maximum sequence length d_2 available if the transform kernel a = 2. Fermat number transforms based on a = 2 are discussed in detail by Rader [8] and are often termed Rader transforms.

Table 2.

Maximum transform length for the first six Fermat numbers F_t

t	F_t	d_{max}	d_2
0	2^1+1	2	2
1	2^2+1	4	4
2	2^4+1	16	8
3	2^8+1	256	16
4	$2^{16}+1$	65536	32
5	$2^{32}+1$	128	64
6	$2^{64}+1$	256	128

5. CONCLUSIONS

An introduction to the theory of Number Theoretic Transforms with circular convolution properties has been given. Following this introduction the realisation of Number Theoretic Transform operations is discussed briefly. Since this paper is only a short survey of the subject, many details have not been mentioned. The references listed are not all referred to in the text but they do represent a reasonably complete bibliography and the interested reader should consult these references for a more detailed study of Number Theoretic Transforms.

6. References

1. Bellman, R., Introduction to Matrix Analysis, McGraw-Hill Book Company, 1970, Chapter 12.

2. Mathews, G.B., Theory of Numbers, Chelsea Publishing Co., New York.

3. Agnew, J., Explorations in Number Theory, Brooks/Cole Publishing Co.

4. Abramowitz, M. AND Stegun, I.A., Handbook of Mathematical Functions, Dover Publications, Inc., New York, Chapter 24.

5. Pollard, J.M., 'The Fast Fourier Transform in a Finite Field', Math.Comput., vol.25, pp.365-374, Apr. 1971.

6. Agarwal, R.C. and Burrus, C.S., 'Fast Convolution using Fermat Number Transforms with Applications to Digital Filtering', I.E.E.E. Trans.Acoustics, Speech, and Signal Processing, vol.ASSP-22, pp.87-97, Apr. 1974.

7. Agarwal, R.C. and Burrus, C.S., 'Number Theoretic Transforms to Independent Fast Digital Convolution', Proc. I.E.E.E., vol.63, pp.550-560, Apr. 1975.

8. Rader, C.M., 'Discrete Convolution via Mersenne Transforms', I.E.E.E. Trans.Comput., vol.C-21, pp.1269-1273, Dec. 1972.

9. Dickson, L.E., History of the Theory of Numbers, vol.1, Washington, D.C., Carnegie Institute, 1919.

10. Nicholson, P.J., 'Algebraic Theory of Finite Fourier Transforms', J.Comput.Syst.Sci., vol.5, pp.524-547, 1971.

11. Good, I.J., 'The Relation Between Two Fast Fourier Transforms', I.E.E.E. Trans.Comput., vol.C-20, pp.310-317, Mar. 1971.

12. Rader, C.M., 'A Note on Exact Discrete Fourier Transforms', I.E.E.E. Trans.Audio and Electroacoust. (Corresp.), vol.AU-21, pp.558-559, Dec. 1973.

13. Reed, I.S. and Truong, T.K., 'The Use of Finite Fields to Compute Convolutions', I.E.E.E. Trans. Info. Theory, vol. IT-21, pp. 208-213, March 1975.

14. Reed, I.S. and Truong, T.K., 'Complex Integer Convolutions over a Direct Sum of Galois Fields', I.E.E.E. Trans. Info. Theory, vol. IT-21, pp. 657-661, Nov. 1975.

15. Brute, J.D., 'Fast Convolution with Finite Field Fast Transforms', I.E.E.E. Trans. Acoust., Speech and Signal Processing, vol. ASSP-23, p. 240, Apr. 1975

16. McClellan, J.H., 'Hardware Realization of a Fermat Number Transform', I.E.E.E. Trans. Acoust., Speech and Signal Processing, vol. ASSP-24, pp. 216-225, June 1976.

17. Nussbaumer, H.J., 'Complex Convolutions via Fermat Number Transforms', IBMJ. Res. Develop., vol. 20, pp. 282-284, May 1976.

18. Pollard, J.M., 'Implementation of Number-Theoretic Transforms', Electronics Letters, vol. 12, pp. 378-379, No. 15, July 1976.

STATIONARY AND NONSTATIONARY LEARNING CHARACTERISTICS OF THE LMS ADAPTIVE FILTER*

Bernard Widrow, John McCool, Michael G. Larimore, C. Richard Johnson, Jr.

Information Systems Laboratory, Stanford University, Stanford, California; Fleet Engineering Department, Naval Undersea Center, San Diego, California

ABSTRACT. This paper describes the performance characteristics of the LMS adaptive filter, a digital filter composed of a tapped delay line and adjustable weights, whose impulse response is controlled by an adaptive algorithm. For stationary stochastic inputs, the mean-square error, the difference between the filter output and an externally supplied input called the "desired response," is a quadratic function of the weights, a paraboloid with a single fixed minimum point that can be sought by gradient techniques. The gradient estimation process is shown to introduce noise into the weight vector that is proportional to the speed of adaptation and number of weights. The effect of this noise is expressed in terms of a dimensionless quantity "misadjustment" that is a measure of the deviation from optimal Wiener performance. Analysis of a simple nonstationary case, in which the minimum point of the error surface is moving according to an assumed first-order Markov process, shows that an additional contribution to misadjustment arises from "lag" of the adaptive process in tracking the moving minimum point. This contribution, which is additive, is proportional to the number of weights but inversely proportional to the speed of adaptation. The sum of the misadjustments can be minimized by choosing the speed of adaptation to make equal the two contributions. It is further

 This work was supported in part by the National Science Foundation under Grant ENGR 74-21752.

G. Tacconi (ed.), Aspects of Signal Processing, Part 1, 355-393.

shown, in Appendix A, that for stationary inputs the LMS adaptive algorithm, based on the method of steepest descent, approaches the theoretical limit of efficiency in terms of misadjustment and speed of adaptation when the eigenvalues of the input correlation matrix are equal or close in value. When the eigenvalues are highly disparate ($\lambda_{max}/\lambda_{min} > 10$), an algorithm similar to LMS but based on Newton's method would approach this theoretical limit very closely.

1. INTRODUCTION

Our purpose is to derive relationships between speed of adaptation and performance of adaptive systems. In general, faster adaptation leads to more noisy adaptive processes. When the input environment of an adaptive system is statistically stationary, best steady-state performance results from slow adaptation. However, when the input statistics are time variable, best performance is obtained by a compromise between fast adaptation (necessary to track variations in input statistics) and slow adaptation (necessary to contain the noise in the adaptive process). These issues will be studied both analytically and by computer simulation. The context of this study will be restricted to adaptive digital filters "driven" by the LMS adaptation algorithm of Widrow and Hoff [1], [2]. This algorithm and similar algorithms have been used for many years in a wide variety of practical applications [3]-[26].

We are attempting to formulate a "statistical theory of adaptation." This is a very difficult subject and the present work should be regarded as only a beginning. Stability and rate of convergence are analyzed first, then gradient noise and its effects upon performance are assessed. The concept of "misadjustment" is defined and used to establish design criteria for an adaptive predictor. Extension of the concept to the analysis of a useful but relatively simple form of nonstationary adaptation leads to criteria governing optimal choice of speed of adaptation.

The results reported here have been gradually developed in our laboratory during the past 15 years and are being extended and applied by ongoing research.

2. AN ADAPTIVE FILTER

The filter considered here comprises a tapped delay line, variable weights (variable gains) whose input signals are the signals at the delay-line taps, a summer to add the weighted signals, and an adaptation process that automatically seeks an optimal impulse response by adjusting the weights. Fig. 1 illustrates the adaptive filter as used in modeling an unknown dynamic system.

In addition to the usual input signals, another input signal, the "desired response," must be supplied to the adaptive filter during the adaptation process. In Fig. 1, essentially the same input is applied to the adaptive filter as to the unknown system to be modeled. The output of this system provides the desired response for the adaptive filter. In other applications, considerable ingenuity may be required to obtain a suitable desired response for an adaptive process.

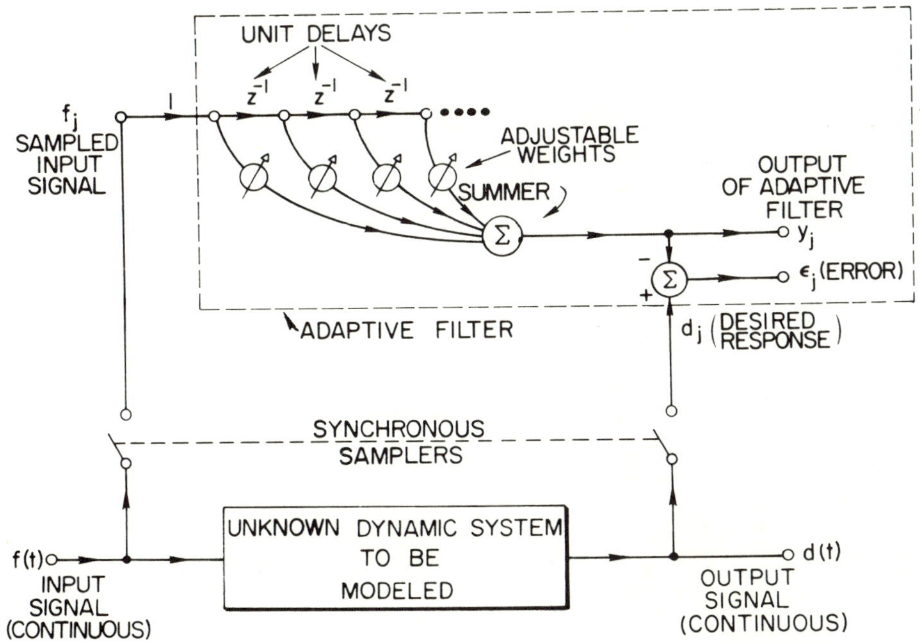

Figure 1. Modeling an unknown system by a discrete adaptive filter.

3. THE PERFORMANCE SURFACE

The analysis of the adaptive filter is developed by considering the "adaptive linear combiner" of Fig. 2, a subsystem of the adaptive filter of Fig. 1, comprising its most significant part.[1]

In Fig. 2, a set of input signals is weighted and summed to form an output signal. The inputs occur simultaneously and discretely in time. The jth input vector is

$$\underset{\sim}{X}_j = [x_{1j}, x_{2j}, \cdots, x_{\ell j}, \cdots, x_{nj}]^T.$$

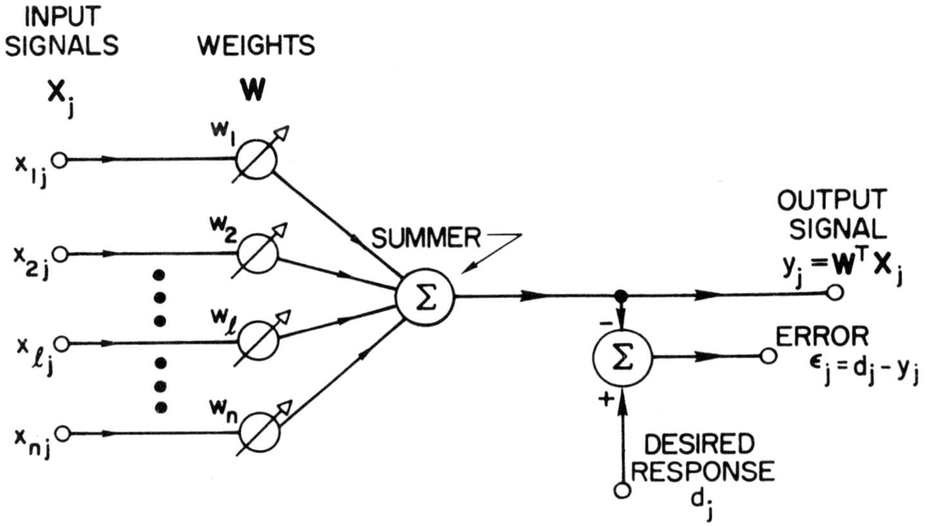

Figure 2. Adaptive linear combiner.

---

[1] This combinational system can be connected to the elements of a phased array antenna to make an adaptive antenna [5]-[9], or to a quantizer to form an adaptive threshold element ("Adaline" [1], [3] or TLU [2]) for use in adaptive logic and pattern-recognition systems. It can also be used as the adaptive portion of certain learning control systems [10], [11]; as a key portion of adaptive filters for channel equalization [12]-[16]; for adaptive noise cancelling [17], [18]; or for adaptive systems identification [19]-[26].

The set of weights is designated by the vector $\underset{\sim}{W}^T = [w_1, w_2,\cdots, w_\ell,\cdots, w_n]$. The jth output signal is

$$y_j = \sum_{\ell=1}^{n} w_\ell x_{\ell j} = \underset{\sim}{W}^T \underset{\sim}{X}_j = \underset{\sim}{X}_j^T \underset{\sim}{W}. \tag{1}$$

The input signals and desired response are assumed to be stationary ergodic processes. Denoting the desired response as d_j, the error at the jth time is

$$\varepsilon_j = d_j - y_j = d_j - \underset{\sim}{W}^T \underset{\sim}{X}_j = d_j - \underset{\sim}{X}_j^T \underset{\sim}{W}. \tag{2}$$

The square of this error is

$$\varepsilon_j^2 = d_j^2 - 2d_j \underset{\sim}{X}_j^T \underset{\sim}{W} + \underset{\sim}{W}^T \underset{\sim}{X}_j \underset{\sim}{X}_j^T \underset{\sim}{W}. \tag{3}$$

The mean-square error ξ, the expected value of ε_j^2, is

$$\begin{aligned} \xi \triangleq E[\varepsilon_j^2] &= E[d_j^2] - 2E[d_j \underset{\sim}{X}_j^T]\underset{\sim}{W} + \underset{\sim}{W}^T\, E[\underset{\sim}{X}_j \underset{\sim}{X}_j^T]\underset{\sim}{W} \\ &= E[d_j^2] - 2\underset{\sim}{P}^T \underset{\sim}{W} + \underset{\sim}{W}^T\, \underset{\sim}{R}\underset{\sim}{W} \end{aligned} \tag{4}$$

where the cross correlation vector between the input signals and the desired response is defined as

$$E[d_j \underset{\sim}{X}_j] = E \begin{bmatrix} d_j x_{1j} \\ d_j x_{2j} \\ \cdot \\ \cdot \\ \cdot \\ d_j x_{nj} \end{bmatrix} \triangleq \underset{\sim}{P} \tag{5}$$

and where the symmetric and positive definite input correlation matrix $\underset{\sim}{R}$ of the x-input signals is defined as

$$E[\underset{\sim}{X}_j \underset{\sim}{X}_j^T] = E \begin{bmatrix} x_{1j}x_{1j} & x_{1j}x_{2j} & \cdots \\ x_{2j}x_{1j} & x_{2j}x_{2j} & \cdots \\ \vdots & \vdots & \\ & \cdots & x_{nj}x_{nj} \end{bmatrix} \triangleq \underset{\sim}{R}. \tag{6}$$

It may be observed from (4) that the mean-square-error (mse) performance function is a quadratic function of the weights, a "bowl-shaped" surface; the adaptive process will be continuously adjusting the weights, seeking the bottom of the bowl. This may be accomplished by steepest descent methods [27], [28] discussed below.

In the nonstationary case, the adaptive process must track the bottom of the bowl, which may be moving. An analysis of a simple nonstationary case is presented in Section 11.

4. THE GRADIENT AND THE WIENER SOLUTION

The method of steepest descent uses gradients of the performance surface in seeking its minimum. The gradient at any point on the performance surface may be obtained by differentiating the mse function, (4), with respect to the weight vector. The gradient vector is

$$\underset{\sim}{\nabla} = -2\underset{\sim}{P} + 2\underset{\sim}{R}\underset{\sim}{W}. \tag{7}$$

Set the gradient to zero to find the optimal weight vector W*.

$$\underset{\sim}{W}^* = \underset{\sim}{R}^{-1} \underset{\sim}{P} \tag{8}$$

which is the Wiener-Hopf equation in matrix form.

The minimum mse is obtained from (8) and (4).

$$\xi_{min} = E[d_j^2] - \underset{\sim}{P}^T\underset{\sim}{W}^*. \tag{9}$$

Substituting (9) into (4) yields a useful formula for mse:

$$\xi = \xi_{min} + (W - W^*)^T R(W - W^*). \tag{10}$$

Define V as the difference between W and the Wiener solution W^*.

$$V \triangleq (W - W^*). \tag{11}$$

Therefore,

$$\xi = \xi_{min} + V^T R V. \tag{12}$$

Differentiation of (12) yields another form for the gradient:

$$\nabla = 2RV. \tag{13}$$

The input correlation matrix, being symmetric and positive definite, may be represented as

$$R = Q\Lambda Q^{-1} = Q\Lambda Q^T \tag{14}$$

where Q is the orthonormal modal matrix of R and Λ is its diagonal matrix of eigenvalues:

$$\Lambda = \text{diag}\ [\lambda_1, \lambda_2, \cdots, \lambda_p, \cdots, \lambda_n]. \tag{15}$$

Equation (12) may be reexpressed as

$$\xi = \xi_{min} + V^T\ Q\Lambda Q^{-1}\ V. \tag{16}$$

Define a transformed version of V as

$$V' \triangleq Q^{-1}\ V \text{ and } V = QV'.$$

Accordingly, equation (12) may be put in normal form as

$$\xi = \xi_{min} + V'^T\ \Lambda V'. \tag{18}$$

The primed coordinates are therefore the principal axes of the quadratic surface. Transformation (17) may be applied to the weight vector itself,

$$W' = Q^{-1}W \text{ and } W = QW'.$$

5. THE METHOD OF STEEPEST DESCENT

The method of steepest descent makes each change in the weight vector proportional to the negative of the gradient vector:

$$\underset{\sim}{W}_{j+1} = \underset{\sim}{W}_j + \mu(-\underset{\sim}{\nabla}_j). \tag{20}$$

The scalar parameter μ is a convergence factor that controls stability and rate of adaptation. The gradient at the jth iteration is $\underset{\sim}{\nabla}_j$. Using (13), (14), and (17), equation (20) becomes

$$\underset{\sim}{V}'_{j+1} - (\underset{\sim}{I} - 2\mu\underset{\sim}{\Lambda})\,\underset{\sim}{V}'_j = 0. \tag{21}$$

This homogeneous vector difference equation is uncoupled. It has a simple geometric solution in the primed coordinates [5].

$$\underset{\sim}{V}'_j = (\underset{\sim}{I} - 2\mu\underset{\sim}{\Lambda})^j\,\underset{\sim}{V}_0 \tag{22}$$

where $\underset{\sim}{V}'_0$ is an initial condition:

$$\underset{\sim}{V}'_0 = \underset{\sim}{W}'_0 - \underset{\sim}{W}^{*\prime}. \tag{23}$$

For convergence, it is necessary that

$$1/\lambda_{max} > \mu > 0 \tag{24}$$

where λ_{max} is the largest eigenvalue of $\underset{\sim}{R}$. From (22), we see that transients in the primed coordinates will be geometric; the geometric ratio of the pth coordinate is

$$r_p = (1 - 2\mu\lambda_p). \tag{25}$$

An exponential envelope can be fitted to a geometric sequence. If the basic unit of time is considered to be the iteration cycle, time constant τ_p can be determined as follows:

$$r_p = \exp\left(-\frac{1}{\tau_p}\right) = 1 - \frac{1}{\tau_p} + \frac{1}{2!\tau_p^2} - \cdots. \tag{26}$$

The case of general interest is slow adaptation, i.e. large τ_p. Therefore,

$$r_p = (1 - 2\mu\lambda_p) \simeq 1 - \frac{1}{\tau_p}$$

or

$$\tau_p \simeq \frac{1}{2\mu\lambda_p}. \tag{27}$$

Equation (27) gives the time constant of the pth mode.

Steepest descent can be regarded as a feedback process where the gradient plays the role of vector error signal. The process, if stable, tends to bring the gradient to zero.[2] Fig. 3 shows a feedback model for a stationary quadratic mse surface being searched by the method of steepest descent. The model is equivalent to the following set of relations.

$$\underset{\sim}{W}_j = \underset{\sim}{W}_{j+1} \text{ delayed one iteration}$$

$$\underset{\sim}{W}_{j+1} = \underset{\sim}{W}_j + \mu(-\underset{\sim}{\nabla}_j)$$

$$\underset{\sim}{\nabla}_j = 2\underset{\sim}{R}(\underset{\sim}{W}_j - \underset{\sim}{W}^*) = 2\underset{\sim}{R}\underset{\sim}{V}. \tag{28}$$

This feedback model is used subsequently in a study of non-stationary adaptation. Notice an input not mentioned earlier, "gradient noise." Because gradients are estimated at each iteration cycle with finite amounts of input data, they will be imperfect or noisy.

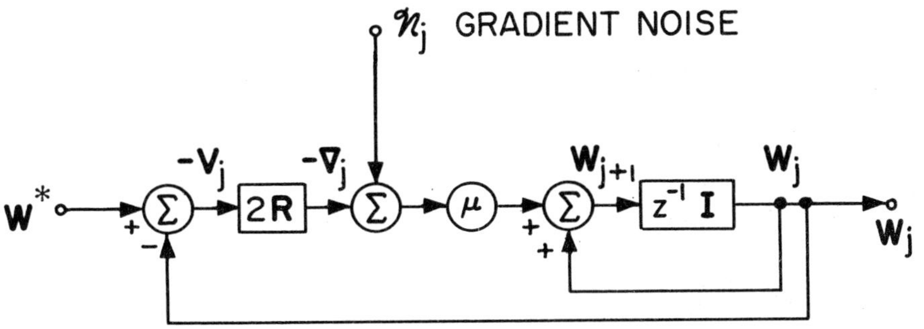

Figure 3. Feedback model of steepest descent.

---
[2]This has been called performance feedback [1], [29].

6. THE LMS ALGORITHM

The LMS algorithm is an implementation of steepest descent using measured or estimated gradients:

$$W_{j+1} = W_j + \mu(-\hat{\nabla}_j). \tag{29}$$

The estimate of the true gradient is $\hat{\nabla}$.

The gradient estimate used by LMS takes the gradient of the square of a single error sample. Thus

$$\hat{\nabla}_j = -2\varepsilon_j X_j. \tag{30}$$

The LMS algorithm can be written as

$$W_{j+1} = W_j + 2\mu\varepsilon_j X_j. \tag{31}$$

If we assume that X_j is uncorrelated over time (i.e., that $E[X_j X_{j+1}^T] = 0, \forall \ell \neq 0$), an assumption common in the field of stochastic approximation [30], [31], then the expected value of the gradient estimate equals the true gradient, and the weight-vector mean is convergent to the Wiener solution of (8), as shown in [4] and [5].

Condition (24) is necessary and sufficient for convergence of the LMS algorithm. However, in practice, the individual eigenvalues are rarely known so that (24) is not always easy to apply. Since tr R is the total input power to the weights, a generally known quantity, and since tr $R > \lambda_{max}$ as R is positive definite, a sufficient condition for convergence is

$$1/\text{tr } R > \mu > 0. \tag{32}$$

7. THE LEARNING CURVE AND ITS TIME CONSTANTS

During adaptation, the error ε_j is nonstationary as the weight vector adapts toward W^*. The mse can be defined only on the basis of ensemble averages. From (18), we obtain

$$\xi_j = \xi_{min} + V_j'^T \Lambda V_j'. \tag{33}$$

Imagine an ensemble of adaptive processes, each having individual stationary ergodic inputs drawn from the same statistical population, with all initial weight vectors equal. The mse ξ_j is a function of iteration number j, obtained by averaging over the ensemble at iteration j.

Using (22), but assuming no noise in the weight vector, equation (33) becomes

$$\xi_j = \xi_{min} + \underset{\sim}{V_0'}^T \underset{\sim}{\Lambda}(\underset{\sim}{I} - 2\mu\underset{\sim}{\Lambda})^{2j} \underset{\sim}{V_0'}$$

$$= \xi_{min} + \underset{\sim}{V_0}^T(\underset{\sim}{I} - 2\mu\underset{\sim}{R})^j \underset{\sim}{R}(\underset{\sim}{I} - 2\mu\underset{\sim}{R})^j \underset{\sim}{V_0}. \tag{34}$$

When the adaptive process is convergent, it is clear from (34) that

$$\lim_{j\to\infty} \xi_j = \xi_{min}$$

and that the geometric decay in ξ_j going from ξ_0 to ξ_{min} will, for the pth mode, have a geometric ratio of r_p^2 and a time constant

$$\tau_{p_{mse}} \triangleq \frac{1}{2}\tau_p = \frac{1}{4\mu\lambda_p}. \tag{35}$$

The result obtained by plotting mse against number of iterations is called the "learning curve." Due to noise in the weight vector, actual practice will show ξ_j to be higher than indicated by (34).

8. GRADIENT AND WEIGHT-VECTOR NOISE

Gradient noise will affect the adaptive process both during initial transients and in steady state. The latter condition is of particular interest here.

Assume that the weight vector is close to the Wiener solution. Assume, as before, that $\underset{\sim}{X}_j$ and d_j are stationary and ergodic and that $\underset{\sim}{X}_j$ is uncorrelated over time; i.e.,

$$E[\underset{\sim}{X}_j\underset{\sim}{X}_{j+k}] = 0, \quad k \neq 0. \tag{36}$$

The LMS algorithm uses a gradient estimate

$$\hat{\underset{\sim}{\nabla}} = -2\epsilon_j\underset{\sim}{X}_j = \underset{\sim}{\nabla}_j + \underset{\sim}{N}_j \tag{37}$$

where $\underset{\sim}{\nabla}_j$ is the true gradient and $\underset{\sim}{N}_j$ is a zero-mean gradient estimation noise vector. When $\underset{\sim}{W}_j = \underset{\sim}{W}^*$, the true gradient is zero, but the gradient would be estimated according to (30) and is equal to the gradient noise:

$$\underset{\sim}{N}_j = -2\varepsilon_j \underset{\sim}{X}_j. \tag{38}$$

According to Wiener filter theory, when $\underset{\sim}{W}_j = \underset{\sim}{W}^*$, ε_j and $\underset{\sim}{X}_j$ are uncorrelated. If they are assumed zero-mean Gaussian, ε_j and $\underset{\sim}{X}_j$ are statistically independent. As such, the covariance of $\underset{\sim}{N}_j$ is

$$\text{cov } [\underset{\sim}{N}_j] = E[\underset{\sim}{N}_j \underset{\sim}{N}_j^T] = 4E[\varepsilon_j^2 \underset{\sim}{X}_j \underset{\sim}{X}_j^T]$$

$$= 4E[\varepsilon_j^2] \, E[\underset{\sim}{X}_j \underset{\sim}{X}_j^T]$$

$$= 4E[\varepsilon_j^2] \, \underset{\sim}{R}. \tag{39}$$

When $\underset{\sim}{W}_j = \underset{\sim}{W}^*$, $E[\varepsilon_j^2] = \xi_{min}$. Accordingly,

$$\text{cov } [\underset{\sim}{N}_j] = 4 \, \xi_{min} \, \underset{\sim}{R}. \tag{40}$$

As long as $\underset{\sim}{W}_j \simeq \underset{\sim}{W}^*$, we assume that the gradient noise covariance is given by (40) and that this noise is stationary and uncorrelated over time. The latter assumption is based on (36) and (38).

Projecting the gradient noise,

$$\underset{\sim}{N}_j' = \underset{\sim}{Q}^{-1} \underset{\sim}{N}_j \tag{41}$$

its covariance becomes

$$\text{cov } [\underset{\sim}{N}_j'] = E[\underset{\sim}{N}_j' \underset{\sim}{N}_j'^T] = E[\underset{\sim}{Q}^{-1} \underset{\sim}{N}_j \underset{\sim}{N}_j^T \underset{\sim}{Q}] = \underset{\sim}{Q}^{-1} \text{cov } [\underset{\sim}{N}_j] \underset{\sim}{Q}$$

$$= 4\xi_{min} \, \underset{\sim}{Q}^{-1} \, \underset{\sim}{R}\underset{\sim}{Q}$$

$$= 4\xi_{min}\underset{\sim}{\Lambda}. \tag{42}$$

Although the components of $\underset{\sim}{N}_j$ are correlated with each other, those of $\underset{\sim}{N}'_j$ are mutually uncorrelated and can, therefore, be handled more easily.

Gradient noise propagates and causes noise in the weight vector. Accounting for gradient noise, the LMS algorithm can be expressed as

$$\underset{\sim}{W}'_{j+1} = \underset{\sim}{W}'_j + \mu(-\hat{\underset{\sim}{\nabla}}'_j) = \underset{\sim}{W}'_j + \mu(-\underset{\sim}{\nabla}'_j + \underset{\sim}{N}'_j). \tag{43}$$

This equation can be written in terms of $\underset{\sim}{V}'_j$ as

$$\underset{\sim}{V}'_{j+1} = \underset{\sim}{V}'_j + \mu(-2\underset{\sim}{\Lambda}\underset{\sim}{V}'_j + \underset{\sim}{N}'_j). \tag{44}$$

Near the minimum point of the error surface in steady-state, the mean of $\underset{\sim}{V}'_j$ is zero and the covariance of the weight-vector noise is [18, appendix D, section B]

$$\text{cov}\,[\underset{\sim}{V}'_j] = \mu\xi_{\min}\underset{\sim}{I} \tag{45}$$

where the components of the weight-vector noise are of equal variance and are mutually uncorrelated. It has been found, however, that (45) closely approximates measured weight-vector covariances under a considerably wider range of conditions than the assumptions above imply.

9. MISADJUSTMENT DUE TO GRADIENT NOISE

Random noise in the weight vector causes an excess mse. If the weight vector were noise free and converged such that $\underset{\sim}{W}_j = \underset{\sim}{W}^*$, then the mse would be $\xi_{\min}$. However, this does not occur in actual practice so that the weight vector is on the average "misadjusted" from its optimal setting.

An expression for mse in terms of $\underset{\sim}{V}'_j$ is given by (33), from which we obtain an expression for excess mse

$$(\text{excess mse}) = \underset{\sim}{V}'^{T}_j\,\underset{\sim}{\Lambda}\underset{\sim}{V}'_j. \tag{46}$$

The average excess mse is an important quantity.

$$E[\underset{\sim}{V}'^{T}_j\,\underset{\sim}{\Lambda}\underset{\sim}{V}'_j] = \sum_{p=1}^{n} \lambda_p\, E[(\upsilon'_{pj})^2] \tag{47}$$

where υ'_{pj} is the pth component of $\underset{\sim}{V}'_j$. After adaptive transients die out, $E[\underset{\sim}{V}'_j] = 0$. Therefore, from (45) we have

$$E[(\upsilon'_{pj})^2] = \mu\xi_{min}, \forall p. \tag{48}$$

Substitution into (47) yields the average excess mse,

$$E[\underset{\sim}{V}'^T_j \underset{\sim}{\Lambda}\underset{\sim}{V}'_j] = \mu\xi_{min} \sum_{p=1}^{n} \lambda_p = \mu\xi_{min} \operatorname{tr} \underset{\sim}{R}. \tag{49}$$

We define the "misadjustment" due to gradient noise as the dimensionless ratio of the average excess mse to the minimum mse,

$$M \triangleq \frac{\text{average excess mse}}{\xi_{min}}. \tag{50}$$

For the LMS algorithm, under the conditions assumed above,

$$M = \mu \operatorname{tr} \underset{\sim}{R}. \tag{51}$$

This formula works well for small values of misadjustment, 25 percent or less, so that the assumption

$$\underset{\sim}{W}_j \simeq \underset{\sim}{W}^* \tag{52}$$

is satisfied. The misadjustment is a useful measure of the cost of adaptability. A value of M = 10 percent means that the adaptive system has a mse only 10 percent greater than ξ_{min}.

It is useful to relate misadjustment to the speed of adaptation and the number of weights being adapted. Since tr $\underset{\sim}{R}$ equals the sum of the eigenvalues,

$$M = \mu \sum_{p=1}^{n} \lambda_p = \mu n \lambda_{ave} \tag{53}$$

where λ_{ave} is the average of the eigenvalues. From (35),

$$\lambda_p = \frac{1}{4\mu}\left(\frac{1}{\tau_{p_{mse}}}\right) \quad \text{or} \quad \lambda_{ave} = \frac{1}{4\mu}\left(\frac{1}{\tau_{p_{mse}}}\right)_{ave}. \tag{54}$$

Substituting into (53) yields

$$M = \frac{n}{4}\left(\frac{1}{\tau_{p_{mse}}}\right)_{ave} . \tag{55}$$

The special case where all eigenvalues are equal is an important one. The learning curve has only one time constant τ_{mse}, and the misadjustment is given by

$$M = \frac{n}{4\tau_{mse}}. \tag{56}$$

When the eigenvalues are sufficiently similar for the learning curve to be approximately fitted by a single exponential, its time constant may be applied to (56) to give an approximate value of M.

Since transients settle in about four time constants, equation (56) leads to an approximate "rule of thumb": the misadjustment equals the number of weights divided by the settling time. A 10-percent misadjustment would be satisfactory for many engineering designs. Operation with 10-percent misadjustment can generally be achieved with an adaptive settling time equal to ten times the memory time span of the adaptive transversal filter.

10. A DESIGN EXAMPLE: CHOOSING NUMBER OF FILTER WEIGHTS FOR AN ADAPTIVE PREDICTOR

Fig. 4 is a block diagram of an adaptive predictor.[3] Its adaptive filter converts the delayed input $x_{j-\Delta}$ into x_j as best possible. If the adaptive-filter weights are copied into an auxiliary filter having a tapped delay-line structure identical to that of the adaptive filter and the input x_j is applied to this auxiliary filter, the resulting output will be a linear least squares estimate of $x_{j+\Delta}$ (limited by finite filter length and misadjustment).

A computer implementation of the adaptive predictor was made using a simulated input signal x_j obtained by bandpass filtering a white Gaussian signal and adding this to another independent white Gaussian signal. Prediction was one time sample in the future, i.e., $\Delta = 1$, using an adaptive filter with five weights, all initially set to zero.

---

[3]This same predictor was described by Widrow in [5]; it has been used for data compression and speech encoding [32] and for "maximum entropy" spectral estimation [33].

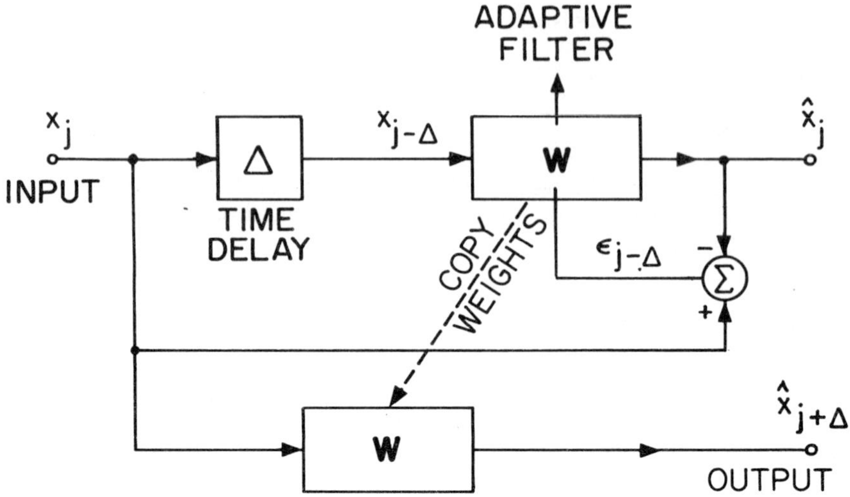

Figure 4. An adaptive predictor.

Fig. 5 depicts three learning curves. For each adaptive step, the mse ξ_j corresponding to the current weight vector $\underset{\sim}{W}_j$ was calculated from (10) using known values of $\underset{\sim}{R}$ and ξ_{min}, giving the "individual learning curve." The smooth "ensemble average learning curve" is simply the average of 200 such individual curves and approximates the adaptive behavior in the mean. The third curve calculated from (34) shows how the process would evolve if perfect knowledge of the gradient were available at each step. It is a noiseless "steepest descent learning curve."

Of particular interest is the residual difference between the ensemble learning curve and the steepest descent learning curve after convergence. The latter, of course, converges to ξ_{min}. The difference is the excess mse due to gradient noise, in this case, giving a measured misadjustment of 3 percent. The theoretical misadjustment was M = 2.5 percent. The minor discrepancy is due mainly to the fact that the input samples are highly correlated in violation of the assumption that $E[\underset{\sim}{X}_j \underset{\sim}{X}^T_{j+k}] = 0, \forall k \neq 0$, used in the derivation of misadjustment formula (56).

The ensemble average learning curve has an effective measured time constant τ_{mse} of about 50 iterations since it falls to within 2 percent of its converged value at around iteration 200. When all eigenvalues are equal, equation (35) becomes

$$\tau_{mse} = \frac{1}{4\mu\lambda} = \frac{n}{4\mu \text{ tr } \underset{\sim}{R}}. \tag{57}$$

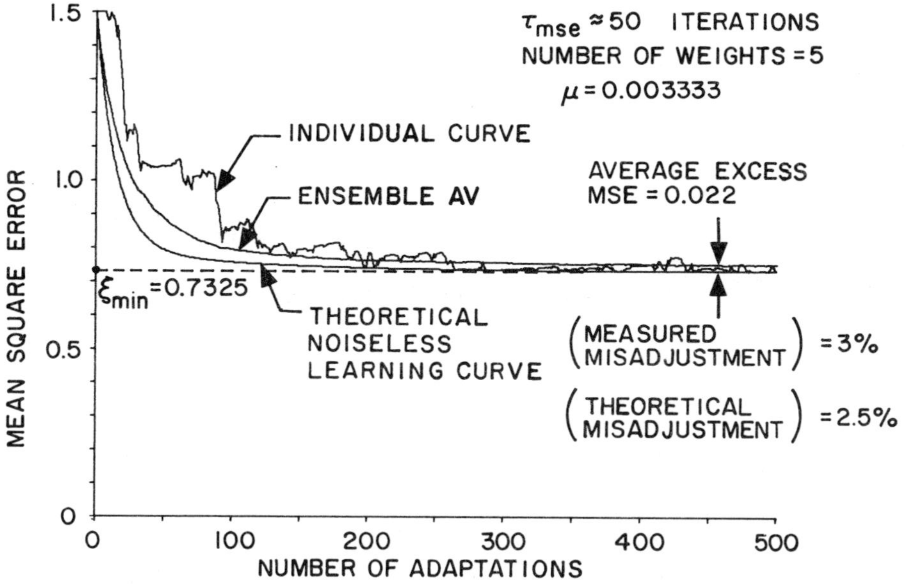

Figure 5. Learning curves for adaptive predictor.

Using (57) in the present case (although the eigenvalues range over a 10 to 1 ratio) yields $\tau_{mse} = 50$, which agrees with experiment. Equation (57) gives a formula for an "effective time constant," useful even when the eigenvalues are highly disparate.

The performance of the adaptive filter may improve with an increase in the number of weights. However, for a fixed rate of convergence, larger numbers of weights increase misadjustment. Fig. 6 shows these conflicting effects. The lowest curve for $\tau_{mse} = \infty$ represents idealized noise-free adaptation, providing the minimum mse $\xi_{min}(n)$ for each value of n. The other curves include average excess mse due to gradient noise. We define the "average mse" to be the sum of the minimum mse and the average excess mse. Thus

$$(\text{average mse}) = [1 + M]\xi_{min}(n). \tag{58}$$

Using this formula, theoretical curves have been plotted in Fig. 6 for approximate values of τ_{mse} of 100, 50, 25, and 15 iterations. It is apparent from these curves that increasing the number of weights does not always guarantee improved system performance. Experimental points derived by computer simulation have compared very well with theoretical values predicted by (58). Typical results are summarized in Table I.

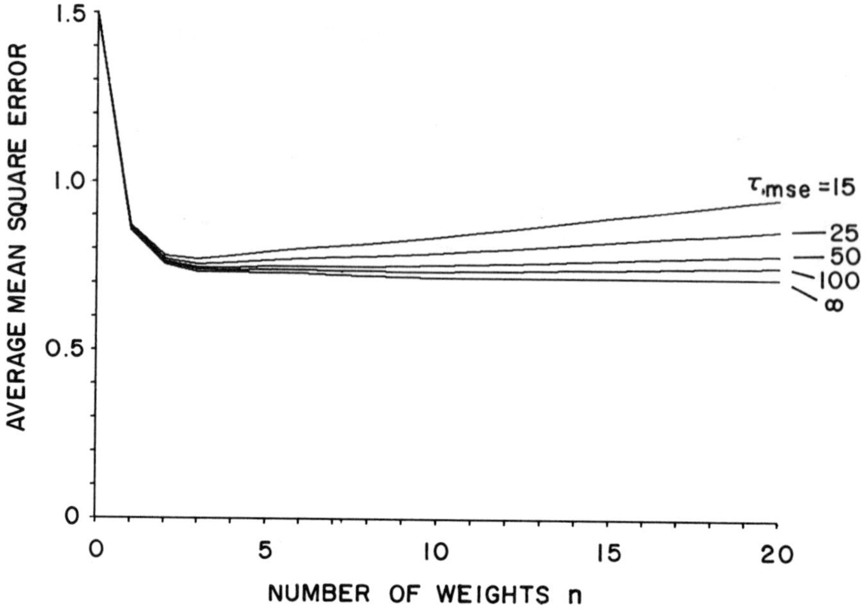

Figure 6. Performance versus number of weights and adaptive predictor time constant.

Table I. Comparison of theoretical and experimental adaptive predictor performance.

Number of Weights n	Approx. Time Constant τ_{mse}	Average mse		Misadjustment	
		Theoretical	Experimental	Theoretical	Experimental
5	100	.742	.751	1.3%	2.5%
5	50	.751	.754	2.5%	3.0%
5	25	.769	.781	5.0%	6.6%
5	15	.794	.824	8.3%	12.6%
10	100	.737	.745	2.5%	3.5%
10	50	.755	.764	5.0%	6.2%

11. RESPONSE OF THE LMS ADAPTIVE FILTER IN A NONSTATIONARY ENVIRONMENT

Filtering nonstationary signals is a major area of application for adaptive techniques, especially when the stochastic properties of the signals are unknown a priori. Although the utility of adaptive filters with nonstationary inputs has been demonstrated experimentally, very little of this work has been published, perhaps due to the inherently complex mathematics associated with such problems [34], [35]. The nonstationary situations to be studied here are highly simplified, but they retain the essence of the problem that is common to more complicated and realistic situations.

The example considered here involves modeling or identifying an unknown time-variable system by an adaptive LMS transversal filter. The unknown system is assumed to be a transversal filter of the same length n whose weights (impulse response values) undergo independent stationary ergodic first-order Markov processes, as indicated in Fig. 7. The input signal x_j is assumed to be stationary and ergodic. Additive output noise, assumed to be stationary, of mean zero, and of variance ξ_{min}, prevents a perfect match between the unknown system and the adaptive system. The minimum mse is, therefore, ξ_{min}, achieved whenever the weights of

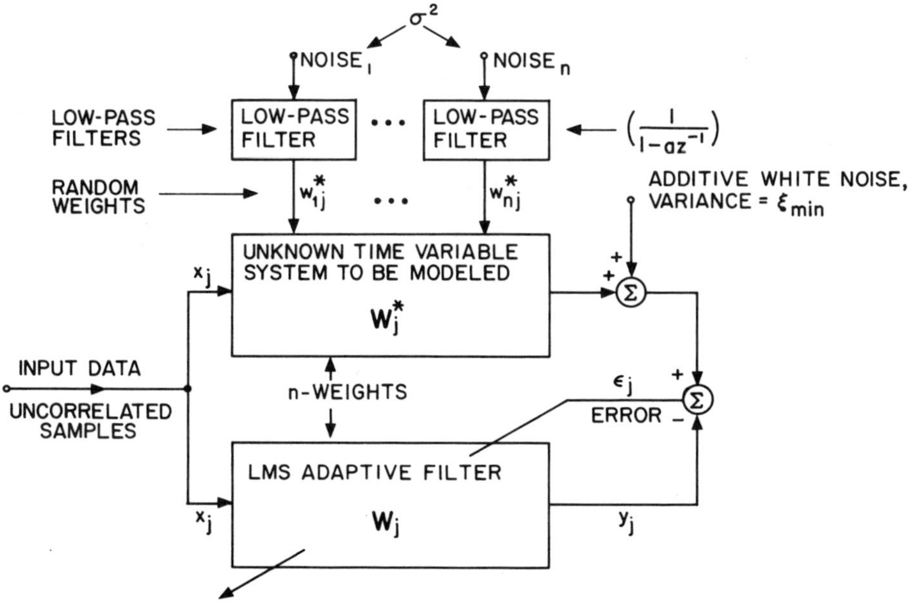

Figure 7. Modeling an unknown time-variable system.

the adaptive filter $\underset{\sim}{W}_j$ match those of the unknown system. The latter are at every instant the optimal values for the corresponding weights of the adaptive filter and are designated $\underset{\sim}{W}_j^*$, the subscript indicating that the unknown "target" to be tracked is time variable.

According to the scheme of Fig. 7, minimizing mse causes the adaptive weight vector $\underset{\sim}{W}_j$ to attempt to best match the unknown $\underset{\sim}{W}_j^*$ on a continual basis. The $\underset{\sim}{R}$ matrix, dependent only on the statistics of x_j, is constant even as $\underset{\sim}{W}_j^*$ varies. The desired response of the adaptive filter d_j is nonstationary, being the output of a time-variable system. The minimum mse ξ_{min} is constant. Thus the mse function, a quadratic bowl, varies in position while its eigenvalues, eigenvectors, and ξ_{min} remain constant.

In order to study this form of nonstationary adaptation both analytically and by computer simulation, a model comprising an ensemble of nonstationary adaptive processes has been defined and constructed as illustrated in Fig. 8. The unknown filters to be modeled are all identical and have the same time-variable weight vector $\underset{\sim}{W}_j^*$ throughout the ensemble. Each ensemble member has its own independent input signal going to both the unknown system and the corresponding adaptive system. The effect of output noise in the unknown systems is obtained by the addition of independent noise of variance ξ_{min}. All of the adaptive filters are assumed to start with the same initial weight vector $\underset{\sim}{W}_0$; each develops its own weight vector over time in attempting to pursue the Markovian target $\underset{\sim}{W}_j^*$.

For a given adaptive filter, the weight-vector tracking error at the jth instant is $(\underset{\sim}{W}_j - \underset{\sim}{W}_j^*)$. This error is due to both the effects of gradient noise and weight-vector lag, and may be expressed as

$$\begin{aligned}(\text{weight-vector error})_j &= (\underset{\sim}{W}_j - \underset{\sim}{W}_j^*) \\ &\equiv \underbrace{(\underset{\sim}{W}_j - E[\underset{\sim}{W}_j])}_{\substack{\text{weight-vector}\\ \text{noise}}} + \underbrace{(E[\underset{\sim}{W}_j] - \underset{\sim}{W}_j^*)}_{\substack{\text{weight-vector}\\ \text{lag}}}. \qquad (59)\end{aligned}$$

The expectations are averages over the ensemble. The components of error are identified in (59). Any difference between the ensemble mean of the adaptive weight vectors and the target value

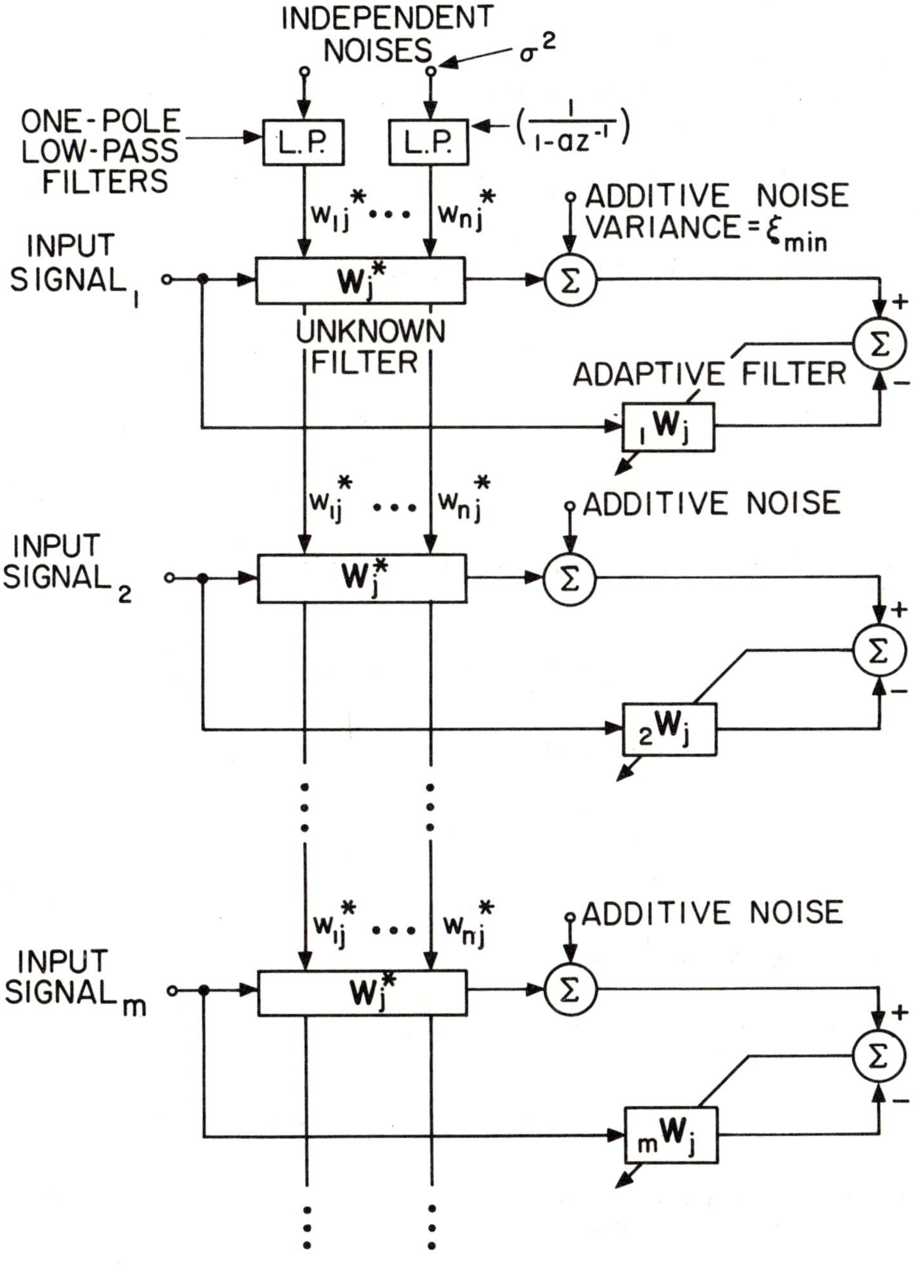

Figure 8. An ensemble of nonstationary adaptive processes.

W_j^* is due to lag in the adaptive process, while the deviation of the individual adaptive weight vectors about the ensemble mean is due to gradient noise.

Weight-vector error causes an excess mse. The ensemble average excess mse at the jth instant is

$$\begin{pmatrix}\text{average excess}\\ \text{mse}\end{pmatrix}_j = E[(W_j - W_j^*)^T R(W_j - W_j^*)]. \qquad (60)$$

Using (59), this can be expanded as follows:

$$\begin{aligned}\begin{pmatrix}\text{average excess}\\ \text{mse}\end{pmatrix}_j &= E[(W_j - E[W_j])^T R(W_j - E[W_j])]\\ &\quad + E[(E[W_j] - W_j^*)^T R(E[W_j] - W_j^*)]\\ &\quad + 2E[(W_j - E[W_j]^T R(E[W_j] - W_j^*)]. \qquad (61)\end{aligned}$$

Expanding the last term of (61) and simplifying since W_j^* is constant over the ensemble,

$$\begin{aligned}&2E[W_j^T RE[W_j] - W_j^T RW_j^* - E[W_j]^T RE[W_j] + E[W_j]^T RW_j^*]\\ &= 2[E[W_j]^T RE[W_j] - E[W_j]^T RE[W_j]\\ &\quad - E[W_j]^T RW_j^* + E[W_j]^T RW_j^*]\\ &= 0 \qquad (62)\end{aligned}$$

Therefore, (61) becomes

$$\begin{aligned}\begin{pmatrix}\text{average excess}\\ \text{mse}\end{pmatrix}_j &= E[(W_j - E[W_j])^T R(W_j - E[W_j])]\\ &\quad + E[(E[W_j] - W_j^*)^T R(E[W_j] - W_j^*)]. \qquad (63)\end{aligned}$$

The average excess mse is thus a sum of components due to both gradient noise and lag.

$$\begin{pmatrix}\text{average excess}\\ \text{mse due to lag}\end{pmatrix}_j = E[(E[\mathbf{W}_j] - \mathbf{W}^*_j)^T \mathbf{R}(E[\mathbf{W}_j] - \mathbf{W}^*_j)]$$

$$= E[(E[\mathbf{W}'_j] - \mathbf{W}^{*\prime}_j)^T \mathbf{\Lambda}(E[\mathbf{W}_j] - \mathbf{W}^{*\prime}_j)]. \quad (64)$$

$$\begin{pmatrix}\text{average excess}\\ \text{mse due to}\\ \text{gradient noise}\end{pmatrix}_j = E[(\mathbf{W}_j - E[\mathbf{W}_j])^T \mathbf{R}(\mathbf{W}_j - E[\mathbf{W}_j])]$$

$$= E[(\mathbf{W}'_j - E[\mathbf{W}'_j])^T \mathbf{\Lambda}(\mathbf{W}'_j - E[\mathbf{W}'_j])]. \quad (65)$$

Fig. 9 is a feedback diagram adapted from Fig. 3, illustrating the two sources of weight-vector error. From the feedback diagram, it can be seen that the "output" $\mathbf{W}_j$ attempts to track the time variable "input" $\mathbf{W}^*_j$. Tracking error $(\mathbf{W}_j - \mathbf{W}^*_j)$ is caused by the propagation of gradient noise and by the response of the adaptive process to the random variations of $\mathbf{W}^*_j$. It will be shown that increasing the time constant of the adaptive process diminishes the propagation of gradient noise but simultaneously increases the lag error that results from the random changes in $\mathbf{W}^*_j$.

The gradient-noise covariance for the stationary case (40) is a function of $\mathbf{R}$. Since $\mathbf{R}$ is constant, equation (40) is a good representation of covariance for the type of nonstationarity under study. Furthermore, Fig. 9 shows that the propagation of gradient noise in the linear feedback system representing the adaptive process is not affected by variability of $\mathbf{W}^*_j$. Therefore, equation (49) can be used to provide an evaluation of (65),

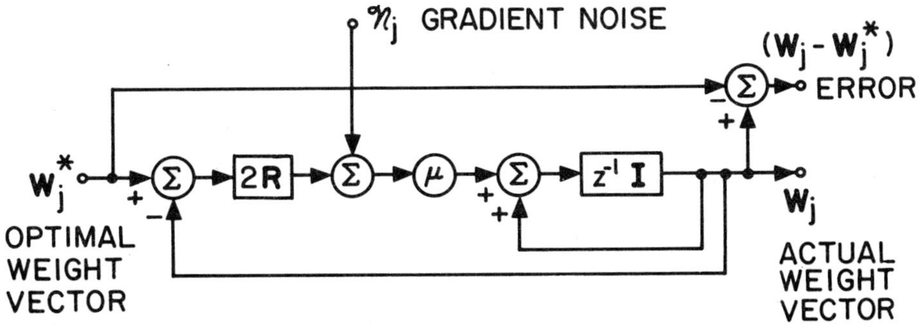

Figure 9. Feedback diagram of steepest descent showing sources of weight tracking error.

the excess mse from gradient noise. The next step is an evaluation of (64), the excess mse due to lag. Statistical knowledge of $(E[\underset{\sim}{W}'_j] - \underset{\sim}{W}^{*\prime}_j)$ will be required. In finding lag effects, we may eliminate gradient noise from consideration so that $E[\underset{\sim}{W}'_j] = \underset{\sim}{W}'_j$. Knowledge of $(\underset{\sim}{W}'_j - \underset{\sim}{W}^{*\prime}_j)$ will be sufficient.

Without gradient noise, the method of steepest descent and the LMS algorithm are represented by (13) and (20). With variable $\underset{\sim}{W}^*_j$, they become

$$\underset{\sim}{W}_{j+1} - (\underset{\sim}{I} - 2\mu \underset{\sim}{R}) \underset{\sim}{W}_j = 2\mu \underset{\sim}{R}\underset{\sim}{W}^*_j. \tag{66}$$

Premultiplying both sides by $\underset{\sim}{Q}^{-1}$ transforms (66) into the primed coordinates,

$$\underset{\sim}{W}'_{j+1} - (\underset{\sim}{I} - 2\mu\underset{\sim}{\Lambda})\underset{\sim}{W}'_j = 2\mu\underset{\sim}{\Lambda}\underset{\sim}{W}^{*\prime}_j. \tag{67}$$

We have assumed for our present study that all components of $\underset{\sim}{W}^*_j$ are stationary, ergodic, independent, and first-order Markov; they all have the same variances and the same autocorrelation functions. Since $\underset{\sim}{W}^{*\prime}_j = \underset{\sim}{Q}^{-1} \underset{\sim}{W}^*_j$ and $\underset{\sim}{Q}^{-1}$ is orthonormal, all components of $\underset{\sim}{W}^{*\prime}_j$ are independent and have the same autocorrelation functions as the components of $\underset{\sim}{W}^*_j$. Therefore, equation (67), being in diagonal form and having a driving function whose components are independent, may be treated as an array of n independent first-order linear difference equations.

Let the z-transform of $\underset{\sim}{W}'_j$ be $\underset{\sim}{W}'(z)$. The z-transform of (67) is then

$$z\underset{\sim}{W}'(z) - (\underset{\sim}{I} - 2\mu\underset{\sim}{\Lambda})\underset{\sim}{W}'(z) = 2\mu\underset{\sim}{\Lambda}\underset{\sim}{W}^{*\prime}(z). \tag{68}$$

Solving (68) yields the transform of $\underset{\sim}{W}'_j$:

$$\underset{\sim}{W}'(z) = 2\mu\underset{\sim}{\Lambda}(z\underset{\sim}{I} - \underset{\sim}{I} + 2\mu\underset{\sim}{\Lambda})^{-1} \underset{\sim}{W}^{*\prime}(z). \tag{69}$$

The weight tracking error $(\underset{\sim}{W}'_j - \underset{\sim}{W}^{*\prime}_j)$ is of direct interest. Its transform is obtained from (69) as

$$\underset{\sim}{W}'(z) - \underset{\sim}{W}^{*\prime}(z) = [2\mu\underset{\sim}{\Lambda}(z\underset{\sim}{I} - \underset{\sim}{I} + 2\mu\underset{\sim}{\Lambda})^{-1} - \underset{\sim}{I}]\, \underset{\sim}{W}^{*\prime}(z). \tag{70}$$

The transfer function connecting $\underset{\sim}{W}^{*\prime}_j$ to the weight tracking error is thus

$$2\mu\underset{\sim}{\Lambda}(z\underset{\sim}{I} - \underset{\sim}{I} + 2\mu\underset{\sim}{\Lambda})^{-1} - \underset{\sim}{I}. \tag{71}$$

Since (71) is diagonal, the scalar transfer function of its pth diagonal element may be written as

$$2\mu\lambda_p(z - 1 + 2\mu\lambda_p)^{-1} - 1 = \frac{(z^{-1} - 1)}{1 - (1 - 2\mu\lambda_p)z^{-1}}. \tag{72}$$

This transfer function has a zero at $z = 1$ and a pole whose impulse response has a geometric ratio of $(1 - 2\mu\lambda_p) = r_p$.

Fig. 10(a) shows the origin of the vector $\underset{\sim}{W}^*_j$ as a first-order Markov process and its propagation into the weight tracking error. $\underset{\sim}{W}^*_j$ is assumed to originate from independent stationary ergodic white-noise excitation (of variance σ^2) to a bank of one-pole filters, all having transfer function $1/(1 - az^{-1})$. The pth channel of this process is shown in Fig. 10(b). Its scalar transfer function is

$$\frac{(z^{-1} - 1)}{(1 - az^{-1})(1 - (1 - 2\mu\lambda_p)z^{-1})} = \frac{(z^{-1} - 1)}{(1 - az^{-1})(1 - r_p z^{-1})}$$

$$= \frac{\left(\frac{1 - a}{a - r_p}\right)}{(1 - az^{-1})} + \frac{\left(\frac{r_p - 1}{a - r_p}\right)}{(1 - r_p z^{-1})}. \tag{73}$$

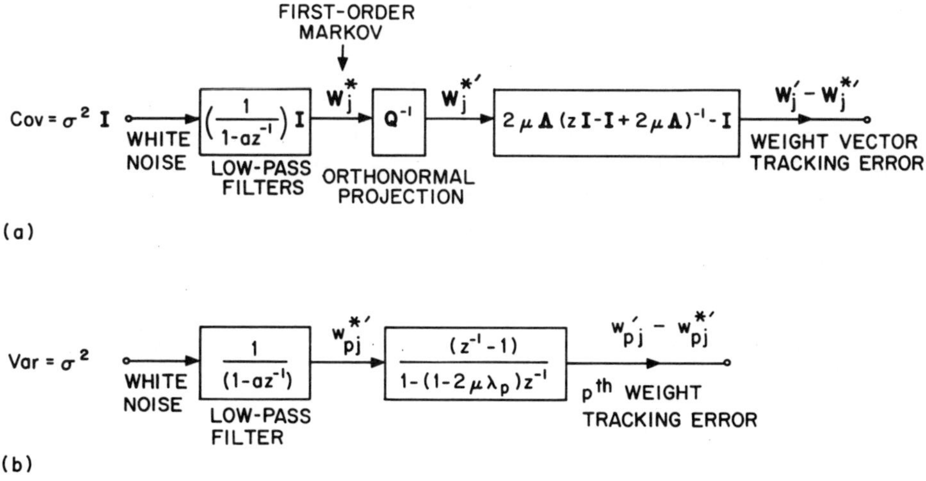

Figure 10. Origin of $\underset{\sim}{W}^*_j$ and its propagation into weight-lag error. (a) All channels. (b) pth channel.

The sampled impulse response of this transfer function is obtained by inversion of (73) into the time domain. From it, the variance of the lag error of the pth component of the primed weight vector can be computed as the sum of the squares of the samples of the impulse response multiplied by σ^2. The sum of squares is given by

$$\text{sum squares} = \sum_{j=0}^{\infty} \left[\left(\frac{1-a}{a-r_p}\right) a^j + \left(\frac{r_p - 1}{a - r_p}\right) r_p^j\right]^2$$

$$= \left(\frac{1}{a - r_p}\right)^2 \left[\left(\frac{1-a}{1+a}\right) + \left(\frac{1-r_p}{1+r_p}\right) + \frac{2(1-a)(r_p - 1)}{(1 - ar_p)}\right]. \quad (74)$$

In cases of interest, τ_p is large so that $r_p \lesssim 1$. From (27),

$$\tau_p = \frac{1}{1 - r_p} = \frac{1}{2\mu\lambda_p}. \quad (75)$$

Furthermore, we assume that the time constant of nonstationarity $\tau_{\underset{\sim}{W}*}$ is also large, so that $a \lesssim 1$.

$$\tau_{\underset{\sim}{W}*} = \frac{1}{1 - a}. \quad (76)$$

A common operating region would be where

$$\tau_{\underset{\sim}{W}*} >> \tau_p, \forall p. \quad (77)$$

The value of μ is set so that the response times of the adaptive weights are short compared to the time constant of nonstationarity. Under these conditions, (74) reduces to

$$(\text{sum squares})_{\tau_{\underset{\sim}{W}*} >> \tau_p} = \frac{1}{2}\tau_p = \frac{1}{4\mu\lambda_p}. \quad (78)$$

Using this relation, the covariance of the lag error is obtained as

$$\text{cov}\,[\underset{\sim}{W}'_j - \underset{\sim}{W}^{*\prime}_j]\Big|_{\substack{N=0 \\ \tau_{\underset{\sim}{W}*} >> \tau_p}} = \frac{\sigma^2}{2}\begin{bmatrix} \tau_1 & & & & 0 \\ & \ddots & & & \\ & & \tau_p & & \\ & & & \ddots & \\ 0 & & & & \tau_n \end{bmatrix} = \frac{\sigma^2}{4\mu}\underset{\sim}{\Lambda}^{-1}. \quad (79)$$

Making use of (64),

$$\text{(average excess mse due to lag)} = \frac{\sigma^2}{2}\sum_{p=1}^{n}\tau_p\lambda_p = \frac{n\sigma^2}{4\mu}. \tag{80}$$

Because of the ergodic properties of $\underset{\sim}{W}^*_j$, this average is not time variable. The misadjustment due to lag is

$$(M_L)_{\tau_{\underset{\sim}{W}^*} \gg \tau_p} = \frac{\sigma^2}{2\xi_{min}}\sum_{p=1}^{n}\tau_p\lambda_p = \left(\frac{n\sigma^2}{4\xi_{min}}\right)\frac{1}{\mu}. \tag{81}$$

Under usual operating conditions, the misadjustment due to lag is inversely proportional to μ.

Set μ to a very small value so that the adaptive weight vector $\underset{\sim}{W}_j$ does not track $\underset{\sim}{W}^*_j$ but merely assumes the value of its time average. As $r_p \to 1$, equation (74) reduces to

$$\text{(sum squares)}_{\mu\approx 0} = \frac{1}{2}\tau_{\underset{\sim}{W}^*}. \tag{82}$$

The misadjustment due to lag turns out to be

$$\text{(NS)} \triangleq (M_L)_{\mu\approx 0} = \frac{\sigma^2}{2\xi_{min}}\tau_{\underset{\sim}{W}^*}\ \text{tr}\ \underset{\sim}{R}. \tag{83}$$

Since there is no tracking, the misadjustment for this case is a measure of the "nonstationarity," NS, of the randomly moving hyperparaboloidal bowl.

An interesting special case is that of all equal eigenvalues. Combining (81) with (83),

$$(M_L)_{\tau_{\underset{\sim}{W}^*} \gg \tau_p} = \text{(NS)}\left[\frac{\tau}{\tau_{\underset{\sim}{W}^*}}\right] = \text{(NS)}\left[\frac{2\tau_{mse}}{\tau_{\underset{\sim}{W}^*}}\right]. \tag{84}$$

This result has intuitive appeal. The misadjustment equals the product of nonstationarity and the ratio of the adaptive time constant to the time constant of nonstationarity.

From (63), the average excess mse is the sum of components due to gradient noise and lag. The total misadjustment is, therefore, the sum of two misadjustment components. Making use of (51) and (81),

$$(M_{sum})_{\tau_{\tilde{W}^*} >> \tau_p} = (\mu)\ \mathrm{tr}\ \tilde{R} + \left(\frac{1}{\mu}\right)\frac{n\sigma^2}{4\xi_{min}}. \tag{85}$$

Optimizing the choice of μ results in minimum M_{sum} when the two right-hand terms are equal. The speed of adaptation is optimized when the loss of performance due to gradient noise equals the loss in performance due to weight-vector lag.[4] The optimal μ is

$$\mu^*\Big|_{\tau_{\tilde{W}^*} >> \tau_p} = \left[\frac{n\sigma^2}{4\xi_{min}\ (\mathrm{tr}\ \tilde{R})}\right]^{1/2} \tag{86}$$

A typical plot of M_{sum} versus μ is shown in Fig. 11, indicating the tradeoffs involved in adjusting μ for minimization of M_{sum}.

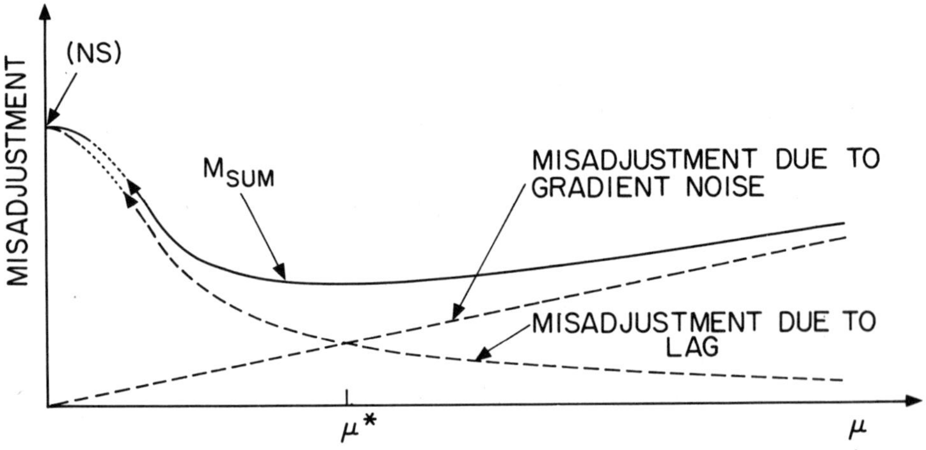

Figure 11. Net misadjustment versus LMS convergence factor μ.

---

[4] Another case has been analyzed by Widrow [29] where the fluctuation of $\tilde{W}^*_j$ has a uniform low-pass power spectrum. In this case, the misadjustment due to lag is proportional to the square of μ; the speed of adaptation is optimized when the gradient-noise loss equals twice the loss due to lag. The misadjustment due to lag turns out to be quite sensitive to the spectral characteristics of the fluctuation of $\tilde{W}^*_j$.

In practice, μ^* might need to be approximated by trial and error, particularly when data are unavailable for application of (86).

The theory developed in this section has been tested extensively by computer simulation based on an ensemble of adaptive processes, as illustrated in Fig. 8. Every mathematical quantity discussed in this section has been measured. Typical experimental results are presented below.

Fig. 12 illustrates weight tracking and the associated errors. The adaptive filter had four weights. Responses are shown only for weight number one. The effects of weight lag are demonstrated by comparing the ensemble average of weight number one plotted over time against weight number one of $\underset{\sim}{W}_j^*$. Averages were taken over 128 ensemble members. The lag effect is highly evident in the first experiment with $\mu = 0.003125$. In the third experiment, with $\mu = 0.05$, the lag is quite small decreasing in proportion to μ.

The effects of gradient noise are demonstrated with the same experiment. The ensemble mean of weight number one is plotted as a function of time j. Theoretical one-standard-deviation lines for weight noise are shown about this mean. In addition, weight number one of $\underset{\sim}{W}_j$ of a single ensemble member is plotted to indicate what occurred in an individual situation. It is clear that weight-noise power increases in proportion to μ.

In these experiments, the inputs were white and of unit power, so that $\underset{\sim}{R} = \underset{\sim}{I}$. The additive output noise power was $\xi_{min} = 1$. Equation (85) has been used to obtain theoretical values of misadjustment and its components. Tables II, III, and IV summarize the results of three experiments, comparing theory and experiments for three values of μ, and fixing everything else. The input data were the same for all three experiments. Initial transients were allowed to die out before measurements were taken. Experimental values of misadjustment and its components were obtained by ensemble average measurements using (60), (64), and (65), normalizing with respect to ξ_{min}, which in this case was 1. Theoretical and experimental results compared well, except for lag misadjustment in the first experiment. In this case, where μ is very small, equation (78) is inaccurate since $\tau_{\underset{\sim}{W}*}$ is no longer much larger than τ_p.

Much more work needs to be done in the study of nonstationary adaptive behavior. We have presented a simplistic but meaningful beginning.

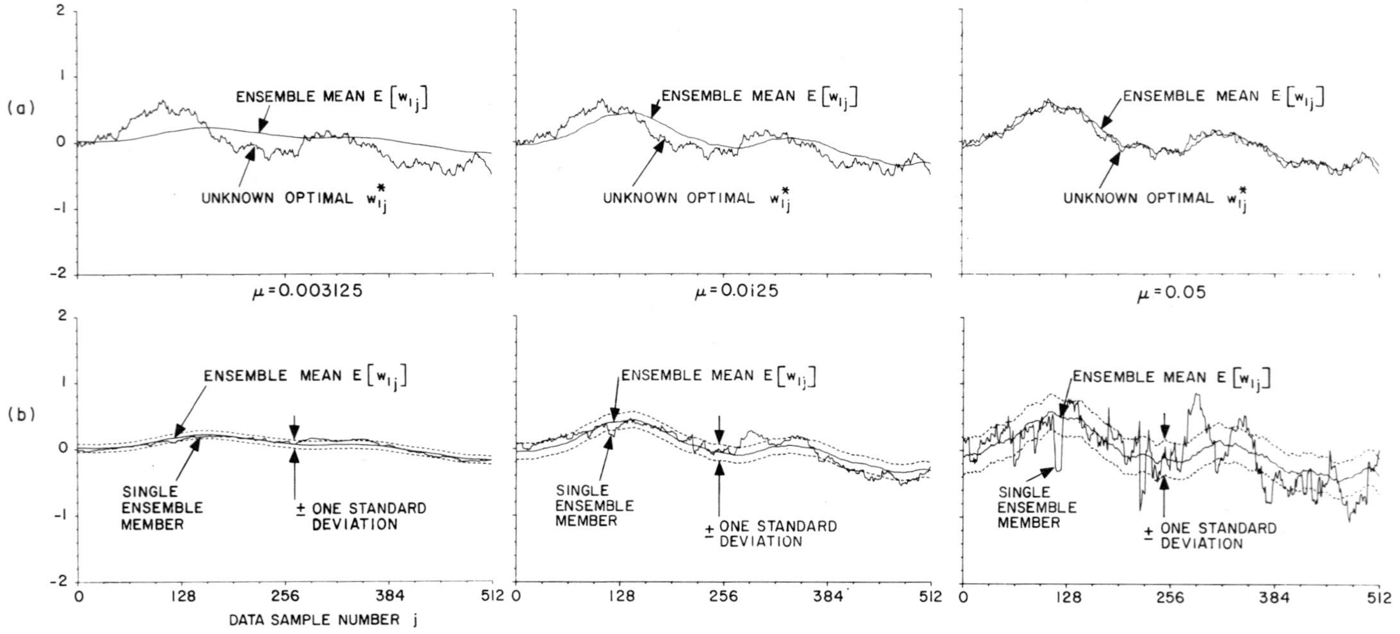

Figure 12. Weight tracking in a nonstationary environment. (a) Plots demonstrate weight lag as a function of μ. (b) Plots demonstrate weight noise as a function of μ.

Table II. First experiment, $\mu = 0.003125$, $n = 4$ weights, $\tau_{mse} = 80$ data samples, $\tau_{\underset{\sim}{W}*} = 125$ data samples, (NS) = 24.9%.

	Misadjustment	
	Due to Weight Lag	Due to Gradient Noise
Theoretical	32.0%	1.25%
Experimental	13.5%	1.5%

Table III. Second experiment, $\mu = 0.0125$, $n = 4$ weights, $\tau_{mse} = 20$ data samples, $\tau_{\underset{\sim}{W}*} = 125$ data samples, (NS) = 24.9%.

	Misadjustment	
	Due to Weight Lag	Due to Gradient Noise
Theoretical	8.0%	5.0%
Experimental	5.6%	5.7%

Table IV. Third experiment, $\mu = 0.05$, $n = 4$ weights, $\tau_{mse} = 5$ data samples, $\tau_{\underset{\sim}{W}*} = 125$ data samples, (NS) = 24.9%.

	Misadjustment	
	Due to Weight Lag	Due to Gradient Noise
Theoretical	2.0%	20.0%
Experimental	1.8%	28.3%

APPENDIX A. EFFICIENCY OF ADAPTIVE ALGORITHMS

We have analyzed the efficiency of the LMS algorithm from the point of view of misadjustment versus rate of adaptation. The question arises, could another algorithm be devised that would produce less misadjustment for the same rate of adaptation?

Suppose that an adaptive linear combiner is fed N independent input n x 1 data vectors $\underset{\sim}{X}_1, \underset{\sim}{X}_2, \cdots, \underset{\sim}{X}_N$ drawn from a stationary ergodic process. Associated with these input vectors are their scalar desired responses $d_1, d_2, \cdots, d_N$, also drawn from a stationary ergodic process. Keeping the weights fixed, a set of N error equations can be written as

$$\varepsilon_i = d_i - \underset{\sim}{W}^T \underset{\sim}{X}_i, \qquad i = 1, 2, \cdots, N. \tag{A.1}$$

Let the objective be to find a weight vector that minimizes the sum of the squares of the error values based on a sample of N items of data.

Equation (A.1) can be written in matrix form as

$$\underset{\sim}{E} = \underset{\sim}{D} - \underset{\sim}{X}\underset{\sim}{W} \tag{A.2}$$

where $\underset{\sim}{X}$ is an N x n rectangular matrix

$$\underset{\sim}{X} \triangleq [\underset{\sim}{X}_1 \underset{\sim}{X}_2 \cdots \underset{\sim}{X}_N]^T \tag{A.3}$$

and where $\underset{\sim}{E}$ is an N element error vector

$$\underset{\sim}{E} \triangleq [\varepsilon_1 \varepsilon_2 \cdots \varepsilon_N]^T. \tag{A.4}$$

A unique solution of (A.1), bringing $\underset{\sim}{E}$ to zero, exists only if $\underset{\sim}{X}$ is square and nonsingular. However, the case of greatest interest is that of N >> n. The sum of the squares of the errors is

$$\underset{\sim}{E}^T\underset{\sim}{E} = \underset{\sim}{D}^T\underset{\sim}{D} + \underset{\sim}{W}^T\underset{\sim}{X}^T\underset{\sim}{X}\underset{\sim}{W} - 2\underset{\sim}{D}^T\underset{\sim}{X}\underset{\sim}{W}. \tag{A.5}$$

This sum multiplied by 1/N is an estimate $\hat{\xi}$ of the mse ξ. Thus

$$\hat{\xi} = \frac{1}{N}\underset{\sim}{E}^T\underset{\sim}{E} \qquad \text{and} \qquad \lim_{N\to\infty} \hat{\xi} = \xi. \tag{A.6}$$

Note that $\hat{\xi}$ is a quadratic function of the weights, the parameters of the quadratic form being related to properties of the N data samples. $(\underset{\sim}{X}^T \underset{\sim}{X})$ is square and positive semidefinite. $\hat{\xi}_{min}$ is the small-sample-size mse function, while ξ is the large-sample-size mse function. These functions are sketeched in Fig. 13.

The function $\hat{\xi}$ is minimized by setting its gradient to zero:

$$\nabla \hat{\xi} = 2\underset{\sim}{X}^T \underset{\sim}{X} \underset{\sim}{W} - 2\underset{\sim}{X}^T \underset{\sim}{D}. \tag{A.7}$$

The "optimal" weight vector based only on the N data samples is

$$\hat{\underset{\sim}{W}}^* \triangleq (\underset{\sim}{X}^T \underset{\sim}{X})^{-1} \underset{\sim}{X}^T \underset{\sim}{D}. \tag{A.8}$$

This formula gives the position of the minimum of the small-sample-size bowl. The corresponding formula for the large-sample-size bowl is the Wiener-Hopf equation (8).

We could calculate $\hat{\underset{\sim}{W}}^*$ by a training process, a regression process, LMS, or some other optimization procedure. Taking the first block of N data samples, we obtain a small-sample-size function $\hat{\xi}_1$ whose minimum is at $\hat{\underset{\sim}{W}}^*_1$. This could be repeated with a second data sample, giving a function $\hat{\xi}_2$ whose minimum is at

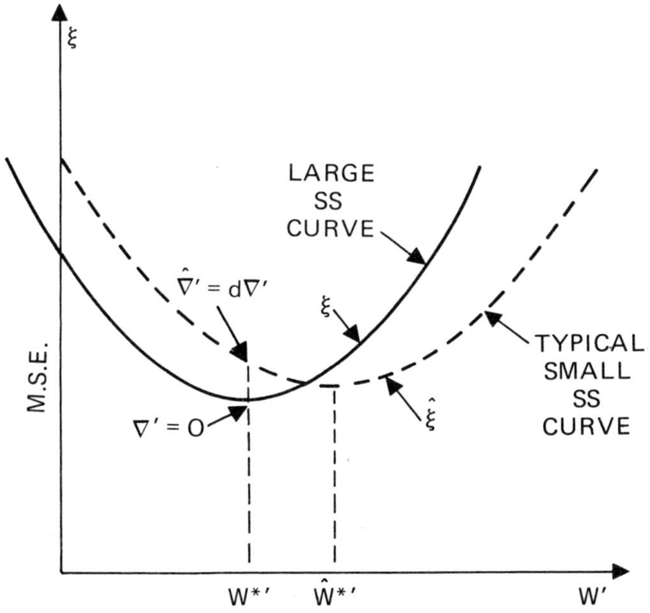

Figure 13. Small- and large-sample-size mse curves.

$\hat{W}^*_2$, etc. Typically, all the values of $\hat{W}^*$ would differ from the true optimum W^* and would, thereby, be misadjusted.

To analyze the misadjustment, assume that N is large and that the typical small-size curve approximately matches the large-sample-size curve. Therefore,

$$\hat{\xi} \approx \xi \quad \text{and} \quad (\xi - \hat{\xi}) \overset{\Delta}{=} d\xi. \tag{A.9}$$

The true large-sample-size function is

$$\xi = \xi_{min} + V'^T \Lambda V'.$$

The gradient of this function expressed in the primed coordinates is

$$\nabla' = 2\Lambda V'.$$

A differential deviation in the gradient is

$$(d\nabla') = 2\Lambda(dV') + 2\,(d\Lambda)V'. \tag{A.10}$$

This deviation could represent the difference in gradients between small- and large-sample-size curves.

Refer to Fig. 13. Let $W' = W^{*\prime}$, then $V' = 0$. The gradient of ξ is zero, while the gradient of $\hat{\xi}$ is $\hat{\nabla}' = d\hat{\nabla}$. Using (A.10),

$$(d\nabla') = 2\Lambda(dV'). \tag{A.11}$$

From (A.11), the deviation in gradient can be linked to the deviation in position of the small-sample-size curve minimum since $(dV') = (W^{*\prime} - \hat{W}^{*\prime})$. Taking averages of (A.11) over an ensemble of small-sample-size curves,

$$\text{cov}\,[d\nabla'] = 4\Lambda\,\text{cov}\,[dV']\Lambda. \tag{A.12}$$

Equation (42) indicates that the covariance of the gradient noise when $W' = W^{*\prime}$ is given by $4\xi_{min}\,\Lambda$. If the gradient were estimated under the same conditions but using N independent error samples,

$$\text{cov}\,[d\nabla'] = \frac{4}{N}\,\xi_{min}\,\Lambda. \tag{A.13}$$

Substituting this into (A.12) yields

$$\mathrm{cov}\,[d\underset{\sim}{V}'] = \frac{1}{N}\,\xi_{min}\,\underset{\sim}{\Lambda}^{-1}. \tag{A.14}$$

The average excess mse, an ensemble average, is

$$\begin{pmatrix}\text{average}\\ \text{excess}\\ \text{mse}\end{pmatrix} = E[(d\underset{\sim}{V}')^T\,\underset{\sim}{\Lambda}(d\underset{\sim}{V}')]. \tag{A.15}$$

Equation (A.14) shows cov $[d\underset{\sim}{V}']$ to be diagonal, so that

$$\begin{pmatrix}\text{average}\\ \text{excess}\\ \text{mse}\end{pmatrix} = \frac{n}{N}\,\xi_{min}. \tag{A.16}$$

The misadjustment is, therefore,

$$M = \frac{n}{N} = \frac{\text{(number of weights)}}{\text{(number of independent training samples)}}. \tag{A.17}$$

This formula was first presented without detailed proof by Widrow and Hoff [1] in 1960. It has been used for many years in pattern recognition studies. For small values of M (less than 25 percent), it has proven to be very useful. A formula similar to (A.17), although based on somewhat different assumptions, was derived by Davisson [36] in 1970.

Although equation (A.17) has been derived for training with finite blocks of data, it can be used to assess the efficiency of steady-flow algorithms. Consider an adaptive transversal filter with stationary stochastic inputs, adapted by the LMS algorithm. For simplicity, let all eigenvalues of $\underset{\sim}{R}$ be equal. As such

$$M = \frac{n}{4\tau_{mse}}. \tag{56}$$

The LMS algorithm exponentially weights its input data over time in determining current weight values. If an equivalent uniform averaging window is assumed equal to the adaptive settling time, approximately four time constants, the equivalent data sample taken at any instant by LMS is essentially $N_{eq} = 4\tau_{mse}$ samples. Accordingly for LMS,

$$M = \frac{n}{N_{eq}}. \tag{A.18}$$

A comparison of (A.18) and (A.17) shows that when eigenvalues are equal, LMS is about as efficient as a least squares algorithm can be.[5] However, with disparate eigenvalues, the misadjustment is primarily determined by the fastest modes while settling time is limited by the slowest modes. To sustain efficiency with disparate eigenvalues, algorithms similar to LMS have been devised based on Newton's method rather than on steepest descent [38], [39]. Such algorithms premultiply the gradient estimate each iteration cycle by an estimate of the inverse of $\underset{\sim}{R}$.

$$\underset{\sim}{W}_{j+1} = \underset{\sim}{W}_j + \mu \hat{\underset{\sim}{R}}^{-1} \hat{\underset{\sim}{\nabla}}_j,$$

or

$$\underset{\sim}{W}_{j+1} = \underset{\sim}{W}_j + 2\mu \hat{\underset{\sim}{R}}^{-1} \varepsilon_j \underset{\sim}{X}_j. \tag{A.19}$$

This process causes all adaptive modes to have essentially the same time constant. Algorithms based on this principle are potentially more efficient that LMS but are typically more difficult to implement.

ACKNOWLEDGMENT

The authors wish to acknowledge the helpful discussions and contributions of C. S. Williams and J. R. Treichler of Stanford University; Dr. O. L. Frost III of Argo Systems, Inc.; and Dr. M. E. Hoff, Jr. of the Intel Corporation. Special thanks are also due Diane Byron, who assisted in editing the paper.

[5]Attempts have been made to devise algorithms more efficient than LMS by using variable μ [37]. Initial values of μ are chosen high for rapid convergence; final values of μ are chosen low for small misadjustment. This works as long as input statistics are stationary. This procedure and the methods of stochastic approximation on which it is based will not perform well in the nonstationary case.

REFERENCES

1. B. Widrow and M. E. Hoff, "Adaptive switching circuits," in 1960 WESCON Conv. Rec., pt. 4, pp. 96-140.
2. N. Nilsson, Learning Machines. New York: McGraw-Hill, 1965.
3. J. Koford and G. Groner, "The use of an adaptive threshold element to design a linear optimal pattern classifier," IEEE Trans. Inform. Theory, vol. IT-12, pp. 42-50, Jan. 1966.
4. B. Widrow, P. Mantey, L. Griffiths, and B. Goode, "Adaptive antenna systems," Proc. IEEE, vol. 55, pp. 2143-2159, Dec. 1967.
5. B. Widrow, "Adaptive Filters," in Aspects of Network and System Theory, R. Kalman and N. DeClaris, Eds. New York: Holt, Rinehart, and Winston, 1971, pp. 563-587.
6. S. P. Applebaum, "Adaptive arrays," Special Projects Lab., Syracuse Univ. Res. Corp., Rep. SPL 769.
7. L. J. Griffiths, "A simple adaptive algorithm for real-time processing in antenna arrays," Proc. IEEE, vol. 57, pp. 1696-1704, Oct. 1969.
8. O. L. Frost III, "An algorithm for linearly constrained adaptive array processing," Proc. IEEE, vol. 60, pp. 926-935, Aug. 1972.
9. W. F. Gabriel, "Adaptive arrays-An introduction," Proc. IEEE, vol. 64, pp. 239-272, Feb. 1976.
10. F. W. Smith, "Design of quasi-optimal minimum-time controllers," IEEE Trans. Automat. Contr., vol. AC-11, pp. 71-77, Jan. 1966.
11. B. Widrow, "Adaptive model control applied to real-time blood-pressure regulation," in Pattern Recognition and Machine Learning, Proc. Japan-U.S. Seminar on the Learning Process in Control Systems, K. S. Fu, Ed. New York: Plenum Press, 1971, pp. 310-324.
12. R. Lucky, "Automatic equalization for digital communication," Bell Syst. Tech. J., vol. 44, pp. 547-588, Apr. 1965.
13. M. DiToro, "A new method of high-speed adaptive serial communication through any time-variable and dispersive transmission medium," in Conf. Record, 1965 IEEE Annual Communications Convention, pp. 763-767.
14. R. Lucky and H. Rudin, "An automatic equalizer for general-purpose communication channels," Bell Syst. Tech. J., vol. 46, pp. 2179-2208, Nov. 1967.
15. R. Lucky et al., Principles of Data Communication. New York: McGraw-Hill, 1968.
16. A. Gersho, "Adaptive equalization of highly dispersive channels for data transmission," Bell Syst. Tech. J., vol. 48, pp. 55-70, Jan. 1969.
17. M. Soudhi, "An adaptive echo canceller," Bell Syst. Tech. J., vol. 46, pp. 497-511, Mar. 1967.
18. B. Widrow et al., "Adaptive noise cancelling: Principles and applications," Proc. IEEE, vol. 63, pp. 1692-1716, Dec. 1975.

19. P. E. Mantey, "Convergent automatic-synthesis procedures for sampled-data networks with feedback," Stanford Electronics Laboratories, Stanford, CA, TR no. 7663-1, Oct. 1964.
20. P. M. Lion, "Rapid identification of linear and nonlinear systems," in Proc. 1966 JACC, Seattle, WA, pp. 605-615, Aug. 1966; also AIAA Journal, vol. 5, pp. 1835-1842, Oct. 1967.
21. R. E. Ross and G. M. Lance, "An approximate steepest descent method for parameter identification," in Proc. 1969 JACC, Boulder, CO, pp. 483-487, Aug. 1969.
22. R. Hastings-James and M. W. Sage, "Recursive generalized-least-squares procedure for online identification of process parameters," Proc. IEE, vol. 116, pp. 2057-2062, Dec. 1969.
23. A. C. Soudack, K. L. Suryanarayanan, and S. G. Rao, "A unified approach to discrete-time systems identification," Int. J. Control, vol. 14, no. 6, pp. 1009-1029, Dec. 1971.
24. W. Schaufelberger, "Der Entwurf adaptiver Systeme nach der direckten Methode von Ljapunov," Nachrichtentechnik, Nr. 5, pp. 151-157, 1972.
25. J. M. Mendel, Discrete Techniques of Parameter Estimation: The Equation Error Formulation. New York: Marcel Dekker, Inc., 1973.
26. S. J. Merhav and E. Gabay, "Convergence properties in linear parameter tracking systems," Identification and System Parameter Estimation-Part 2, Proc. 3rd IFAC Symp., P. Eykhoff, Ed. New York: American Elsevier Publishing Co., Inc., 1973, pp. 745-750.
27. R. V. Southwell, Relaxation Methods in Engineering Science. New York: Oxford, 1940.
28. D. J. Wilde, Optimum Seeking Methods. Englewood Cliffs, N.J.: Prentice-Hall, 1964.
29. B. Widrow, "Adaptive sampled-data systems," in Proc. First Intern. Cong. Intern. Federation of Automatic Control, Moscow, 1960.
30. H. Robbins, and S. Monro, "A stochastic approximation method," Ann. Math. Statist., vol. 22, pp. 400-407, 1951.
31. A. Dvoretzky, "On stochastic approximation," in Proc. Third Berkeley Symp. Math. Statist. and Probability, J. Neyman, Ed. Berkeley, CA: University of California Press, 1956, pp. 39-55.
32. J. Makhoul, "Linear prediction: A tutorial review," Proc. IEEE, vol. 63, pp. 561-580, Apr. 1975.
33. L. J. Griffiths, "Rapid measurement of digital instantaneous frequency," IEEE Trans. Acoust., Speech, Signal Processing, vol. ASSP-23, pp. 207-222, Apr. 1975.
34. Y. T. Chien, K. S. Fu, "Learning in non-stationary environment using dynamic stochastic approximation," in Proc. 5th Allerton Conf. Circuit and Systems Theory, pp. 337-345, 1967.
35. T. P. Daniell and J. E. Brown III, "Adaptation in nonstationary applications," in Proc. 1970 IEEE Symp. Adaptive Processes (9th), Austin, TX, paper no. XXIV-4, Dec. 1970.

36. L. D. Davisson, "Steady-state error in adaptive mean-square minimization," IEEE Trans. Inform. Theory, vol. IT-16, pp. 382-385, July 1970.
37. T. J. Schonfeld and M. Schwartz, "A rapidly converging first-order training algorithm for an adaptive equalizer," IEEE Trans. Inform. Theory, vol. IT-17, pp. 431-439, July 1971.
38. K. H. Mueller, "A new, fast-converging mean-square algorithm for adaptive equalizers with partial-response signaling," Bell Syst. Tech. J., vol. 54, pp. 143-153, Jan. 1975.
39. L. J. Griffiths and P. E. Mantey, "Iterative least-squares algorithm for signal extraction," in Proc. Second Hawaii Int. Conf. System Sciences, Western Periodicals Co., pp. 767-770, 1969.

DISCUSSION

Comment : D.G. PINCOCK

In your adaptive-noise cancelling scheme I can visualize many situations in which there will be some correlation between your "noise source" and the signal. Have you investigated the effect this may have on system performance?

Reply : B. WIDROW

Yes we have. If the "noise source" or, using our own terminology, the "reference input" contains noise correlated with the primary noise and some signal components in addition, the signal-to-noise ratio in any given frequency bin at the system output will be equal to the noise-to-signal ratio of the reference input, you get "signal-to-noise inversion". This is described more fully in a paper by Widrow et al, entitled "Adaptive Noise Cancelling: Principles and Applications", which appeared in the December 1975 issue of the Proceedings of the IEEE.

AN ADAPTIVE APPROACH TO UNDERWATER PASSIVE DETECTION*

G. BIENVENU

Thomson-CSF, DASM Cagnes sur Mer, France

1. INTRODUCTION

The main purpose of underwater passive listening is to detect the presence of noise sources. Since the time structures of the signals emitted by these sources are unknown, the main criterion which can be used is their spatial coherence: these sources produce on the sensors of the receiving antenna signals which are almost identical except for a time delay according to their direction. The detection of the eventual presence of a source in a given direction is blurred by other coherent noise sources which can be targets of interest or merely jammers, and by background noise. In this paper, feasible treatments according to which hypothesis are made on the properties of noisefield are going to be examined. The first hypothesis, which will be used throughout this paper, assumes that the sources produce signals which are quite identical on the sensors except for a time delay between each of them. The signal is then said to have a perfect spatial coherence. Another hypothesis is that the signals emitted by the sources are statistically independent of each other and of the background noise. Hypothesis on the background noise will be made later when necessary.

First, the theory of optimal processing for the detection of an unknown signal of known direction will be briefly reviewed [1]. It assumes that the correlation matrix of the received signals without the signal to be detected s(t) is known. Let $x_k(t)$ be the

* This work has been supported by "Directions des Recherches et Moyens d'Essais.

G. Tacconi (ed.), Aspects of Signal Processing, Part 1, 395-400.

output signal of the kth sensor delayed according to the look direction. When the signal s(t) is present, $x_k(t)$ is equal to $s(t) + x'_k(t)$. s(t) is the same for all the sensors; $x'_k(t)$ is the contribution from other sources and the background noise which are supposed to be zero-mean gaussian processes. The optimum processing is obtained by using the generalized likelihood ratio test criterion. It consists of a spatial processor which gives the maximum likelihood estimate $\hat{s}_{MLE}(t)$ of s(t), followed by a time processor. This estimate is the sum of the signal s(t) and a noise b(t). The time processor is that which is obtained by the generalized likelihood ratio test for the detection of the unknown signal s(t) in the noise b(t). In the case of an infinite observation interval, the transfer function of the spatial processor is given by the filtering vector:

$$\vec{H}_{MLE}(f) = \Gamma_{x'}^{-1}(f)\, \vec{N} \left[\vec{N}^{+} \Gamma_{x'}^{-1}(f)\, \vec{N} \right]^{-1} \qquad (1)$$

and the output of the time processor by:

$$\int \left[\hat{S}_{MLE}(f)\, \hat{S}^{*}_{MLE}(f) \,/\, \gamma_b(f) \right] df \qquad (2)$$

where: $\Gamma_{x'}(f)$ is the cross-spectral density matrix of the signals $x'_k(t)$

$\vec{N}$ is a column vector of K (number of sensors) ones

$\hat{S}_{MLE}(f)$ is the Fourier transform of s(t)

$\gamma_b(f) = \left[\vec{N}^{+} \Gamma_{x'}^{-1}(f)\, \vec{N} \right]^{-1}$ is the spectral density of b(t)

Therefore, the optimum processing uses first the spatial properties of the signals to clean s(t) as best as possible, and then it uses their available time properties to detect.

In practice, $\Gamma_{x'}(f)$ is not known, and the optimum processing cannot be used. Nevertheless, it is possible to obtain an estimation of its spatial processor, but it is impossible to have an estimation of the time processor without making any additional hypothesis on $\Gamma_{x'}(f)$.

2. ADAPTIVE ANTENNA

It can be shown that in relation (1), $\Gamma_{x'}(f)$ can be replaced by $\Gamma_x(f)$, the cross-spectral density matrix of the received signals $x_k(t)$. As it is possible to make an estimation $\hat{\Gamma}_x(f)$ of $\Gamma_x(f)$, an estimation $\vec{H}_A(f)$ of the optimum filtering vector $\vec{H}_{MLE}(f)$ may be obtained by substitution in relation (1) of $\Gamma_{x'}(f)$ by $\hat{\Gamma}_x(f)$. It is the adaptive antenna [2]. This processing can also be built up with wide-band time algorithms [3, 4]. They minimize the output power of the processing under the constraint that the signal be undistorted. If $\vec{W}(n)$ and $\vec{X}(n)$ are respectively filter weights and sampled signals vectors at the nth sampling period, the algorithm will be written as follows:

$$\vec{W}(n + 1) = \vec{W}(n) - \varepsilon P \vec{X}(n) y(n) \quad (3)$$

where P is a projection matrix due to the constraint and y(n) the output sample. A much simpler algorithm can be used if the constraint is realized by a transformation of the input data before minimization:

$$\vec{W}' (n + 1) = \vec{W}'(n) - \varepsilon \vec{X}'(n) y(n) \quad (4)$$

The performance of the adaptive antenna has been measured during experiments carried out on a lake by using a processor which works in the 2-4 KHz bandwidth. The receiving antenna is linear (sensors spacing: 16.3 cm). The sources were simulated by transducers. Some of the results are presented in figure 1: They show output levels of both the classical and the adaptive processors versus bearing.

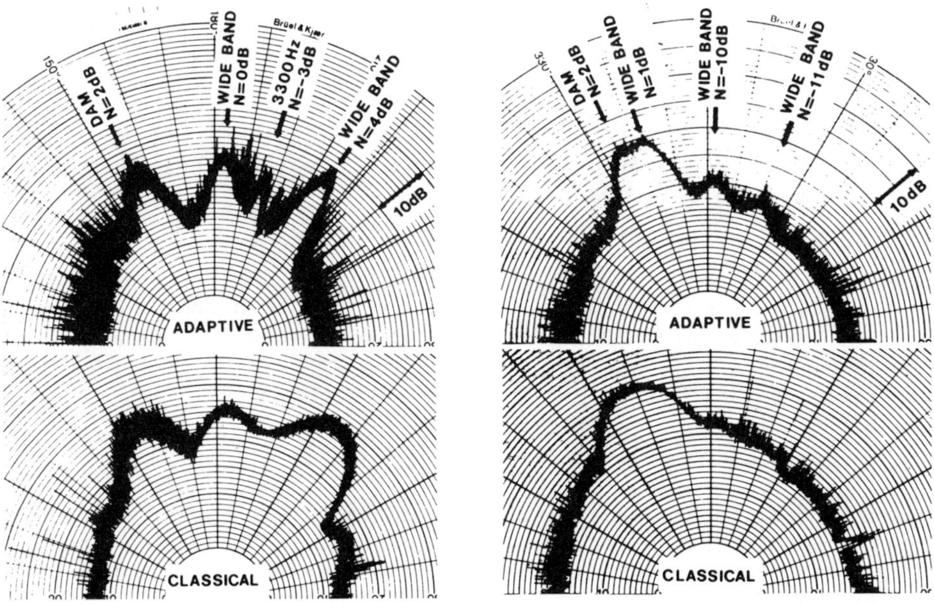

Fig. 1. Experimental comparison of classical and adaptive antenna.

3. DETECTION TEST

3.1. Hypothesis on the background noise

As $\Gamma_{x'}(f)$ is unknown, an attempt can be made to obtain an estimation of it. But if we then apply a generalized likelihood ratio test, this procedure fails, because the contrast that we have used between the signal and the noise is not sufficient. In order to build a detection test, we must define some statistic of the noise which is modified in a known way by the presence of the signal. When an antenna is used, the statistic depends upon two parameters: time and space. As it was seen, in underwater passive listening the time structures of the signals are unknown, and only the spatial properties which are better known can be used. The spatial coherence of the signal is assumed to be perfect: this hypothesis is sufficient for its estimation. For detection purposes, the spatial coherence of the noise must be given. The process of detection consists then in testing if the received signals possess the spatial coherence of the noise, or if they contain some contribution which has the spatial coherence of a signal arriving from the look direction. Obviously, it is impossible to formulate an hypothesis on $\Gamma_{x'}(f)$. The noisefield contains in fact several sources which have the typical spatial coherence of a signal. Thus we can attempt to make an hypothesis only on the spatial coherence of the background noise. In other words, of the two terms which compose $\gamma_b(f)$: $\gamma_c(f)$ and $\gamma_\sigma(f)$ produced respectively by the spatially coherent sources and the background noise, only $\gamma_\sigma(f)$ can be estimated. $\gamma_c(f)$ will give the spatial ambiguity that the detection test must necessarily possess.

In order to show how to obtain an estimation $\hat{\gamma}_\sigma(f)$ of $\gamma_\sigma(f)$, it can be assumed, for example, that the cross-spectrum density matrix of the background noise is:

$$\Gamma_G(f) = \gamma_G(f)\ J(f) \qquad (5)$$

where $\gamma_G(f)$ which depends on the time correlation and on the power of the background noise is unknown and J(f) which represents the spatial coherence is given. Under these conditions, the cross-spectrum density matrix of the received signals is:

$$\Gamma_x(f) = \gamma_G(f)\ J(f) + \sum_{i=1}^{I} \gamma_i(f)\ \vec{D}_i(f)\ \vec{D}_i^{+}(f) \qquad (6)$$

where $\gamma_i(f)$ is the spectral density of the ith coherent source and $\vec{D}_i(f)$, its direction vector. In order that the sources be resolved, it is assumed that their number I is smaller than the number of sensors K. It can be shown that the estimation $\gamma_G(f)$ of $\gamma_G(f)$ is equal to the minimum of the expression:

$$\vec{H}_G^{+}(f)\ \hat{\Gamma}_x(f)\ \vec{H}_G(f) \qquad (7)$$

where $\vec{H}_G(f)$ is a vector such as:

$$\vec{H}_G^{+}(f)\ J(f)\ \vec{H}_G(f) = 1 \tag{8}$$

and $\hat{\Gamma}_x(f)$ is an estimation of $\Gamma_x(f)$ obtained from the signals received on the sensors. $\hat{\gamma}_G(f)$ is consistent. $\hat{\gamma}_\sigma(f)$ is then computed by:

$$\hat{\gamma}_\sigma(f) = \hat{\gamma}_G(f)\ \vec{H}_A^{+}(f)\ J(f)\ \vec{H}_A(f) \tag{9}$$

The vector $\vec{H}_G(f)$ can be obtained by using a constrained minimization algorithm.

3.2. Detection test

If an estimation $\hat{\gamma}_\sigma(f)$ of $\gamma_\sigma(f)$ is known, it is possible to build a detector by following the same proceeding as that shown by the optimum test. First, the signal s(t) is cleaned as best as possible by a spatial processor: this work is carried out by the adaptive antenna which approximates the optimum spatial processor (1) and tends asymptotically towards it. This treatment gives a signal $\hat{s}_A(t)$ which is the sum of s(t) and of the contributions $s'_c(t)$ and $b'_\sigma(t)$ of respectively the other coherent sources and the background noise. The problem is now to detect an unknown signal s(t) corrupted by a noise $b'_\sigma(t)$ and in the presence of terms $s'_c(t)$ resulting from signals to be detected when their direction coincides with the look direction, and which cannot be taken into account in the detection process: their influence have been reduced to a minimum by the adaptive antenna. If $b'_\sigma(t)$ is assumed Gaussian, the detection test should be identical to the optimum test (2) in which $\gamma'_\sigma(f)$ is substituted for $\gamma_b(f)$ and $\hat{S}_A(f)$ for $S_{MLE}(f)$. But as $\gamma'_\sigma(f)$ is unknown, we propose to replace it by its estimation $\hat{\gamma}_\sigma(f)$, and the detection test is therefore:

$$\Lambda = \int \left[\ \hat{S}_A(f)\ \hat{S}_A^{*}(f)\ /\ \hat{\gamma}_\sigma(f)\ \right] df \tag{10}$$

The detection test have been simulated on a computer with narrow band signals (3,000 Hz ± 50 Hz). The background noise was assumed to be isotropic and independent between sensors. Therefore, its cross-spectrum density matrix is: $\gamma_G(f)I$, where I is the identity matrix. The antenna simulated is linear and has 10 sensors with a 16 cm spacing. The results are presented in figure 2. They show, as a function of bearing, the square roots of the test $\Lambda^{\frac{1}{2}}$ (-) and of the output power of the classical $C^{\frac{1}{2}}$ (x) and adaptive $A^{\frac{1}{2}}$ (.) antenna normalized by the power of the background noise. Theoretical asymptotic curves obtained under the assumption that $\Gamma_x(f)$ is known are also presented for the detection test (....) and for the adaptive antenna (----). The synthetized noisefields are composed of background noise and two sinusoidal signals (3,000 Hz

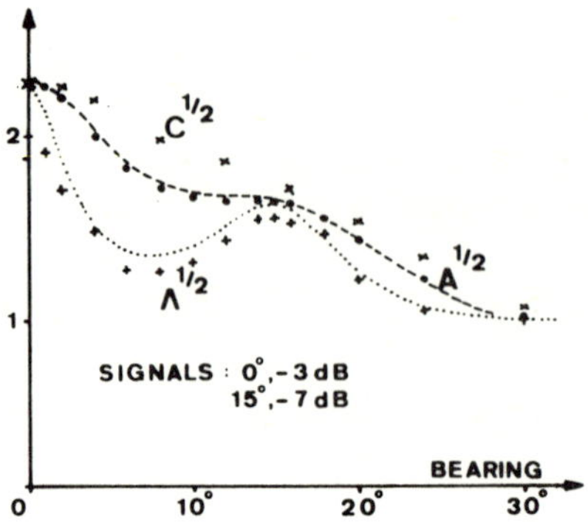

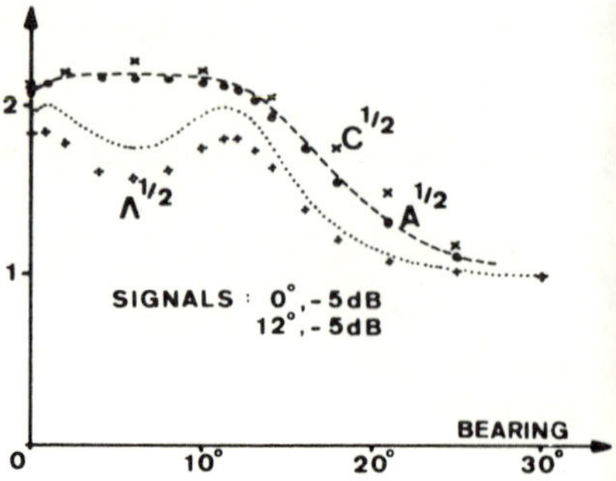

Fig. 2. Simulation results of the detection test.

and 3,030 Hz) whose bearing and level in respect to that of the background noise are indicated in each figure.

The detection test Λ has a better spatial resolution than the adaptive antenna. This is quite natural for the adaptive antenna is an estimator constrained to restitute the signal without distortion, whereas the detection test whose function is to indicate the presence only of the signal is not so constrained.

4. FURTHER WORK

In fact, the actual spatial coherence of the signal is never completely perfect as it was assumed, and that of the background noise is not well known; moreover, both are fluctuating. So, a more fitted test would be obtained if the spatial coherences were defined by statistical classes.

REFERENCES

1. H.L. Van Trees, Optimum Processing for Passive Sonar Arrays, IEEE, Ocean El. Symp., Honolulu, Hawaii, Aug. 1966
2. J. Capon, High-resolution Frequency Wave-number Spectrum Analysis, Proc. IEEE, Vol. 54, n° 8, p. 1408-1418, Aug. 1969.
3. O.L. Frost, An Algorithm for Linearly Constrained Adaptive Array Processing, Proc. IEEE, Vol. 60, n° 6, p. 926-935, Aug. 1972.
4. G. Bienvenu, J.L. Vernet, Enhancement of Antenna Performance by Adaptive Processing, NATO A.S.I. Proc., Loughborough, Aug. 1972.